普通高等教育农业农村部“十三五”规划教材

农药专业英语

（第二版）

骆焱平　主编

化学工业出版社

·北京·

内容简介

本书为农药学科的专业英语教材，以化学农药、生物农药、农药剂型、纳米农药、农药应用、化感作用、基因工程农药、农药生物测定、农药抗药性、农药残留、农药降解、农药毒理、农药风险评估、农药管理、农药设计十五个主题进行详细展开，课后有习题及阅读材料。书后附录包括常见农药专业英语词根、前后缀，农药剂型名称及代码，常见农药中英文名称等。

本书可供农药学专业的本科生和研究生使用，也可作为相关专业人员的教学和参考用书。

图书在版编目（CIP）数据

农药专业英语/骆焱平主编. —2 版. —北京：化学工业出版社，2021.4
ISBN 978-7-122-38643-4

Ⅰ.①农… Ⅱ.①骆… Ⅲ.①农药-英语-高等学校-教材 Ⅳ.①TQ45

中国版本图书馆 CIP 数据核字（2021）第 039166 号

责任编辑：刘 军 冉海滢 装帧设计：关 飞
责任校对：张雨彤

出版发行：化学工业出版社（北京市东城区青年湖南街 13 号 邮政编码 100011）
印 装：大厂聚鑫印刷有限责任公司
787mm×1092mm 1/16 印张 20¾ 字数 526 千字 2021 年 7 月北京第 2 版第 1 次印刷

购书咨询：010-64518888 售后服务：010-64518899
网 址：http://www.cip.com.cn
凡购买本书，如有缺损质量问题，本社销售中心负责调换。

定 价：68.00 元

本书编写人员

主　　编　骆焱平

副 主 编　马志卿　李向阳　曾志刚

编写人员（按姓氏笔画排列）

马志卿　西北农林科技大学

李双梅　海南大学

李向阳　贵州大学

吴友根　海南大学

张　莉　中国农业大学

张云飞　海南大学

邵旭升　华东理工大学

郑　丽　海南大学

骆焱平　海南大学

曾志刚　湖北科技学院

薛超彬　山东农业大学

序

仓廪实，天下安。粮食安全是社会稳定发展的基石，关乎国计民生，关乎千家万户。习近平总书记指出，“要依靠自己保口粮，集中国内资源保重点，做到谷物基本自给、口粮绝对安全。更加注重农产品质量和食品安全，转变农业发展方式，抓好粮食安全保障能力建设”。

我国用占世界7%的耕地，养活了占世界22%的人口。预计到2050年，世界人口将接近100亿，食物需求将翻倍。然而，由于植物病虫草鼠害造成全球重大作物产量损失，严重威胁着全球粮食安全。农药是保障粮食丰收的重要农业生产资料，对维护国家粮食安全起着非常重要的作用。据不完全统计，农药可以减少病虫草鼠等的危害，挽回至少10%～30%的经济损失。

我国是农药生产和出口大国，在全球农药产业链中处于重要位置。与此同时，我国的农药研究和开发，已经从过去的跟踪、模仿，走向了创制和原创，我国已经成为世界上主要的新农药研发大国之一，正在逐步走向强国。据了解，我国每年生产的300多万吨农药，一半以上出口，足迹已遍布“一带一路”沿线国家。随着时代的发展，我国农药产业呈现出转型升级的良好发展态势，农药科学研究蓬勃发展，国际交流日益增强，为国家服务、为全球服务、为绿色生态和社会可持续发展服务，且亟需既懂农药专业知识，又懂农药专业外语的复合型人才。因此，开展农药专业英语教学，对培养农药后备人才，将有积极的促进作用。

《农药专业英语（第二版）》是在2009年出版的第一版基础上，新增纳米农药、化感作用、基因工程农药、农药风险评估、农药设计等内容编写而成。本书基本覆盖了现有农药学科的相关专题，内容丰富而全面；同时附有农药专业词汇，可供农药学本科生、研究生及相关研究人员使用。

作为农药学科的首本专业英语教材，本书入围农业农村部“十三五”规划教材，是编者辛勤劳动的成果，也是我国大力支持农药发展的结果，为此作序，以兹鼓励。

钱旭红

中国工程院院士

华东师范大学校长

2020年12月

前言

随着科学技术不断发展，人们生活水平不断提高，对物质生活的追求也出现了质的飞跃。在作物保产增收的过程中，农药发挥了积极作用。现如今，农药已非传统意义上的“农药”，钱旭红院士提出的“绿色农药”，宋宝安院士提出的植物诱抗激活剂等，新的技术如RNAi技术的出现，为农药赋予了新的内涵。因此，2009年出版的《农药专业英语》也要与时俱进。

为此，本书在第一版的基础上，由10个部分增加到15个部分，由20个单元40篇课文增加到30个单元60篇课文。分别介绍化学农药、生物农药、农药剂型、纳米农药、农药应用、化感作用、基因工程农药、农药生物测定、农药抗药性、农药残留、农药降解、农药毒理、农药风险评估、农药管理、农药设计等内容。在第一版的基础上新增纳米农药、化感作用、基因工程农药、农药风险评估、农药设计等内容，使知识面更加丰富，专业词汇更加完备。同时，每个单元后面均列有详细注释，并提供一定量的习题供读者进行练习和巩固。书后附录包含常见农药专业英语词根、前后缀，农药剂型名称及代码，常见农药中英文名称等。

本书内容选自外文书籍或相关文献，部分章节内容经过删减、加工等，力求内容全面，通俗易懂。除编写人员外，本书的编写还得到刘健、王兰英、张淑静、杨育红等的帮助，在此一并表示感谢。

本书出版得到了海南大学教材出版基金项目、国家自然科学基金项目（31860513）和海南大学教学名师工作室（hdms202013）的资助。

由于编者水平有限，难免存在不足之处，敬请同行专家及广大师生提出宝贵的意见和建议。

编者

2020年11月

目录

PART I

CHEMICAL PESTICIDES

Unit 1 Chemistry of Pesticides

History

Although the use of chemicals to combat agricultural pests dates from antiquity, the large scale utilization of chemicals as major components of pest management systems is a twentieth century development. However, types of chemicals in use have changed substantially in response to environmental concerns that have arisen since their introduction. As late as 1950, many inorganic chemicals were still in use, including calcium arsenate, copper sulfate, lead arsenate, and sulfur, but, with the exception of sulfur, these materials were almost completely displaced by synthetic organic pesticides in subsequent years.

The 1940s and 1950s were productive years in terms of new synthetic organic chemistry. The chemical industry faced a major challenge in its efforts to synthesize and manufacture replacements for materials that were critically needed to protect crops from insect pests and protect personnel in tropical areas from malaria and other insect-borne diseases. The discovery of the insecticidal properties of hexachlorocyclohexane almost simultaneously in France and England in 1940 was one of the first successes. The discovery of the insecticidal activity of Lindane (the gamma isomer of hexachlorocyclohexane), followed by the widely acclaimed successes of DDT in controlling vector-borne diseases, reinforced efforts to discover and commercialize new synthetic insecticides. At the end of World War Ⅱ, newly developed chemical technologies became the basis for the manufacture of a number of new insecticides, particularly the application of the Diels-Alder reaction to synthesize cyclodiene insecticides. Because their acute mammalian toxicity was generally low and their spectrum of activity was very broad, such insecticides could be used to control many agricultural insect pests. The organochlorine insecticides were described as "nerve poisons".

Another structural lead emerged from studies of the pharmacological properties of the alkaloid physostigmine, the toxico-principle of the beans of the plant *Physostigma venenosum*, which is used as an ordeal poison in West Africa. Extensive research on the structure and activity of physostigmine showed that it was a potent inhibitor of cholinesterase, the enzyme responsible for degrading acetylcholine, a substance involved in neural transmission. Synthetic analogs used medicinally are ionizable compounds; nonionizable analogs were shown to have insecticidal activity. In 1947, several *N*-methylcarbamates that possessed significant acetylcholinesterase inhibitory activity were synthesized in Switzerland by the Syngenta Company and developed as insecticides. One of the best known of the carbamate insecticides is carbaryl (*N*-methylnaphthyl carbamate).

In Germany in 1937, in the course of investigations to find substitutes for nicotine, Schrader noted the insecticidal activity of organophosphorus compounds. Investigations were diverted from pesticide development to the development of potential agents for use in chemical warfare because some organophosphates were powerful inhibitors of cholinesterase and were very toxic to mammals. The first practical organophosphorus insecticide was Bladan, which contained tetraethyl pyrophophosphate. It was marketed in Germany in 1944. Subsequently, the discovery of parathion by Schrader in 1944 was followed by the synthesis of many related compounds. The high mammalian toxicity of many organophosphates calls for extreme caution in their practical application, but by varying substituent patterns, many less toxic analogs were manufactured and approved for use as insecticides, fungicides, and plant growth regulators, among other agricultural uses.

These new developments in the control of agricultural insect pests were paralleled in the search for chemical agents for control of weeds and plant diseases. Research into the nature of plant growth regulators led to the identification of indoleacetic acid as the first of the plant growth hormones. This compound and a variety of analogs were shown to elicit a variety of responses in plants, but it was not until the 1940s that these compounds were applied to weed control. Description of the growth regulating activity of 2,4-dichlorophenoxyacetic acid (2, 4-D) in 1942 was followed by field trials in which it was shown to kill weeds selectively. Subsequently 2, 4-D was developed for use as a major herbicide for control of weeds in corn and other cereals. It was widely used to control annual and perennial broadleaf weeds in tolerant crops and on noncrop areas. Although dinitrophenols had been used in the 1930s, the scale of herbicide use in agriculture expanded after the introduction of 2,4-D and this was followed by the introduction of atrazine, the first of the triazine herbicides, in 1958. Since then many, new herbicides that represent a wide variety of chemical classes have been commercialized to improve environmental safety, selectivity, and weed control at low rates of application.

The introduction of newer synthetic techniques, such as combinatorial chemistry, which can generate large numbers of new compounds, made it is possible to increase the throughput of compounds. Although there is a constant flow of new compounds through the developmental stages, industrial resources dedicated to the search for improved chemical controls are currently shifting to biotechnological approaches. One application of biotechnology is to increase herbicide tolerance in existing crops by genetic modification. Seeds of crop plants that are resistant to environmentally safe herbicides have been produced by genetic manipulation. Weeds then can be eliminated by conventional herbicides without damage to the growing crop.

Classification of Pesticides

1. Nomenclature

Active ingredients of pesticides represent a very diverse array of chemical structures including many biological agents. Many pesticide structures are very complex and cannot be categorized simply; therefore, classification systems in use have evolved to accommodate the increasing diversity of chemical and biological agents used in pest control or management.

One convenient basis for classification is the target organism. Thus pesticides may be classified, for example, as herbicides, insecticides, fungicides, nematicides and rodenticides. These classes then may be conveniently subdivided into smaller classes based on chemical structure, but a classification system based on mode of action may be equally valid. From the point of view of the toxicologist or the applicator, classification systems based on hazard are important. The terminology and classification systems adopted often may be quite arbitrary and occasionally are misleading.

From a legal standpoint, if it is claimed that a substance may be used to control, mitigate, or repel pests, it must be regarded as a pesticide. "Active ingredient" as defined in the data requirements for USEPA registration means " any substance (or group of structurally similar substances) that will prevent, destroy, repel or mitigate any pest, or that functions as a plant regulator, desiccant, or defoliant within the meaning of FIFRA". Thus, substances such as repellents and attractants are covered by pesticide legislation and are included in discussions of pesticides.

Assessments of toxicity must be based on material of clearly defined composition. Data obtained from any studies or tests conducted on pesticides should be preceded by a clear statement of the identity of the test material. Pesticides are sold to the user as formulated products, which contain, in addition to the active ingredient, a variety of additives and diluents. "Formulation" in the legal sense means: (1) the process of mixing, blending or dilution of one or more active ingredients, without an intended chemical reaction, to obtain a manufacturing use product or an end use product, or (2) the repackaging of any registered product.

The purpose of formulation is to improve the efficacy of the pesticide, facilitate its application, and ensure that it remains stable in storage. Thus, additives include inert ingredients, such as wetting agents, stickers, emulsifiers, antioxidants, and diluents. The label names the active ingredient and defines the amount contained in the formulation. Most pesticides have common names agreed upon by the International Organization for Standardization and it is usual to refer to the active ingredient by its common name. Occasionally, the common name may refer to a product of defined composition that is not homogeneous, but may contain isomers or closely related compounds. Chemical compounds of known structure may be described uniquely by systematic chemical names, which are given according to the rules of the International Union of Pure and Applied Chemistry (IUPAC) or the Chemical Abstracts nomenclature.

2. Structures and Stereochemistry

The conventional structural representation of a chemical compound uniquely defines its attributes. It represents the spatial arrangement of the component atoms and depicts bond types. To the chemist, it indicates potential reactivity and it can be used as a basis for estimation of physical properties and environmental behavior.

Details of spatial arrangements are an important factor in defining the biological activity of a compound. Carbon atoms in organic compounds are tetravalent. A molecule that contains a carbon atom with four different substituents lacks a center of symmetry and a molecule that contains one or more such asymmetric carbon atoms is termed a chiral molecule. If a molecule has one or more chiral centers, nonsuperimposable structures that are mirror images are termed enantiomers. Molecules that have non-superimposable structures that are not mirror images are called diastereoisomers. Diastereoisomers differ in their physical and chemical properties, whereas enantiomers have identical physical and chemical properties, and differ only in their ability to rotate the plane of polarized light to different extents.

Biological processes such as those that occur at the surface of an enzyme or protein are chiral. They involve interactions between molecules that have a three-dimensional arrangement that is defined by energy relationships. The chirality of a molecule will, therefore, determine its behavior in a biological environment. Processes at the molecular level may be triggered or inhibited by interactions between chiral biological receptors and biologically active molecules. Most natural organic compounds are chiral and occur in a single enantiomeric form.

Chirality is an increasingly important consideration in the design of a pesticide. For a pesticide to function effectively, it must reach the receptor site, where its effectiveness may depend on the geometry and dimensions of that portion of the molecule that interacts with the receptor. In nature, biological activity is generally a property that is restricted to a single enantiomer and only one isomer of a pair of optically active synthetic pesticides or pharmaceuticals will be effective. The use of screening systems, which are effective at the molecular level, is likely to reveal the superior activity of a particular enantiomer. Other isomers may react with a different receptor and cause undesirable side effects. If they are ineffective, they represent a wasteful production cost or possibly an undesirable load on the environment. In case of molecules that have more than one chiral center, the situation is more complex, and a molecule that contains n chiral centers may have 2^n stereoisomers.

Selected from: Krieger R I. Handbook of Pesticide Toxicology: Vol 1, Principles, 2nd. California: Academic Press, 2001: 95-98.

Words and Expressions

pesticide /'pestɪsaɪd/ *n*. 农药

pest /pest/ *n*. 害虫，有害之物，讨厌的人

utilization /ˌjuːtəlaɪ'zeɪʃn/ *n*. 利用，使用

calcium arsenate　砷酸钙
lead arsenate　砷酸铅
sulfur　/'sʌlfər/ *n*.［化］硫黄；*vt*.用硫黄处理
synthetic　/sɪn'θetɪk/ *adj*.合成的，人造的，综合的
organic　/ɔː'gænɪk/ *adj*.［有化］有机的，组织的，器官的，根本的
insecticide　/ɪn'sektɪsaɪd/ *n*.杀虫剂
pyrethrum　/paɪ'riːθrəm/ *n*.［植］除虫菊，［药］除虫菊杀虫剂
synthesize　/'sɪnθəsaɪz/ *v*.综合，合成
tropical　/'trɒpɪkl/ *adj*.热带的，热情的
malaria　/mə'leəriə/ *n*.疟疾，瘴气
hexachlorocyclohexane　/'heksəˌklɔːrə/ *n*.［化］六氯环己烷，六六六
gamma　/'gæmə/ *n*.伽玛（希腊字母），微克
isomer　/'aɪsəmər/ *n*.异构体
cyclodiene　/ˌsaikləu'daiiːn/ *n*.环戊二烯类杀虫剂
organochlorine　/ˌɔːgənəu'klɔːriːn/ *n*. & *adj*.有机氯（的），有机氯杀虫剂（的）
pharmacological　/ˌfɑːməkə'lɒdʒɪkl/ *adj*.药理学的
alkaloid　/'ælkəlɔɪd/ *n*.［化］生物碱，植物碱基
physostigmine　/ˌfaɪsəu'stɪgmiːn/ *n*.［药］毒扁豆碱（一种眼科缩瞳药）
Physostigma venenosum　［医］毒扁豆
ordeal　/ɔː'diːl/ *n*.严酷的考验，痛苦的经验，折磨
cholinesterase　/ˌkəulɪ'nestəreɪz/ *n*.［生化］胆碱酯酶
enzyme　/'enzaɪm/ *n*.［生化］酶
ionizable　/'aɪənɪzəbl/ *adj*.电离的（被离子化的）
acetylcholinesterase　/əˌsitlˌkolə'nɛstəˌres/ *n*.［生化］乙酰胆碱酯酶
carbamate　/'kɑːbəmeɪt/ *n*.［化］氨基甲酸盐
carbaryl　/'kɑːbəraɪl/ *n*.［化］甲萘威
nicotine　/'nɪkətiːn/ *n*.［有化］尼古丁，［有化］烟碱
organophosphorus　/ˌɔːgənəu'fɒsfərəs/ *adj*.［化］有机磷的
tetraethyl pyrophophosphate　焦磷酸四乙酯
parathion　/ˌpærə'θaɪən/ *n*.［化］对硫磷
mammalian　/mə'meɪliən/ *n*.哺乳动物；*adj*.哺乳动物的
toxicity　/tɒk'sɪsəti/ *n*.毒性
substituent　/səb'stɪtjuənt/ *n*.取代（基）；*adj*.取代的
fungicide　/'fʌŋgɪsaɪd/ *n*.杀真菌剂
parallel　/'pærəlel/ *adj*.平行的，并联的；*n*.平行线，相似物；*v*.平行
indoleacetic acid　［生化］吲哚乙酸，异植物生长素
perennial　/pə'reniəl/ *adj*.四季不断的，长期的，（植物）多年生的
field trial　田间试验
nomenclature　/nə'meŋklətʃər/ *n*.命名法，术语
herbicide　/'hɜːbɪsaɪd/ *n*.除草剂
tolerant　/'tɒlərənt/ *adj*.容忍的，宽恕的，有耐药力的

dinitrophenol /dai,naitrəu'fi:nɔl/ *n.* 二硝基酚

atrazine /'ætrəzi:n/ *n.* [化] 莠去津

triazine /'traɪəzi:n/ *n.* [有化] 三嗪

manipulation /mə,nɪpju'leɪʃn/ *n.* 处理，操纵，被操纵

eliminate /ɪ'lɪmɪneɪt/ *vt.* 排除，消除；*v.* 除去

categorize /'kætəgəraɪz/ *v.* 加以类别，分类

commercialize /kə'mɜ:ʃəlaɪz/ *v.* 使商业化，使商品化

nematicide /nɪ'mætɪsaɪd/ *n. adj.* 杀线虫剂（的）

rodenticide /rəu'dɛntɪ,saɪd/ *n.* 灭鼠剂

terminology /,tɜ:mɪ'nɒlədʒi/ *n.* 术语，术语学；用辞

mitigate /'mɪtɪgeɪt/ *vt.* 使缓和，使减轻

desiccant /'desɪkənt/ *n.* 干燥剂；*adj.* 使干燥的，去湿的

defoliant /di:'fəuliənt/ *n.* 脱叶剂，落叶剂

repellent /rɪ'pelənt/ *n. adj.* 排斥（的）

attractant /ə'træktənt/ *n.* 引诱剂，引诱物

antioxidant /,ænti'ɒksɪdənt/ *n.* [化] 抗氧化剂，硬化防止剂

diluent /'dɪljuənt/ *adj.* 冲淡的，稀释的；*n.* [医] 稀释液，稀释药

formulation /,fɔ:mju'leɪʃn/ *n.* 用公式表示，明确地表达，剂型

stereochemistry /,sterɪə'kemɪstrɪ/ *n.* [化] 立体化学

tetravalent /,tetrə'veɪlənt/ *adj.* [化] 四价的；*n.* 四价染色体

symmetry /'sɪmətri/ *n.* 对称，匀称

asymmetric /,eɪsɪ'metrɪk/ *adj.* 不均匀的，不对称的

chiral /'tʃɪrəl/ *adj.* [化] [物] 手（征）性的

non-superimposable *adj.* 非重叠的

enantiomer /ɪ'næntiəmə/ *n.* [化] 对映（结构）体

diastereoisomer /'daiə,steriəu'aisəmə/ *n.* [化] 非对映异构体

stereoisomer /,sterɪəu'aɪsəmə/ [化] 立体异构体

Notes

[1] insect-borne disease：虫源性疾病。由昆虫传播病原菌引起的疾病。

[2] lindane：林丹。有机氯类农药，六六六的 γ 异构体，目前我国已禁止使用。

[3] vector-borne disease：媒介传播疾病。由媒介生物直接或间接传播的疾病。

[4] Diels-Alder reaction：狄尔斯-阿尔德反应。指含有双键或叁键的不饱和化合物与链状或环状含共轭双键体系化合物发生 1,4-加成环化，生成六元碳环的氢化芳香族化合物的反应。

[5] acute mammalian toxicity：哺乳类动物的急性毒性。

[6] Syngenta（先正达）：2000 年 11 月 13 日，捷利康农化公司以及诺华的作物保护和种子业务分别从原公司中独立出来，合并组建全球最具实力的专注于农业科技的企业——先正达。2017 年 4 月 4 日，美国联邦贸易委员会发布声明批准中国化工收购先正达的交易。

[7] carbaryl (*N*-naphthyl methylcarbamate)：甲萘威。化学名称：甲氨基酸-1-萘酯。纯品为白色晶体。微溶于水，溶于大多数有机溶剂。

[8] USEPA：美国国家环境保护局。成立于1970年12月，致力于发展和加强环境保护方面的政策法规管理，署长由美国总统直接任命。

[9] FIFRA：《联邦杀虫剂、杀菌剂、灭鼠剂法案》。

[10] combinatorial chemistry：组合化学。组合化学是一门将化学合成、组合理论、计算机辅助设计及机械手结合于一体，在短时间内将不同构建模块用巧妙构思，根据组合原理，系统反复连接，从而产生大批的分子多样性群体，形成化合物库，然后运用组合原理，以巧妙的手段对库成分进行筛选优化，得到可能的有目标性能的化合物结构的科学。

Exercises

Ⅰ Answer the following questions according to the text.

1. How many chemical pesticides do you know? Please give a list as possible as you can. And classify all of them according to your understanding.

2. What can be regarded as a pesticide? How to understandc "Active ingredient" of a pesticide?

3. What does formulation mean? What is the purpose of formulation?

4. What is the meaning of a chiral molecule? Why is chirality an increasingly important consideration in the design of a pesticide?

Ⅱ Translate the following English phrases into Chinese.

organic pesticide	nerve poison	receptor site	screening system
insecticidal activity	practical application	physical property	copper sulfate
genetic modification	spatial arrangement	target organism	mirror image

Ⅲ Translate the following Chinese phrases into English.

除草剂	杀虫剂	杀菌剂	杀线虫剂	灭鼠剂	副作用
添加剂	稀释剂	抗氧化剂	组合化学	驱避剂	活性成分

Ⅳ Choose the best answer for each of the following questions according to the text.

1. Which were described as "nerve poisons"?

A. fungicides　　B. nematicides

C. organochlorine insecticides　　D. herbicides

2. Which of the following is not an inorganic chemical?

A. calcium arsenate　　B. copper sulfate

C. lead arsenate　　D. parathion

3. Which of the following pesticide is acclaimed success in controlling vector-borne diseases?

A. lindane　　B. DDT　　C. carbaryl　　D. paraoxon

4. Which of the following belong to pesticides?

A. herbicides and insecticides　　B. fungicides and nematicides

C. rodenticides　　D. all of the above

5. The purpose of formulation is to ________.

A. improve the efficacy of the pesticide

B. facilitate application of the pesticide

C. ensure that the pesticides remains stable in storage

D. all of the above

Ⅴ **Translate the following short passages into Chinese.**

1. As late as 1950, many inorganic chemicals were still in use, including calcium arsenate, copper sulfate, lead arsenate, and sulfur, but, with the exception of sulfur, these materials were almost completely displaced by synthetic organic pesticides in subsequent years.

2. " Active ingredient " means any substance (or group of structurally similar substances) that will prevent, destroy, repel or mitigate any pest, or that functions as a plant regulator, desiccant, or defoliant within the meaning of FIFRA.

3. The purpose of formulation is to improve the efficacy of the pesticide, facilitate its application, and ensure that it remains stable in storage. Thus, additives include inert ingredients, such as wetting agents, stickers, emulsifiers, antioxidants, and diluents.

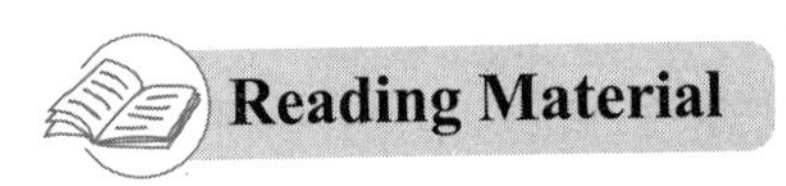

Pesticides, Common Names, Chemical Names and Trade Names

Pesticides are chemicals specifically developed and produced for use in the control of agricultural and public health pests, to increase production of food and fiber, and to facilitate modern agricultural methods. Antibiotics to control microorganisms are not included. They are usually classified according to the type of pest (fungicides, algicides, herbicides, insecticides, nematicides, and molluscicides) they are used to control. When the word *pesticide* is used without modification, it implies a material synthesized by humans. *Plant pesticide* is a substance produced naturally by plants that defend against insects and pathogenic microbes—and the genetic material required for production.

The term *biocide* is not used much in the scientific literature. It may be used for a substance that is toxic and kills several different life-forms. Mercury salts (Hg^{2+}) may be called biocides because they are toxic for microorganisms, animals, and many other organisms, whereas DDT is not a biocide because of its specificity toward organisms with a nervous system. The word is also sometimes used as a collective term for substances intentionally developed for use against harmful organisms. In a directive from the European Community, we find the following definition:

> The new Directive describes biocides as chemical preparations containing one or more active substances that are intended to control harmful organisms by either chemical or biological, but by implication, not physical means. The classification of biocides is broken down into four main groups—disinfectants and general biocides, preservatives, pest control and other biocides and these are further broken down into 23

separate categories.

Pesticides have one *standard name* and one *chemical name*. The different companies make products with registered *trade names*. They should be different from the standard names, but also have to be approved. The chemical industry also frequently uses a code number for its products. In Germany, for instance, old farmers still know parathion by the number *E*-605, which was used by Bayer Chemie before a standard name and a trade name were given to *O*,*O*′-diethyl paranitrophenyl phosphorothioate. The chemical name is often very complicated and even difficult to interpret for a chemist. The chemical formula, however, is often much simpler and may tell something about the property of the compound even to a person with moderate knowledge of chemistry.

One or more national standardization organizations and the International Organization of Standardization approve standard names. The chemical names are either according to the rules of the International Union of Pure and Applied Chemistry or according to Chemical Abstracts. The so-called Chemical Abstracts Services Registry Number (CAS-RN) is a number that makes it easy to find the product or chemical in databases from Chemical Abstracts. The standard names are regarded as ordinary nouns, but the pesticide products are sold under a trade name that is treated as a proper name with a capital initial letter. An insecticide was provided as an example:

Common Names: chlorantraniliprole

Chemical Names: 3-bromo-*N*-(4-chloro-2-methyl-6-(methylcarbamoyl) phenyl)-1-(3-chloropyridin-2-yl)-1*H*-pyrazole-5-carboxamide

Trade Names: Coragen; Rynaxpyr

Registry Number (CAS-RN): CAS No. 500008-45-7

Chemical Structure:

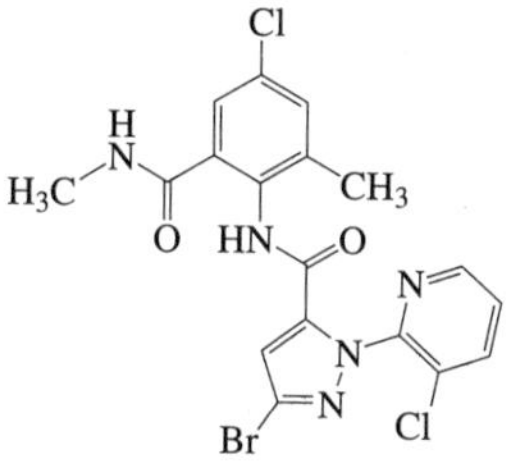

Selected from: Stenersen J. Chemical Pesticides: Mode of Action and Toxicology. Florida: CRC Press LLC, 2004: 10-11.

Words and Expressions

facilitate /fə'sɪlɪteɪt/ *vt*.(不以人作主语的)使容易,使便利,推动,促进

microorganism /ˌmaɪkrəʊ'ɔːgənɪzəm/ *n*.[微生]微生物,微小动植物

algicide /'ældʒɪsaɪd/ *n*.[化]灭藻剂,杀藻剂,除海藻的药

molluscicide /mə'lʌskɪsaɪd/ *n*.软体动物杀灭剂,灭螺剂(亦作 molluscacide)

pathogenic /ˌpæθə'dʒenɪk/ *adj*.致病的,病原的,发病的

disinfectant /ˌdɪsɪnˈfektənt/ *n*. 消毒剂；*adj*. 消毒的

preservative /prɪˈzɜːvətɪv/ *n*. 防腐剂，预防法，防护层；*adj*. 防腐的，有保护性的

O,*O*′-diethyl paranitrophenyl phosphorothioate *O*,*O*′-二乙基对硝基苯基硫逐磷酸酯

the International Union of Pure and Applied Chemistry 国际纯粹与应用化学联合会

Chemical Abstract （美国）化学文摘

chlorantraniliprole /klɔːrænt'rænɪlɪprəʊl/ 氯虫苯甲酰胺，氯虫酰胺

Unit 2 Chemical Insecticides

Insecticides predominated in the early years of synthetic organic pesticide, but total usage decreased as pest management systems developed alternative practices to control insects in which chemicals were used more effectively. Amounts applied per acre were also reduced by the introduction of chemicals that were more effective.

Organochlorines

Organochlorine insecticides were heavily used after their introduction, but the disadvantages of using compounds that do not readily degrade under environmental conditions soon became apparent. As mentioned earlier, implementation of national environmental policies, starting in 1970, became a major driving force in the choice of molecules that were appropriate for development as pesticides.

Organochlorine compounds affect neural transmission. Although carbamate and organophosphate insecticides were much more susceptible to environmental degradation than the majority of organochlorine pesticides, they were both inhibitors of acetylcholinesterase.

Organophosphates

The term organophosphates are used generically to include organic compounds that contain a phosphorus atom. These compounds are widely used as insecticides and large quantities are used to control agricultural pests and disease vectors.

Compounds are named as derivatives of the corresponding parent phosphorus compounds in which phosphorus may be trivalent or pentavalent. They owe their pesticidal activity to their ability to inhibit cholinesterase, which involves phosphorylation at the biochemical target site. This may involve a stage termed activation in which the insecticide is converted by an enzymic or chemical process from a rather weak inhibitor of cholinesterase, as determined *in vitro*, to a highly active compound. An example is the conversion of parathion to paraoxon. Parathion is a phosphorothionate that contains a P=S linkage and this is converted to a phosphate-containing or P=O compound.

Nonenzymic reactions of organophosphates are important. Although generally of higher mammalian toxicities than organochlorine insecticides, they do not present the same problems of environmental persistence and biological uptake. Organophosphates are reactive toward nucleophiles. They generally undergo rapid hydrolysis in the presence of a base and are also hydrolyzed by acids. Thus they may be readily detoxified. The process of conversion of phosphorothionates to phosphates also may take place chemically or photolytically and paraoxon has been detected as a residue in fog.

Isomerization of organophosphates is also important in relationship to toxicity and it is important that the isomer content of a product be established. The P =S linkage may isomerize to the P-S-form and the product may be substantially more toxic to mammals. The toxicity of malathion to humans is quitc low, but the presence of more than 2% of isomalathion, the *S*-methyl isomer, in a manufactured product used for malaria control led to an outbreak of poisoning among 7500 workers. In the case of parathion, the product is an *S*-ethyl derivative of paraoxon. Organophosphates are normally stable at ambient temperatures, but at elevated temperatures, good yields of isomers may be obtained.

Carbamates

Carbamates are esters of carbamic acids NHRCOOR'. The substituent R in the majority of insecticidal carbamates is a methyl group. One of the best known is carbaryl in which R' is a 1-naphthyl group. Like the organophosphates, carbamates are potent inhibitors of cholinesterase. Their mammalian toxicity is generally lower than that of the organophosphates.

However, aldicarb [2-methyl-2 (methylthio) propionaldehyde *O*-methylcarbamoyl oxime] is an exception, having a somewhat higher toxicity than most members of this class (LD_{50} = 0. 93 mg/kg, acute oral in rats). The carbamates are degraded in soils, but in sandy soils, aldicarb sulfoxide and aldicarb sulfone, the principal metabolites of aldicarb, leached sufficiently to enter around water.

Pyrethroids

Pyrethroid insecticides have their origins in the naturally occurring insecticides in pyrethrum extracts. The name pyrethrins are applied to the crude mixture of naturally occurring pyrethroids obtained from pyrethrum extracts. The compounds currently in use owe their origin to the ingenuity of synthesis chemists whose structural variations on the theme of naturally occurring pyrethrins led to extremely effective insecticides, which are widely used in crop protection.

Pyrethrum extracts are obtained from the flowers of chrysanthemum species. Pyrethrum has rapid knockdown action and causes paralysis in insects, but has low toxicity for mammals. Its use in agriculture is limited because it readily degrades in sunlight. A major objective of the search for synthetic replacements was to obtain agriculturally useful analogs that had greater photochemical stability and a longer field life.

Pyrethrum extract is a mixture of three closely related insecticidal esters of chrysanthemic acid and the three corresponding esters of pyrethric acid formed by the alcohols cinerolone or pyrethrolone. Structural elucidation work during the 1940s was followed by the synthesis of many analogs by modifying the acid moiety and the alcohol moiety. For example, a number of newer analogs contain 3-phenoxybenzyl alcohol or α-cyano-3-phenoxybenzyl alcohol as the alcohol component esterified with an acid component in which halogenated alkyl or alkenyl residues are linked to the cyclopropanecarboxylic acid moiety.

Modifications in structure may not only enhance activity, but may also facilitate production on a large scale. Some analogs bear only a vestigial resemblance to the parent compounds and recent research has delineated the critical features that are essential for analogs to manifest biological activity. The activity of pyrethroids is often enhanced by the addition of a synergist such as piperonyl butoxide, which acts by inhibiting mixed function oxidase activity and thus reduces the rate of detoxication of pyrethroids.

The goal of environmental stability without loss of insecticidal activity has been achieved in the synthetic pyrethroids, and those that have entered the market are superior in insecticidal activity to the natural products and are active at extremely low rates of application. Permethrin was introduced in 1973. Its success was accompanied by an increase of industrial interest in this class: currently, more than 30 pyrethroids are in the market or in development.

Neonicotinoids

Neonicotinoid insecticides have been developed since the late 1980s, principally as systemic insecticides with a similar mode of action to nicotine. They selectively bind and interact with the insect nicotinic acetylcholine receptor site and cause paralysis, which leads to death, often within a few hours, but are much less toxic to mammals. Rapidly, they became the main commercial insecticide where insect pests were resistant to the pyrethroids.

As with other main groups of insecticides, there are considerable variations between individual neonicotinoids. The group can be divided into four types based on their chemistry: chloropyridyl (e. g. imidacloprid), thiazoyl (e. g. thiamethoxam), furanyl (e. g. dinotefuran) and sulfoximine (e. g. sulfoxaflor). Thiacloprid is the first chloronicotinyl insecticide to show activity against weevils, leafminers and various species of beetles, as well as against sucking insects such as aphids, whiteflies and some jassids. It has also shown good control of pests within the upper plant foliage after soil application.

Neonicotinoids were developed due to their low mammalian toxicity and good systemic activity, which, theoretically, makes them far superior to many of the insecticides previously developed. However, the incident reported above demonstrates that the whole process of delivery of an insecticide in the field has to be designed so that adverse effects on non-target species are avoided. Recently, it has been pointed out that the sensitivity of honey bees to neonicotinoids varies by orders of magnitude, with bees being less sensitive to *N*-cyanoamidine compounds, such as thiacloprid.

Phenyl-pyrazole Insecticides

Fipronil, discovered in 1987, was developed originally by Rhône-Poulenc and initially promoted as an alternative to control the desert locust. Trials were carried out in west Africa and it was shown to be very effective against acridids. FAO discontinued recommending blanket treatments, even at a dose of 4 g (a. i.) /ha, as its use in Madagascar over an extensive area had a devastating impact on the harvester termites (*Coarctotermes clepsydra*) and other non-target organisms. However, its use continued in Australia where ULV spraying at very low dose rates was used to control the mobile bands of plague locust Chortoicetes terminifera hoppers. Barrier treatments applied swaths at 300 m intervals for an overall dose of 0. 33 g (a. i.) /ha. Studies indicated no impact on common wood-eating *Microcerotermes* spp..

Anthranilamides/Diamides

These insecticides are refinements of the botanical insecticide ryania and function by activating the ryanodine receptors in intracellular calcium channels in larvae and adult insects that ingest them. This causes a large release of calcium ions and, consequently, muscle contraction so that the insect dies.

Chlorantraniliprole, introduced in 2007, has low mammalian toxicity and is especially effective against lepidopteran pests, but also against beetles, including the Colorado beetle.

Cyantraniliprole is recommended in the UK, except when crops are flowering, as part of an IPM programme, e. g. on cabbage root fly on broccoli, brussels sprouts, cabbages and cauliflowers. It is highly toxic to bees.

Flubendiamide, originally reported in 1979, was registered in the USA for use on over 200 crops with some crops having as many as six applications per year, but in 2016 the EPA concluded that the continued use of flubendiamide would result in unreasonable adverse effects on the environment, particularly benthic invertebrates, which are an important part of the aquatic food chain, particularly for fish.

Insect Growth Regulators (IGRs)

There are hormonal IGRs that mimic or inhibit the juvenile hormone that affects the moulting of the insect as it progresses from a young larva towards pupation, for example causing premature moulting, or preventing a pupa becoming an adult. Other IGRs act by inhibiting another hormone, ecdysone, thus causing the larva to moult into a larger larva and not into a pupa. They have been effective against the nymphal stages of sucking pests such as whiteflies. Other compounds, the benzoylureas, impair the formation of new cuticles at moulting. IGRs are of low mammalian toxicity and are generally slow-acting, which makes some users apprehensive as to whether they will be able to control a pest.

Methoprene, developed from 1973 as a juvenile hormone insecticide, has been used for control of mosquito larvae. Pyriproxifen, a juvenile hormone analogue, is a pyridine insecticide, developed in 1989, that has been very successful for controlling mosquito larvae at very low rates of application. Studies have shown that adult mosquitoes inside houses can pick up this insecticide and redistribute it while ovipositing to control larvae.

Diflubenzuron, a chitin deposition inhibitor, which was developed in 1972, has been applied in many different crops and in forestry against mostly lepidopteran larvae. It has also been recommended to control desert locust hoppers and immature stages of mosquitoes. Although it does not kill adult insects, there is evidence that oviposition is reduced.

Other benzoylurea insecticides are chlorfluazuron, flufenoxuron (now banned within the EU due to potential of bioaccumulation in the food chain), hexaflumuron and triflumuron. Lufenuron was developed and is used mainly against fleas on pets. Teflubenzuron, developed in 1983, is an acyl-urea insecticide that has also been used to control lice on fish. Tebufenozide, developed in 1992, is unusual as it acts as an anti-juvenile hormone, causing larvae to try to moult into precocious adults. It has been used against forest pests.

Selected from: Matthews G A. A History of Pesticides. Boston: CABI Press, 2018: 86-93; Krieger R I. Handbook of Pesticide Toxicology: Vol 1, Principles, 2nd. California: Academic Press, 2001: 101-103.

Words and Expressions

predominate /prɪ'dɒmɪneɪt/ *vt.* 掌握，支配；*vi.* 统治，占优势
degrade /dɪ'greɪd/ *v.*（使）降级，（使）堕落，（使）退化
implementation /ˌɪmplɪmen'teɪʃn/ *n.* [计] 实现，履行，安装启用
organophosphate /ˌɔːgənəu'fɒsfeɪt/ *n.* & *adj.* [生化] 有机磷酸酯（的）
phosphorus /'fɒsfərəs/ *n.* 磷
derivative /dɪ'rɪvətɪv/ *adj.* 引出的，系出的；*n.* 派生的事物，派生词
trivalent /traɪ'veɪlənt/ *adj.* [化] 三价的
pentavalent /ˌpentə'veɪlənt/ *adj.* [化] 五价的，五种价的
phosphorylation /fɔˌsfɔri'leiʃen/ *n.* [化] 磷酸化作用
paraoxon /ˌpærə'ɔksɔn/ 对氧磷，磷酸二乙基对硝基苯基酯
linkage /'lɪŋkɪdʒ/ *n.* 连接，结合，联接，联动装置
nonenzymic /nɒ 'nenziːmɪk/ *adj.* [生化] 非酶的，不涉及酶作用的
nucleophile /'njuːklɪəfaɪl/ *n.* [化] 亲核试剂
hydrolysis /haɪ'drɒlɪsɪs/ *n.* 水解
detoxify /diː'tɒksɪfaɪ/ *vt.* 使解毒
residue /'rezɪdjuː/ *n.* 残余，渣滓，滤渣，残数，剩余物
malathion /ˌmælə'θaɪən/ 马拉硫磷
ester /'estər/ *n.* [化] 酯
carbamic /kɑr'bæmɪk/ *adj.* [化] 氨基甲酸的，由氨基甲酸获得的
naphthyl /'næfθɪl/ *n.* [化] 萘基
aldicarb /'ældɪˌkɑrb/ [化] 涕灭威，丁醛肟威
sulfoxide /sʌlf'ɒksaɪd/ *n.* [化] 亚砜
sulfone /'sʌlfəun/ *n.*（=sulphone）砜
leach /liːtʃ/ *v.* 滤去
pyrethroid /paɪ'riːθrɒɪd/ *n.* [化] 拟除虫菊酯，合成除虫菊酯
pyrethrin /paɪ'riːθrɪn/ *n.* [化] 除虫菊酯
chrysanthemum /krɪ'zænθəməm/ *n.* 菊花
chrysanthemic acid　菊酸
pyrethric acid　第二菊酸
cinerolone　瓜菊醇酮，瓜叶除虫菊醇酮
pyrethrolone /paɪriːθrəʊ'ləʊn/ [医] 除虫菊醇酮
elucidation /ɪˌluːsɪ'deɪʃn/ *n.* *vt.* 阐明，说明
esterify /e'sterɪfaɪ/ *v.*（使）酯化
halogenate /hə'lɑdʒɪneɪt/ *v.* [化] 卤化
alkyl /'ælkaɪl/ *n.* [化] 烷基，烃基；*adj.* 烷基的，烃基的

alkenyl /'ɔːlkənɪl/ *n.* [化] 烯基，链烯基
cyclopropanecarboxylic acid 环丙基甲酸
neonicotinoid 新烟碱，烟碱类农药
systemic insecticide 内吸性杀虫剂
nicotinic acetylcholine receptor 烟碱型乙酰胆碱受体
paralysis /pə'raləsɪs/ *n.* 麻痹，无力，停顿
chloropyridyl 氯吡啶基
imidacloprid /ɪmɪdæk'lɒprɪd/ *n.* 吡虫啉
thiazoyl 噻唑基
thiamethoxam /θaɪəmɪðɒk'sæm/ 噻虫嗪
furanyl 呋喃基
dinotefuran /dɪ'nɒtiːfrən/ 呋虫胺
sulfoximine /sʌlfɒk'sɪmaɪn/ 砜亚胺
sulfoxaflor 氟啶虫胺腈，砜虫啶
chloronicotinyl 氯化烟酰
weevil /'wiːvl/ *n.* [昆] 象鼻虫
leafminer /liːf'maɪnər/ 潜叶蝇，斑潜蝇
beetle /'biːtl/ *n.* 甲虫
sucking insect 吸啜式昆虫
aphid /'eɪfɪd/ *n.* [昆] 蚜虫
whitefly /'waɪtflaɪ/ *n.* [昆] 粉虱
jassid /'dʒæsɪd/ *n.* 浮尘子科动物，叶蝉
thiacloprid /θɪæk'ləʊprɪd/ 噻虫啉
phenyl-pyrazole 苯基吡唑
fipronil /fɪp'rɒnɪl/ 氟虫腈
acridid 蝗科，蝗虫
desert locust 荒地蚱蜢，沙漠蝗虫
plague locust 蝗灾
Chortoicetes terminifera 澳洲疫蝗
anthranilamide 邻氨基苯甲酰胺
diamide /'daiəmaid/ *n.* [有化] 二酰胺，肼
intracellular /ˌɪntrə'seljulə/ *adj.* 细胞内的
larvae /'lɑrvi/ *n.* 幼虫，幼体（larva 的复数）
lepidopteran /ˌlepɪ'dɒptərən/ *n.* 鳞翅目昆虫
chlorantraniliprole /klɔːrænt'rænɪlɪprəʊl/ 氯虫苯甲酰胺
cyantraniliprole 氰虫酰胺
broccoli /'brɒkəli/ *n.* 花椰菜，西兰花
Brussels sprouts 球芽甘蓝
cauliflower /'kɒliflaʊə(r)/ *n.* 花椰菜，菜花
flubendiamide /fluːbenda'ɪæmaɪd/ 氟虫双酰胺，氟虫酰胺
juvenile hormone [生理] 保幼激素

moulting /məultiŋ/ *n*. 脱毛，蜕皮

pupation /pjuːˈpeiʃn/ *n*. 蛹化

ecdysone /ˈekdɪsəʊn/ *n*. 蜕化素，蜕皮激素

pupa /ˈpjuːpə/ *n*. [昆] 蛹

nymphal /ˈnɪmfəl/ *adj*. 蛹的，幼虫的

benzoylurea 苯基脲

cuticle /ˈkjuːtɪkl/ *n*. 角质层，表皮，护膜

methoprene /ˈmiːθəpriːn/ 烯虫酯

pyriproxifen 吡丙醚

pyridine /ˈpɪrɪdiːn/ *n*. [有化] 吡啶

oviposit /ˌəjvɪˈpɒzɪt/ *v*. 产卵，排卵

diflubenzuron /daɪfluːˈbenzuərɒn/ *n*. 除虫脲，二氟脲

chlorfluazuron /klɔːfljuəzˈjuərɒn/ 氟啶脲

flufenoxuron /fluːfeˈnɒksjʊrɒn/ 氟虫脲

bioaccumulation /ˌbaɪoəˌkjʊmjəˈleʃən/ 生物累积，生物富集，生物积聚

hexaflumuron /heksɑːflˈjuːmuərɒn/ 氟铃脲

triflumuron /traɪflˈjuːmuərɒn/ 杀铃脲

lufenuron /luːˈfenerɒn/ 虱螨脲

teflubenzuron /tefluːbenˈzuerɒn/ 氟苯脲

acyl-urea 酰基脲

lice /laɪs/ *n*. 虱子 (louse 的复数)

tebufenozide /tɪbjuːfeˈnɒzaɪd/ 虫酰肼

Notes

[1] As mentioned earlier, implementation of national environmental policies, starting in 1970, became a major driving force in the choice of molecules that were appropriate for development as pesticides.

本句中，starting in 1970 是插入语，用来修饰 national environmental policies，that 引导定语从句修饰 molecules。全句可翻译为：正如前面所提到的，从 1970 年开始，国家就实施了环境保护政策。这进一步鼓励了科研人员更加注重选择适用于农药的分子（化合物）。

[2] The goal of environmental stability without loss of insecticidal activity has been achieved in the synthetic pyrethroids, and those that have entered the market are superior in insecticidal activity to the natural products and are active at extremely low rates of application.

本句中，定语从句 that have entered the market 修饰 those；be superior to 是一个表示比较关系的短语，意思是：优于……，高于……。本句翻译为：合成杀虫菊酯在不丧失其杀虫活性的前提下，还能达到维持环境稳定性的目标。目前市场上销售的合成杀虫菊酯，不但杀虫活性优于天然产品，而且在用量极低的情况下，仍然能够保持其应有的活性。

[3] Rhône-Poulenc：罗纳-普朗克化工集团。法国化工公司，总部在巴黎。该公司成立于 1895 年，原名罗纳化工公司，于 1928 年与普朗克兄弟公司合并。主营产品有基础化学

品、专用化学品、精细化学品，纺织品，农业化学品，医药用品，信息记录材料和薄膜5大类。1985年在国内外有84个分公司、子公司和合股公司，有150余座工厂，除法国外分布于英国、瑞士、巴西、西班牙、阿根廷等地。该公司是以石油化工为基础的联合企业，生产和经营的产品达3000多种。

[4] ULV：超低容量喷雾剂。是供超低容量喷雾（指喷到靶标作物上的药液，以极细的雾滴、极低的用量喷出）施用的一种专剂型。

[5] juvenile hormone：保幼激素，又称返细激素。昆虫在发育过程中，由咽侧体分泌的一种内激素。在幼虫期能抑制成虫特征的出现，使幼虫蜕皮后仍保持幼虫形态；在成虫期有控制性的发育、产生性引诱、促进卵子成熟等作用。

[6] pheromone：信息素。信息素由体内腺体制造，直接排出散发到体外，信息素依靠空气、水等传导媒介传给其他个体。信息素主要有性信息素、聚集信息素、告警信息素、示踪信息素、标记信息素等。

Exercises

Ⅰ Answer the following questions after reading the text.

1. How many types of compounds can be used as chemical pesticides according to the text? What are they? And make brief comparison with each other.

2. Why does aldicarb have a somewhat higher toxicity than most members of carbamates?

3. What are the advantages and disadvantages of pyrethrum? What was the major objective of the search for its synthetic replacements?

4. You know about tobacco containing nicotine, but do you know the origin of neonicotinoid insecticides?

5. What are advantages of JHs as pesticides?

Ⅱ Translate the following English phrases into Chinese.

functional group	3rd instar larvae	structural variation
knockdown action	biological activity	metabolic pathway
neural transmission	ambient temperature	mammalian toxicity
photochemical stability	*in vitro*	environmental degradation

Ⅲ Translate the following Chinese phrases into English.

解毒	乙醛	异构化	信息素	二氧化碳	保幼激素
水解	降解	磷酸化	亲核试剂	作用机理	有机磷酸酯

Ⅳ Choose the best answer for each of the following questions according to the text.

1. Which is one of best known carbamates?

A. methoprene B. carbayl C. hydroprene D. parathoxon

2. The mammalian toxicity of carbamates is generally ________ than that of the organophosphates.

A. higher B. lower C. same D. uncertain

3. The activity of pyrethroids is often enhanced by the addition of a (an)________.

A. stimulant B. synergist C. attractant D. repllent

4. Which of the following are not semiochemicals?

A. pheromones B. kairomones

C. synthetic attractants D. wet agents

5. One of the drawbacks of JHs for agricultural pest control lies in ________.

A. its very rapid toxic action B. its very slow toxic action

C. its persistence D. high residual activity

Ⅴ Translate the following short passages into English.

1. The term organophosphates are used generically to include organic compounds that contain a phosphorus atom. These compounds are widely used as insecticides and large quantities are used to control agricultural pests and disease vectors.

2. Pyrethrum extracts are obtained from the flowers of chrysanthemum species. Pyrethrum has rapid knockdown action and causes paralysis in insects, but has low toxicity for mammals.

3. Neonicotinoids were developed due to their low mammalian toxicity and good systemic activity, which, theoretically, makes them far superior to many of the insecticides previously developed.

4. Insect growth regulators (IGRs) are of low mammalian toxicity and are generally slow-acting, which makes some users apprehensive as to whether they will be able to control a pest.

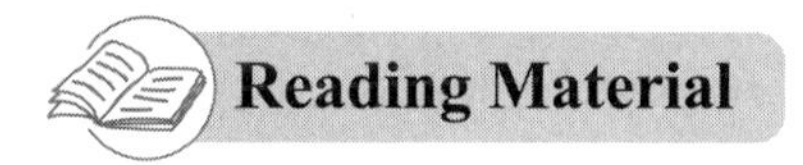

Chemical Herbicides

Herbicides can be classified in different ways. Selective herbicides were developed to kill weeds without damaging the crop, while non-selective herbicides kill or injure all plants within an area being treated.

Non-selective Herbicides

Apart from agriculture, non-selective herbicides are used in very diverse situations, including forestry and amenity areas, alongside roads and railway tracks and in urban and industrial areas on pavements.

1. Paraquat

Paraquat, a bipyridylium, is a non-selective foliar contact herbicide manufactured and marketed by ICI from 1962. It was promoted as an ideal herbicide to kill both grasses and broad-leaved weeds in no-till agriculture. Spray deposits were rainfast as soon as the spray had dried and rapidly stopped further growth of the weeds, while spray reaching the soil was not active against weeds.

2. Diquat

Developed in the late 1950s, diquat is a non-residual bipyridyl contact herbicide to control broad-leaved weeds and grasses. It has also been used as a pre-harvest desiccant, especially on potatoes after the use of sulfuric acid was banned. It was the only herbicide to be approved in New Zealand to control unwanted target weed species in freshwater as it is rapidly removed from the water and deactivated by adsorption on sediments and various compounds in the water.

3. Glyphosate

This is a phosphono amino acid that, when sprayed on plants, is translocated downwards as well as into the foliage. It is non-selective but is generally more effective against grasses compared with broad-leaved weeds. It is not effective in soil. Formulations usually contain a surfactant to increase spread over foliage and to speed up penetration into leaves, an important factor in areas where rainfall can occur within two hours of a spray application. Compared with paraquat, it is slower in action so it may take ten days before weeds are obviously affected.

4. Glufosinate

In the 1970s, a racemic mixture of phosphinothricin was synthesized and marketed as glufosinate, a glutamine synthetase inhibitor that controlled a range of weed species. Later, in 1995, canola was the first crop genetically modified to be tolerant of glufosinate, followed by maize, cotton and soybean. The tolerance to glufosinate was achieved by inserting bar or pat genes from *Streptomyces* into these crops, so that they can detoxify phosphinothricin and prevent it doing damage to the plants. Previously, the amino acid phosphinothricin had been isolated from species of *Streptomyces*. As pointed out earlier, an alternative herbicide was needed to control glyphosate-resistant weeds, hence the significance of having glufosinate-tolerant crops.

5. Imazapyr

Imazapyr is a non-selective herbicide, which can control a broad range of weeds including terrestrial annual and perennial grasses and broadleaved herbs, woody species and riparian and emergent aquatic species. It was introduced in 1985 but is no longer registered within the EU.

Selective Herbicides

1. Sulfonylureas

The first sulfonylurea herbicide was discovered in 1975 and introduced a unique mode of action, by inhibiting the synthesis of amino acids, such as valine, isoleucine and leucine, and aceto-lactase inhibitors (ALS). Young weeds starve as they cannot produce proteins needed for growth. Crops such as rice, wheat, barley, soybean and maize can metabolyse sulfonylureas safely, so this group of herbicides has been used extensively and is applied at very low doses of active ingredient. This required a change for farmers as they were used to applying herbicides in much larger doses and not in g/ha. They now form one of the largest groups of herbicides with many different actives registered across the world.

Chlorsulfuron was first registered in 1980 and used the following year on small-grain crops. Subsequently, other similar herbicides were developed including metsulfuron (1983), primisulfuron (1987), rimsulfuron (1989), nicosulfuron (1990), amidosulfuron (1992), sulfosulfuron (1995), iodosulfuron (1999) and mesosulfuron (2001). Some of these are used as mixtures, thus iodosulfuron and mesosulfuron are registered for use in the UK as a mixture on winter wheat, with an extension of use on some similar crops, such as rye.

The different sulfonylureas offer weed control as a pre- or post-emergence treatment, with choice of length of residual control. They are not volatile, so there is no off-target movement from spray deposits.

Another herbicide with a similar mode of action is propoxycarbazone-sodium, a triazolone introduced around 2000 as a post-emergent treatment to control blackgrass in winter wheat. In the UK, the maximum total dose is one full-dose treatment and not in a programme where other ALS inhibitors are used. Thiencarbazone-methyl is another new herbicide that is often used in mixtures with isoxaflutole for pre-emergence control of grassy and broad-leaved weeds in maize, soybeans, wheat, turf and ornamentals. A product with the mixture is marketed and claimed to be reactivated following rainfall to control late weeds.

Other ALS inhibitors are the imidazolinone herbicides. Imazamethabenz was recorded in 1982, followed by imazaquin (1983), imazethapyr (1984), imazamox (1995) and imazapic (1997). Some are used as selective herbicides, thus imazamethabenz provides good control of wild oats and volunteer canola in spring-sown barley, but it is not registered within the EU. Imazaquin is used to control weeds in grass and turf. Imazathapyr controls grasses in legume crops. Imazamox is used to control reeds and grasses in aquatic areas but is also used with metazachlor to control broad-leaved weeds and annual grasses in canola or with pendimethalin in legume crops, respectively. Imazapic is also selective and can be used in legume crops, but is not registered in the EU.

2. Propionates

Diclofop-methyl, introduced in 1975, is applied as a post-emergence herbicide with MCPA and mecoprop-*P* to control annual grasses including wild oats in wheat, and is also used with ferrous sulfate and MCPA in managed amenity turf production. Fluazifop-*P*-butyl was first marketed in 1981 and applied as a post-emergence herbicide to control annual grasses including wild oats in wheat and broad-leaved crops and ornamentals.

Quizalofop-*P*-ethyl, introduced in 1989, is another post-emergence herbicide used for grass weed control alongside quizalofop-*P*-tefuryl as an alternative to others with the same mode of action. Fenoxaprop-*P*-ethyl, introduced in 1989, is a post-emergence herbicide to control annual and perennial grasses and has been used in rice crops. In 1990, clodinafop-propargyl was first marketed as an alternative propionate post-emergent herbicide to control annual grasses and can be applied mixed with cloquintocet-mexyl, which is a herbicide safener that accelerates the herbicide detoxification in the treated crop. Pinoxaden, another post-emergence herbicide, was introduced in 2006.

3. N-phenyl phthalimides

Flumioxazin is for pre-emergence broad-spectrum control of weeds near or in water to control algae and pond weeds. It was introduced in 1994, and in the UK it is registered for one early post-emergence treatment per crop in winter wheat and oats, provided plants have been hardened by cool weather and are not lush with soft growth. Flumiclorac-pentyl, released in 1995, is a dicarboximide, which has been used to control certain problematic weeds.

A number of other herbicides with different chemical structures were introduced in the second half of the 20th century. Fluometuron (a phenylurea) has a similar mode of action as mesotrione and was developed in 1964. It was used as a soil-applied selective herbicide to control annual grasses and broad-leaved weeds in cotton and sugar cane. Oxadiazon, an oxadiazole, is a pre- and early post-emergent herbicide developed in 1969 and is used to control bindweed and many annual broad-leaved weeds. It acts by inhibiting protoporphyrinogen oxidase (PPO).

Butafenacil, a pyrimidindione, was introduced in 2000 to control annual and perennial broad-leaved weeds in fruit and other crops. It is not registered in the EU. Also propoxycarbazone-sodium, a triazolone, was introduced in 2000 as a new residual grass weed herbicide aimed at controlling black grass. A triketone, mesotrione, introduced in 2001, is a pre- and post-emergent herbicide used mostly in maize crops to control some grass and broad-leaved weeds. It acts as a 4-hydroxyphenylpyruvate dioxygenase inhibitor. In the UK, it is also registered for use in a mixture with nicosulfuron. Topramezone, a benzoyl pyrazole, introduced in 2006, and tembotrione, introduced in 2007, have the same mode of action and provide similar control to mesotrione, but are not registered in the UK. A benzoylpyrazole released in 2006, is a post-emergence herbicide for broad-leaved weeds and grasses used mainly on maize.

According to Grossman and Ehrhardt (2007), the tolerance of maize to topramezone is due to more rapid metabolism combined with a lower sensitivity of the 4-HPPD target enzyme. Since 2006, pinoxaden is a new phenylpyrazoline post-emergence herbicide to control grass weeds in cereals and is used in turf management. It is also marketed in mixtures with clodinafop-propagyl and florasulam.

Selected from: Matthews G A. A History of Pesticides. Boston: CABI Press, 2018: 99-112.

Words and Expressions

selective herbicide 选择性除草剂

non-selective herbicide 非选择性除草剂

amenity /ə'mi:nəti/ *n.* 舒适，礼仪，愉快，便利设施

paraquat /'pærəkwɒt/ *n.* [农药] 百草枯

bipyridylium *n.* 联吡啶

contact herbicide 触杀型除草剂

broad-leaved weed 阔叶杂草

no-till 免耕

diquat /'daɪkwɒt/ *n.* 敌草快（一种接触性除草剂）

bipyridyl /bɪpɪ'raɪdɪl/ 联吡啶基

pre-harvest /pri:'hɑ:vist/ *adj.* 收获期前的，收获前的

glyphosate /ˈɡlaɪfəseɪt/ *n*.［农药］草甘膦

surfactant /sɜːˈfæktənt/ *n*. 表面活性剂；*adj*. 表面活性剂的

glufosinate /ɡlʌfəʊzɪˈneɪt/ 草铵膦

racemic mixture ［化学］外消旋混合物

phosphinothricin 草丁膦

imazapyr /ɪˈmæzəpiːr/ 灭草烟（除草剂）

sulfonylurea /ˌsʌlfənɪlˈjʊərɪə/ *n*. 磺酰脲

valine /ˈveɪliːn/ *n*.［生化］缬氨酸

isoleucine /ˌaɪsəˈluːsiːn/ *n*.［生化］异亮氨酸

leucine /ˈluːsiːn/ *n*.［生化］亮氨酸；白氨酸

aceto-lactase inhibitor 乙酰乳酸合成酶抑制剂

chlorsulfuron /klɔːsʌlˈfuərɒn/ 氯磺隆

metsulfuron /metˈsʌlfuərɒn/ 甲磺隆

primisulfuron /prɪmaɪsʌlˈfuərɒn/ 氟嘧磺隆

rimsulfuron /rɪmˈsʌlfuərɒn/ 砜嘧磺隆

nicosulfuron /nɪkəʊˈsʌlfuərɒn/ 烟嘧磺隆

amidosulfuron 酰嘧磺隆

sulfosulfuron 磺酰磺隆

iodosulfuron 碘甲磺隆，甲基碘磺隆

mesosulfuron 甲基二磺隆

post-emergence *adj*. 萌发后的，出苗后的

propoxycarbazone-sodium 丙苯磺隆钠盐

thiencarbazone-methyl 甲基噻酮磺隆

isoxaflutole 异噁唑草酮

imidazolinone 咪唑啉酮

imazamethabenz 咪草酸，咪草酯

imazaquin 咪唑喹啉酸

imazethapyr 咪唑乙烟酸

imazamox 甲氧咪草烟

imazapic 甲基咪草烟

pendimethalin /pendɪmeˈθeɪlɪn/ 二甲戊乐灵

legume /ˈleɡjuːm/ *n*. 豆类，豆科植物，豆荚

propionate /ˈprəʊpɪəneɪt/ *n*.［有化］丙酸盐；丙酸酯

diclofop-methyl ［农药］禾草灵

fluazifop-*P*-butyl 精吡氟禾草灵

quizalofop-*P*-ethyl 精喹禾灵

quizalofop-*P*-tefuryl 喹禾糠酯

fenoxaprop-*P*-ethyl 精噁唑禾草灵

cloquintocet-mexyl 解毒喹；解草酯

detoxification /diːˌtɒksɪfɪˈkeɪʃn/ *n*. 解毒，［生化］解毒作用

pinoxaden 唑啉草酯

flumioxazin /fluːmaˈɪɒksɑːzɪn/ 丙炔氟草胺
pre-emergence 萌发前
flumiclorac-pentyl 氟烯草酸
dicarboximide /dɪkɑːbɒkˈsɪmaɪd/ 二甲酰亚胺
fluometuron /fluːəˈmetjuərɒn/ 氟草隆
mesotrione /meˈsətriən/ 硝磺草酮
oxadiazon /ˌɔksəˈdaiəzɔn/ 噁草酮
oxadiazole /ɒkseɪˈdɪætsəʊl/ 噁二唑
bindweed /ˈbaɪndwiːd/ *n*. 旋花类的植物
protoporphyrinogen oxidase 原卟啉原氧化酶
butafenacil 氟丙嘧草酯
pyrimidindione 嘧啶二酮
triketone /traɪˈkiːtəʊn/ 三酮
4-hydroxyphenylpyruvate dioxygenase 羟苯丙酮酸二加氧酶
topramezone 苯吡唑草酮，苯唑草酮
benzoyl pyrazole 苯甲酰吡唑
tembotrione 环磺酮

PART Ⅱ

BIOLOGICAL PESTICIDES

Unit 3　Botanical Pesticides

At least 2000 plants are known to have pesticidal activity. Neem is not the first botanical pesticide. Pyrethrins, which are naturally derived from daisy-like flowers of certain species of Chrysanthemum, have been used for centuries. Almost 2000 years ago the Chinese knew that Chrysanthemum plants had insecticidal value; some 2400 years ago the Persians used them. Not until recent centuries, however, were the potentials of the pyrethrums, extracted from the flowers, fully appreciated. Supposedly, an American trader, who had learned the secret while traveling in the Caucasus, introduced the insecticide into Europe early in the nineteenth century. Last century, Yugoslavia became the center of the world's pyrethrum industry, but after World War I, Japan became the main producer. With supplies cut off during World War II, the Allies began producing the flowers in Kenya. Since the 1960s, pyrethrum production has been established in the New Guinea highlands as well.

Like neem products, pyrethrins are valued for their low toxicity to mammals and birds. However, the ingredients in these insecticidal chrysanthemums are lethal to insects in a different way from those in neem. They are nerve poisons and contact insecticides. Pyrethrum has quick knockdown properties and is the active ingredient in millions of aerosol spray cans people use against flies and mosquitoes.

Despite the development of many synthetic insecticides, this chemical from chrysanthemums has maintained its position as a major commercial product. Its production is more than 10,000 ton world over. Although powerful synthetic analogues have been developed, demand for the natural material has remained high, and for the past several years it has been in short supply. The increasing concern of man for the environment seems likely to result in a rising demand for pesticides from plants rather than from petroleum. Such "soft" pesticides represent the hope that agricultural pests can be controlled while maintaining environmental stability. Now neem, another botanical pesticide, can perhaps step up to take an equally important, but complementary, role in the rising soft market.

Pyrethrins

Pyrethrin is an extract of the dried flowers of the pyrethrum, *C. cinerariaefolium*, commercially grown in Kenya. The active ingredients contained in the plant are various compounds known as pyrethrins. The word "pyrethrum" is the name for the crude flower dust itself, and the term "pyrethrins" refers to the six related insecticidal compounds that occur naturally in the crude material. Pyrethrins have a rapid "knockdown" effect on many insects and are irritating, which has caused them to be used for such purposes as wasp and hornet sprays, household aerosols, or for flushing cockroaches.

Most insects are highly susceptible to low concentrations of pyrethrins. The toxins cause immediate knockdown or paralysis on contact, but insects often metabolize them and recover. Pyrethrins break down quickly and have a short residual and low mammalian toxicity, making them among the safest insecticides in use. Pyrethrins may be used against a broad range of pests including ants, aphids, roaches, fleas, flies, and ticks. They are available in dusts, sprays, and aerosol "bombs". Pyrethrins have very low toxicity to mammals and are rapidly broken down when exposed to light. As a result, certain pyrethrins are the only insecticides registered for use in food handling areas. Pyrethrins are widely labeled for use on most food crops as well.

Labels for pyrethrins list many insects. However, with respect to their use on shade trees and shrubs, they are probably most useful for control of exposed caterpillars, sawfly larvae, leaf beetles, and leafhoppers. Their short persistence can limit effectiveness, yet it also helps minimize impacts on natural enemies.

Neem

Neem insecticides are extracted from the seeds of the neem tree, *A. indica*, that grows in arid tropical and sub-tropical regions on several continents. This plant has long been used in Africa and Southern Asia as a source of pharmaceuticals, such as wound dressings and toothpaste. The active ingredient is used both as feeding deterrent and a growth regulator. The treated insect usually cannot molt to its next life stage and dies. It acts as a repellent when applied to a plant and does not produce a quick knockdown and kill. It has low mammalian toxicity and does not cause skin irritation in most formulations. The neem tree supplies at least two compounds with insecticidal activity (azadirachtin and salanin) and other unknown compounds with fungicidal activity. More recently its ability to control insects has been developed.

Neem seed extracts contain oils and a variety of compounds that can affect insect development. Most important is azadirachtin, which has various effects from inhibition of feeding, interference with molting or egg production, and disruption of hormones important in growth. Treated insects rarely show immediate symptoms, and death may be delayed a week or longer, usually occurring during a molt. Affected insects are often sluggish and feed little. The low toxicity and broad labeling of neem insecticides recommend its use. Furthermore, effects on beneficial species are minimal. Slow action and a limited range of susceptible insect species are the primary limitations of neem insecticides.

Rotenone

Rotenone occurs in the roots of two tropical legume *Lonchocarpus* species in South America, *Derris* species in Asia, and several other related tropical legumes. Insects quickly stop feeding, and death occurs several hours to a few days after exposure. Rotenone degrades rapidly when exposed to air and sunlight. It is not phytotoxic, but it is extremely toxic to fish and moderately toxic to mammals. It may be mixed with pyrethrins or piperonyl butoxide to improve its effectiveness. Rotenone is a broad-spectrum contact and stomach poison that is effective against leaf-feeding insects, such as aphids, certain beetles (asparagus beetle, bean leaf beetle, Colorado potato beetle, cucumber beetle, strawberry leaf beetle, and others), and caterpillars, as well as fleas and lice on animals. It is commonly sold as a 1% dust or a 5% powder for spraying.

Sabadilla

Sabadilla is derived from the ripe seeds of *Schoenocaulon officinale*, a tropical lily plant that grows in Central and South America. The alkaloids in Sabadilla affect insect nerve cells, causing loss of nerve function, paralysis, and death. The dust formulation of Sabadilla is the least toxic of all registered botanical insecticides. However, pure extracts are very toxic if swallowed or absorbed through the skin and mucous membranes. It breaks down rapidly in sunlight and air, leaving no harmful residues.

Sabadilla is a broad-spectrum contact poison but has some activity as a stomach poison. It is commonly used in organic fruit and vegetable production against squash bugs, harlequin bugs, thrips, caterpillars, leafhoppers, and stink bugs. It is highly toxic to honey bees, however, and should only be used in the evening, after bees have returned to their hives. Formulations include baits, dusts, or sprays.

Ryania

Ryania is extracted from stems of a woody South American plant, *R. speciosa*, and causes insects to stop feeding soon after ingestion. It works well in hot weather. Ryania is moderate in acute or chronic oral toxicity in mammals. It is generally not harmful to most natural enemies but may be toxic to certain predatory mites. Ryania has longer residual activity than most other botanicals. It is used commercially in fruit and vegetable production against caterpillars (European corn borer, corn earworm, and others) and thrips. Its effectiveness may be enhanced if mixed with rotenone and pyrethrin.

Nicotine

It is a simple alkaloid derived from tobacco, *N. tabacum*, and other Nicotiana species. It is a fastacting nerve toxin and is highly toxic to mammals. Insecticidal formulations generally contain nicotine in the form of 40% nicotine sulfate, which is diluted in water and applied as a spray. Dusts can irritate skin and are normally not available for garden use. Nicotine is used primarily for insects with piercing-sucking mouth parts such as aphids, whiteflies, leafhoppers, thrips, and mites. Nicotine is more effective when applied during warm weather. It was registered for use on a wide range of vegetables and fruit crops but is no longer registered commercially.

Although many pure chemical compounds were isolated from these plants, entomological and field studies are lacking. The list of such plants include *A. squamosa* (seeds, leaves), *Tephrosia purpurea* (pods, roots), *T. villosa* (pods, roots), *Pongamia pinnata* (leaves and seeds), *Lantana camara* (leaves, stems, and flowers), *Ocimum sanctum* (holy basil, leaves, whole plant), *Vitex negundo* (leaves), *Zingiber officinale* (rhizome), *Curcuma longa* (rhizome), *Allium sativum* (garlic bulbs), and *A. cepa* (onion bulbs). In addition, some botanical insecticides that had enjoyed use in North America and Western Europe have lost their regulatory status as approved products. These include nicotine (from *N. tabacum*), quassin (from *Q. amara* and *P. excelsa*), and ryania (from *R. speciosa*). As a consequence, the only botanicals in wide use in North America and Europe are pyrethrum (from *C. cinerariaefolium*) and rotenone (from *Derris* spp. and *Lonchocarpus* spp.), although neem (*A. indica A. Juss*) is approved for use in the United States and regulatory approval is pending in Canada and Germany.

Botanical pesticides do not produce knockdown effect on insects. The insect population get reduced after spraying. Promising results will be obtained if botanical pesticides are sprayed as prophylactic agents.

Selected from: Nollet L M L, Rathore H S. Handbook of pesticides: methods of pesticide residues analysis. US: Taylor & Francis Group, CRC Press, 2010: 68-72.

Words and Expressions

neem /niːm/ *n*. 印度楝；苦楝树；尼姆树
pyrethrin /paiˈriːθrɪn/ *n*. [农药] 除虫菊酯
chrysanthemum /krɪˈzænθəməm/ *n*. 菊花
C. cinerariaefolium 菊花，野菊花
aphid /ˈeɪfɪd/ *n*. [昆] 蚜虫
flea /fliː/ *n*. 跳蚤，低廉的旅馆，生蚤的动物
tick /tɪk/ *n*. 扁虱，蜱亚目，虱蝇
caterpillar /ˈkætəpɪlə(r)/ *n*. [无脊椎] 毛虫，履带车；*adj*. 有履带装置的
sawfly /ˈsɔːflaɪ/ *n*. 叶蜂，锯蝇
leafhopper /ˈlifˌhɑpə/ *n*. 叶蝉
A. indica 印楝
azadirachtin 印楝素
salanin 茄碱苷
rotenone /ˈrəutənəun/ *n*. [农药] 鱼藤酮
legume /ˈlegjuːm/ *n*. 豆类，豆科植物，豆荚
Lonchocarpus 鱼藤属，尖荚豆属

phytotoxic /ˌfaɪte(ʊ)'tɒksɪk/ *adj*. 植物性毒素的

piperonyl butoxide /ˌpipərənilbju'tɔksaid/ 增效醚，胡椒基丁醚

sabadilla /ˌsæbə'dijə/ *n*. 沙巴藜芦，藜芦碱

Schoenocaulon officinale 沙巴草，沙巴藜芦

alkaloid /'ælkəlɔɪd/ *n*.［有化］生物碱，植物碱基

mucous membrane 黏膜

squash bug 南瓜虫，南瓜椿象

harlequin bug 卷心菜斑色蝽

thrip *n*. 蓟马

stink bug （虫）椿象

R. speciosa 大枫子科灌木尼亚那

N. tabacum 普通烟草，红花烟草，黄花烟草

A. squamosa 鳞叶蜂

Tephrosia purpurea 灰毛豆

T. villosa 绒毛山毛茛

Pongamia pinnata 水黄皮，野豆

Lantana camara 马缨丹

Ocimum sanctum 圣罗勒

Vitex negundo 黄荆

Zingiber officinale 生姜

Curcuma longa 姜黄

Allium sativum 大蒜

A. cepa 顶球洋葱

Q. amara 苏林南苦木

P. excelsa 牙买加苦树

Notes

[1] *Azadirachta indica* (*A. indica*)：印楝，是世界上公认的理想杀虫植物。印楝素是从印楝中提取的一类高度氧化的柠檬素，其活性化合物有印楝素 A 至印楝素 G 7 种化合物，其中印楝素 A 是主要杀虫成分。印楝素及其制剂对昆虫具有拒食、忌避、生长调节、绝育等多种作用。

[2] rotenone：鱼藤酮，从植物的根皮部提取的生物碱，用作杀虫剂，对昆虫尤其是菜粉蝶幼虫、小菜蛾和蚜虫具有强烈的触杀和胃毒作用，也可防治人畜体外寄生虫。作用机制主要是影响昆虫的呼吸作用，使害虫细胞的电子传递链受到抑制，降低生物体内的 ATP 水平，致使害虫得不到能量供应，行动迟滞、麻痹而缓慢死亡。

Exercises

Ⅰ Anawer the following questions according to the text.

1. In the text, how many botanical pesticides are introduced respectively? Please list

3～5 botanical pesticides.

2. What are the advantages of botanical pesticides?

3. Compared with chemical pesticides, what are the disadvantages of botanical pesticides?

4. Are botanical pesticides really safe?

Ⅱ Translate the following English phrases into Chinese.

botanical pesticde	contact insecticide	natural material	nerve poison
botanical insecticide	synthetic analogue	natural enemy	soft market
insect development	prophylactic agent	stomach poison	food crops

Ⅲ Translate the following Chinese phrases into English.

生物农药	气雾喷雾器	虫口密度	原材料	哺乳动物毒性
快速击倒	合成杀虫剂	中等毒性	剧毒的	触杀毒剂
经口毒性	环境稳定性	皮肤刺激	低毒	广谱活性

Ⅳ Choose the best answer for each of the following questions according to the text.

1. Which of the following is not a botanical pesticide?

A. neem B. pyrethrin C. rotenone D. Bt

2. Which is highly toxic to honey bees?

A. neem B. sabadilla C. rotenone D. ryania

3. Which has longer residual activity than most other botanicals?

A. neem B. sabadilla C. rotenone D. ryania

4. Which of the following botanical pesticides belong to light-stable?

A. neem B. sabadilla C. rotenone D. Pyrethrin

5. Which is highly toxic to fishes?

A. neem B. sabadilla C. rotenone D. ryania

6. Disadvantages of botanical pesticides are ________.

A. light-unstable B. slow action

C. raw material limitation D. all of the above

Ⅴ Translate the following sentences into Chinese.

1. Most insects are highly susceptible to low concentrations of pyrethrins. The toxins cause immediate knockdown or paralysis on contact, but insects often metabolize them and recover.

2. Neem insecticides are extracted from the seeds of the neem tree, *A. indica*, that grows in arid tropical and sub-tropical regions on several continents. This plant has long been used in Africa and Southern Asia as a source of pharmaceuticals, such as wound dressings and toothpaste.

3. Rotenone degrades rapidly when exposed to air and sunlight. It is not phytotoxic, but it is extremely toxic to fish and moderately toxic to mammals. It may be mixed with pyrethrins or piperonyl butoxide to improve its effectiveness.

4. Botanical pesticides do not produce knockdown effect on insects. The insect population get reduced after spraying. Promising results will be obtained if botanical pesticides are sprayed as prophylactic agents.

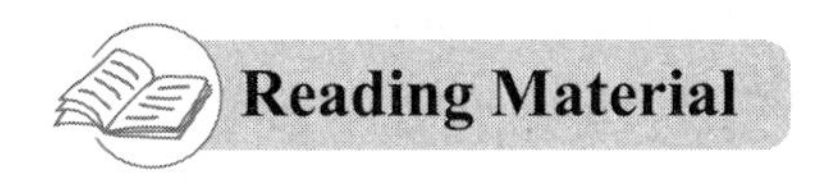

Essential Oils

To defend themselves against herbivores and pathogens, plants naturally release a variety of volatiles including various alcohols, terpenes, and aromatic compounds. These volatiles can deter insects or other herbivores from feeding, can have direct toxic effects, or they may be involved in recruiting predators and parasitoids in response to feeding damage. They may also be used by the plants to attract pollinators, protect plants from disease, or they may be involved in interplant communication. Based on these natural properties, essential oils containing these compounds have been recently touted as potential alternatives to current commercially available insecticides.

Essential oils demonstrate a wide range of bioactivities from direct toxicity to insects, microorganisms and plants, to oviposition and feeding deterrence as well as repellence and attraction. How these effects are mediated is still being elucidated; however, there is a growing set of results which point to membrane disruption (plants, microbes, and possibly insects) and effects on the nervous system of insects. And while individual constituents can mediate some of the effects, it is evident that complete essential oils are more effective than individual constituents or even a combination of constituents.

Insecticidal/Deterrent Effects

Numerous studies have assessed the ability of plant essential oils and their constituents to protect plants or their crops from insect pests with much of this research focusing on controlling stored product pests using essential oils as fumigants and repellents. Essential oils from a wide range of plants have been tested against the rice weevil, the maize weevil, the red-flour beetle, the bean weevil, and other stored product pests. In particular, nutmeg oil has been determined to significantly impact both *S. zeamais* and *T. castaneum* and demonstrates both repellent and fumigant properties (concentration dependent). oil, eucalyptus oil, oils from various mint species, lavender oil, and rosemary oil, although not all essential oils are active against all the insect pests. There is also growing research documenting the effects of these oils as contact toxicants, antifeedant compounds and repellents against a range of other insects.

In research conducted with the two-spotted spider mite *Tetranychus urticae*, rosemary oil has been demonstrated to have contact toxicity while the oils of caraway seed, citronella java, lemon eucalyptus, pennyroyal, and peppermint all exhibit fumigant activity. Perhaps more important, however, is that commercial formulations tested against both *T. urticae* and the predatory mite *Phytoseiulus persimilis* demonstrated high levels of toxicity to *T. urticae* but not *P. persimilis*, suggesting that these commercial formulations may work well in conjunction with an integrated pest management program.

Additional research from our laboratory shows that several essential oils possess antifeedant and oviposition deterrent effects. In particular, thymol, the major constituent of thyme oil, has been shown to be a deterrent to the lepidopteran pest species *Plutella xylostella* and *Pseudaletia unipuncta*. It should be noted that larvae experienced with thymol showed reduced deterrence, suggesting that repeated application of feeding deterrent chemicals in the field could limit their effectiveness. Also, while some essential oils may possess deterrent properties, this deterrence is reduced over time, possibly due to habituation, the volatilization of the essential oils, or a combination of both.

While the mode of action of essential oils is still relatively unknown, new research is providing insights. As previously mentioned, essential oils are likely neurotoxic to insects and mites, and research using individual constituents seems to suggest this. Thus far, evidence has been provided suggesting that some constituents such as eugenol or thymol may work by blocking octopamine (a neurotransmitter in arthropods) receptors or by potentially working through the tyramine receptor cascades. Physical effects such as membrane disruption or blockage of the tracheal systems may also be involved; however, conclusive evidence is still lacking.

Herbicidal Activities

Some essential oils are not only insecticidal but also possess strong phytotoxic effects. In many cases, this would be considered a serious drawback to the use of these essential oils for insect pest control; however, this also opens the door to the use of some oils as herbicides. Although few studies have addressed this herbicidal activity, work completed by Tworkoski demonstrated that the essential oils of red thyme (*Thymus vulgaris*), summer savory (*Satureja hortensis*), cinnamon (*Cinnamomum zeylanicum*) and clove (*Syzyium aromaticum*) are highly phytotoxic. Further analysis of the major constituents of cinnamon oil found that the herbicidal activity was due to eugenol, which is also the major constituent in clove oil.

The herbicidal activity of clove oil and eugenol was further studied in our laboratory using broccoli, common lambsquarters, and redroot pigweed seedlings in an attempt to determine the role of leaf epicuticular wax in susceptibility of these plants to damage. Seedling growth was significantly inhibited by both clove oil and, to a lesser extent, eugenol, while those plants with more epicuticular wax showed reduced electrolyte leakage.

Besides the direct use as herbicides, one specific use of phytotoxic essential oils could be for chemical thinning of fruit trees. In initial trials with a commercially available clove oil-based herbicide (Matran), apple blossom thinning effects were observed; however, extensive leaf and fruit russeting was also observed, with effects dependent on concentration and the apple cultivar tested. While the initial results are promising, further studies will be required before such herbicides can be used as chemical thinning agents.

Antimicrobial Activities

The use of plant-derived compounds to treat infectious diseases or to protect crops dates back several centuries and essential oils are no exception. The essential oils from Ceylon cinnamon, rosemary, thyme, and willow have all been described as possessing activity against a wide range of microbes (e.g. bacteria, fungi, and viruses), and have been suggested to work specifically though membrane disruption. In one recent study assessing the effects of three essential oils against 13 bacterial strains and 6 fungal species, oregano (*Origanum vulgare*) essential oil was effective in reducing bacterial growth by up to 60% using a 20% solution, and was even effective in reducing bacterial growth in penicillin-resistant strains by up to 50%. In another study assessing the ability of clove oil to protect chicken frankfurters from *Listeria monocytogenes*, both 1% and 2% solutions of clove oil were effective in inhibiting growth, with the only major concern being the effect of the clove oil on flavor. The essential oils of oregano and thyme were also found to be active against *Phytophthora infestans*, the fungus causing late blight in tomato and potato crops world wide, while the essential oils from rosemary, lavender, fennel, and laurel also all showed reduced bioactivity.

Selected from: Ohkawa H, Miyagawa H, Lee P W. Pesticide Chemistry. Weinheim: WILEY-VCH Verlag GmbH, 2007: 201-207.

Words and Expressions

herbivore /'hɜːbɪvɔː(r)/ *n*. 草食动物
terpene /'tɜːpiːn/ *n*.［化］萜烯，萜（烃）
aromatic /ˌærə'mætɪk/ *adj*. 芳香的，芬芳的，芳香族的
recruit /rɪ'kruːt/ *n*. 新兵，新会员；*vt*. 使恢复，征募；*vi*. 征募新兵，复原
parasitoid /'pærəsɪtɒɪd/ *n*. 拟寄生物（尤指胡蜂）；*adj*. 拟寄生物的
tout /taut/ *vt*. 兜售，招徕；*vi*. 兜售，拉选票；*n*. 侦查者，兜售者
oviposition /ˌəuvipɔ'ziʃən/ 产卵，下子
deterrence /dɪ'terəns/ *n*. 挽留的事物，妨碍物
weevil /'wiːvl/ *n*.［昆］象鼻虫
beetle /'biːtl/ *n*. 甲虫；*vi*. 悬垂，突出
nutmeg /'nʌtmeg/ *n*. 肉豆蔻，肉豆蔻种子中的核仁
S. zeamais 玉米象
T. castaneum 赤拟谷盗
eucalyptus oil［林］桉叶油，桉油，［林］桉树油
lavender /'lævəndə(r)/ *n*.［植］熏衣草，淡紫色；*adj*. 淡紫色的；*vt*. 用熏衣草熏
rosemary /'rəuzməri/ *n*.［植］迷迭香
caraway /'kærəweɪ/ *n*. 香菜
pennyroyal /penɪ'rɒɪəl/ *n*. 薄荷类，薄荷油
peppermint /'pepəmɪnt/ *n*. 胡椒薄荷，薄荷油
Tetranychus urticae 二斑叶螨
Phytoseiulus persimilis 智利小植绥螨
thyme oil ［油脂］百里香精油

thymol /'θaɪmɒl/ *n*. 百里香酚；[有化] 麝香草酚
lepidopteran /ˌlepɪ'dɒptərən/ *adj*. [昆] 鳞翅类的；*n*. 鳞翅类
Plutella xylostella 小菜蛾
Pseudaletia unipuncta 美洲黏虫
octopamine /ɔk'təupəmiːn/ *n*. [有化] 真蛸胺，章鱼胺
neurotransmitter /'njuərəutrænzmɪtə(r)/ *n*. [生理] 神经递质，[生理] 神经传递素
habituation /həbɪtʃu'eɪʃ(ə)n/ *n*. 习惯，熟习，[生理] 习惯化
red thyme 红色百里香
Thymus vulgaris 银斑百里香
Satureja hortensis 夏季香薄荷，风轮菜
cinnamon /'sɪnəmən/ *n*. 樟属的树，肉桂，肉桂色，肉桂皮
Cinnamomum zeylanicum 锡兰肉桂
Syzyium aromaticum 丁香
broccoli /'brɒkəli/ *n*. 花椰菜，西兰花
epicuticular /ˌepɪkjuː'tɪkjulə/ 上表皮的
electrolyte /ɪ'lektrəlaɪt/ *n*. 电解液，电解质，电解
russet /'rʌsɪt/ *adj*. 黄褐色的，赤褐色的
oregano /ˌɒrɪ'gɑːnəu/ *n*. [植] 牛至
listeria monocytogenes 单核细胞增多性李司忒氏菌
phytophthora infestans 马铃薯晚疫病病菌

Unit 4 Microbial Pesticides

Bacillus thuringiensis

The most successful insect pathogen used for insect control is the bacterium *Bacillus thuringiensis* (Bt), which presently is 2% of the total insecticidal market. Bt is almost exclusively active against larval stages of different insect orders and kills the insect by disruption of the midgut tissue followed by septicemia caused probably not only by Bt but probably also by other bacterial species.

Bt action relies on insecticidal toxins that are active during the pathogenic process but these bacteria also produce an array of virulence factors that contribute to insect killing. Upon sporulation, Bt produces insecticidal crystal inclusions that are formed by a variety of insecticidal proteins called Cry or *Cyt* toxins.

Bacillus thuringiensis is well known for its ability to produce parasporal crystalline protein inclusions, which have attracted worldwide interest for various pest management applications because of their specific pesticidal activities. Since the cloning and sequencing of the first crystal proteins genes in the early-1980s, many others have been characterized and are now classified according to the nomenclature of Crickmore *et al*.

These toxins show a highly selective spectrum of activity killing a narrow range of insect species. The *Cry* and *Cyt* toxins belong to a class of bacterial toxins known as pore forming toxins (PFT) that are secreted as water-soluble proteins that undergo conformational changes in order to insert into the membrane of their hosts. Despite the limited use of Bt products as sprayable insecticides, *Cry* toxins have been introduced into transgenic crops providing a more targeted and effective way to control insect pests in agriculture. Concomitantly, this approach has resulted in significant reduction in the use of chemical insecticides in places where this technology has been embraced.

It is well known that Bt remains active against the pest for only several hours on plant foliage under typical field conditions because of UV degradation, rainfall, and other environmental perturbations. In the early 1980s, Monsanto developed a recombinant plant-colonizing pseudomonad for delivery of Bt genes, with the objective of improving residual activity and efficacy of Bt proteins. This concept was developed by Mycogen into the products MVP™ and M-Trak™. The Bt-bearing pseudomonad is killed (to avoid regulatory hurdles for registering recombinant microorganisms) and sprayed on the crop as other Bt products are. The pseudomonad cell is reported to protect the Bt protein from environmental degradation, thus providing longer residual activity. These Bt products have had modest commercial success. A starch-encapsulationprocedure for virus and Bt designed to improve the survival and efficacy of these microbial products in the field is under development by the US Department of Agriculture (USDA) Agricultural Research Service (ARS) in Peoria, Illinois.

Novo Nordisk discovered an enhancer of Bt, a natural substance produced in the Bt at a very low concentration. The same natural product was isolated previously by University of Wisconsin researchers as a fungicide, which they called zwittermicin A. When the compound is combined at higher concentrations with the Bt protein, efficacy against the most refractory caterpillars, such as *Helicoverpa zea* and *Spodoptera exigua*, is increased substantially in the field. The most successful Bt products are ones that provide efficacy, ease of use, and consistency approaching traditional chemical pesticides. Improved armyworm (*Spodoptera*) and bollworm (*Helicoverpa*) products are the most important developments in the use of Bt microbials in agriculture.

Baculoviruses

Although they have not had the commercial success of Bt, baculoviruses could have important potential for use. They have a number of advantages. Baculoviruses are ideal, because as far as is known, they are safe for nontarget insects, humans, and the environment. Baculoviruses might, in some cases, be the only effective biocontrol agents available for controlling insect species and they provide an avenue for overcoming specific problems, such as resistance. It is important to have a selection of control agents. Because viruses are not likely to elicit cross resistance to chemicals, they should receive more attention from university and industrial researchers. The use of multiple biological products has the advantage of lowering the potential for evolution of pest resistance.

Helicoverpa zea Boddie (cotton bollworm) nuclear polyhedrosis virus (NPV) was the first baculovirus to be marketed in the United States. In Europe, a number of companies have introduced viral products for the insecticide market or are developing them.

Viral products include *Cydia pomonella* L. (codling moth) granulosis virus (GV), *Neodiprion sertifer* (Geoffrey) (European pine sawfly) NPV, *Spodoptera exigua* (Hübner) (beet armyworm) NPV, and *Autographa californica* (Speyer) (alfalfa looper) NPV. The largest use of baculoviruses is in Brazil, where *Anticarsia gemmatalis* Hübner (velvetbean caterpillar) NPV protects 5.9 million hectares of soybeans against the velvetbean caterpillar. The Canadian Forest Service holds registrations for *O. pseudotsugata* NPV and *Neodiprion lecontei* (Fitch) (redheaded pine sawfly) NPV.

Lower production costs are essential for both recombinant and wildtype baculoviruses to compete with classical insecticides. There are active research programs in both *in vivo* and *in vitro* production. Although viruses are less expensive to produce *in vivo* than *in vitro*, the cost still exceeds that of Bt. Viruses are formulated to be applied in the same fashion as Bt strains.

Another limitation of baculoviruses is their host specificity, which can reduce their commercial potential. However, the host specificity is viewed positively from the environmental and IPM standpoints. Two viruses with relatively broad host ranges are *Autographa californica* (alfalfa looper) NPV and *Syngrapha falcifera* (Kirby) (celery looper) NPV, each of which kills over 30 insect species. Host range can be broadened through molecular means or by mixing two viruses.

In the long term, the development of recombinant baculoviruses that can kill rapidly will allow them to compete more effectively with classical pesticides. To increase the ability of baculoviruses to kill early, research to insert specific genes into the baculoviral genome is under way. These genes will serve as toxins or disrupters of larval development. Among the proteins being tested for exploitation are Bt endotoxin (which failed to improve the virus), juvenile hormone esterase, prothoracicotropic hormone (PTTH), melittin, trehalase, scorpion toxin, and mite toxin. The knowledge of the molecular biology of viruses has also promoted interest in modifying and improving baculoviruses with regard to host range and virulence.

Entomopathogenic Fungi

Fungi infect a broader range of insects than do other microorganisms, and infections of lepidopterans (moths and butterflies), homopterans (aphids and scale insects), hymenopterans (bees and wasps), coleopterans (beetles), and dipterans (flies and mosquitoes) are quite common. In fact, some fungi have very broad host ranges that encompass most of those insect groups. That is true of *Beauveria bassiana* (Balsamo) Vuillemin, *Metarhizium anisopliae* (Metschnikoff) Sorokin, *Verticillium lecanii* (Zimmerman) Viegas, and *Paecilomyces* spp., all of which have worldwide distributions; these are the most commonly used insect pathogens developed for commercial pest-management products.

1. Beauveria bassiana

Beauveria bassiana has been identified in many insect species in temperate and tropical regions and is used for pest control on a moderate scale in eastern Europe and China. Mycogen produced a *B. bassiana*-based bioinsecticide, which has been shown to be highly pathogenic in coleopterans. The fungus also is amenable to mass production of conidia by semisolid fermentation. The product has been field-tested against citrus root weevil. Another *B. bassiana* strain, researched at the USDA ARS showsgood control of rasping or sucking insects, such as thrips, whiteflies, and aphids.

2. Metarhizium anisopliae

Metarhizium anisopliae has been most extensively used in Brazil for control of spittlebugs on sugar cane. Using Metarhizium as the control agent, EcoScience Laboratories, Inc. has developed infection chambers in which insects (cockroaches and flies) brush against spores of the pathogen, which later germinate and infect the insect.

3. Verticillium lecanii

Verticillium lecanii is a pathogen that has demonstrated good control of greenhouse pests, such as *Myzus persicae* (Sulcer) aphids, on chrysanthemums. A distinct isolate of *V. lecanii* was obtained from whitefly and provided excellent control of greenhouse whitefly, *Trialeurodes vaporariorum* (Westwood), and of *Thrips tabaci* Lindeman on cucumber. *V. lecanii* was produced commercially as Vertalec for aphid control and Mycotal for control of whitefly from 1982 to 1986, and there is a resurgence of commercial interest in its use for control of aphids, whiteflies, and thrips because these greenhouse pests have developed resistance to chemical pesticides typically used for their control.

Bacillus subtilis

The use of *Bacillus subtilis* as a fungicidal treatment has been demonstrated on a number of diseases, including cornstalk rot (*Fusarium roseum*), onion white rot (*Sclerotium cepivorum*), potato charcoal rot (*Macrophomina phaseolina*), bean rust (*Uromyces phaseoli*), apple blue mold (*Penicillium expansum*), and peach brown rot (*Monilinia fructicola*). The vast majority of the work with *Bacillus subtilis* has concentrated on treatment of seeds or soil to control pathogens; in general, the use of biocontrol as a foliar treatment is much less developed than in the soil-rhizoplane environment.

Selected from: Committee on the Future Role of Pesticides in US Agriculture, Board on Agriculture and Natural Resources and Board on Environmental Studies and Toxicology, Commission on Life Sciences. The future role of pesticides in US agriculture. Washington, D. C.: National Academy Press, 2000: 157-173.

Words and Expressions

Bacillus thuringiensis 苏云金杆菌
sporulation /ˌspɔːrjuˈleɪʃən/ *n.* [生物] 孢子形成
Cyt toxin Cyt 毒素
parasporal 伴孢
pore forming toxin 孔隙形成毒素
concomitantly /kənˈkɒmɪtəntlɪ/ *adv.* 附随地
perturbation /ˌpɜːtəˈbeɪʃn/ *n.* [数] [天] 摄动，不安，扰乱
pseudomonad /ˌpsjudoˈmonæd/ *n.* 假单胞菌
zwittermicin A 兹维特霉素 A
Helicoverpa zea 棉铃虫
Spodoptera exigua 甜菜夜蛾
baculovirus /ˈbækjulə(ʊ)ˌvaɪrəs/ *n.* 杆状病毒
nuclear polyhedrosis virus 核型多角体病毒
Cydia pomonella 苹果小卷蛾，苹果蠹蛾
granulosis virus 颗粒症病毒
Neodiprion sertifer 欧洲松叶蜂，欧洲新松叶蜂
Autographa californica 苜蓿银纹夜蛾
Anticarsia gemmatalis 大豆夜蛾，黎豆夜蛾
O. pseudotsugata 黄杉毒蛾
Neodiprion lecontei 红头松叶蜂
in vivo /inˈviːvəu/ (拉) [生物] 在活的有机体内
in vitro /inˈviːtrəu/ 在体外，在试管内
Syngrapha falcifera 芹菜夜蛾
endotoxin /ˈendəuˌtɒksɪn/ *n.* [病理] 内毒素
prothoracicotropic hormone 促前胸腺激素

melittin /mə'lɪtɪn/ *n*.［生化］蜂毒肽，蜂毒素
trehalase /trɪ'hɑlez/ *n*.［生化］海藻糖酶
scorpion toxin 蝎子毒素
mite toxin 螨毒素
entomopathogenic /'entərəʊ,pæθə'dʒenɪk/ *adj*. 致肠病的
fungi /'fʌndʒaɪ/ *n*. 真菌，菌类，蘑菇（fungus 的复数）
homopteran /hɒ'mɒptərən/［昆］同翅亚目的，［昆］同翅昆虫
hymenopteran /,haɪme'nɒpt(ə)rən/ *adj*. 膜翅目的；*n*. 膜翅目昆虫
coleopteran /,kɒlɪ'ɒpt(ə)rən/ *n*. 鞘翅类昆虫，甲虫类昆虫
dipteran /'dɪpt(ə)r(ə)n/ *adj*. 双翅类昆虫的（等于 dipterous）；*n*. 双翅类昆虫
Beauveria bassiana 白僵菌，球孢白僵菌
Metarhizium anisopliae 绿僵菌，金龟子绿僵菌
Verticillium lecanii 蜡蚧轮枝菌
Paecilomyces spp. *n*. 拟青霉属
bioinsecticide 生物杀虫剂
conidia /kə'nɪdiə/ *n*. 分生孢子（conidium 的复数）
Myzus persicae 桃蚜，烟蚜
Trialeurodes vaporariorum 温室粉虱，温室白粉虱
Thrips tabaci 葱蓟马，烟蓟马
Bacillus subtilis 枯草芽孢杆菌
Fusarium roseum 粉红镰孢，粉红镰刀菌
Sclerotium cepivorum 白腐病，白腐小核菌
Macrophomina phaseolina 壳球孢菌，球孢菌
Uromyces phaseoli 菜豆单胞锈菌
Penicillium expansum 青霉病菌
Monilinia fructicola 褐腐病菌，褐腐病
biocontrol /,baɪokən'trol/ *n*. 生物防除，［生物物理］生物电控制

Notes

［1］Monsanto：孟山都公司，美国的一家跨国农业公司，总部设于美国密苏里州圣路易斯市。其生产的旗舰产品是全球知名的草甘膦除草剂。2018 年 6 月 7 日起，拜耳成为孟山都公司的唯一股东。

［2］Novo Nordisk：诺和诺德公司，总部位于丹麦首都哥本哈根，是一家致力于保护人类健康的生物制药公司。在行业内拥有广泛的糖尿病治疗产品。此外，止血管理、生长保健激素以及激素替代疗法等多方面研究居世界领先地位。

Exercises

Ⅰ **Answer the following questions according to the text.**

1. In this text, how many microbial pesticides are introduced respectively? Please give

examples.

2. What are the advantages and disadvantages of microbial pesticides?

3. Please compare chemical pesticides, botanical pesticides, and microbial pesticides.

4. Which species of the baculoviruses are listed in this text?

5. Which species of the entomopathogenic fungi are listed in this text?

Ⅱ Translate the following English phrases into Chinese.

insecticidal protein　insecticidal toxin　UV degradation　semisolid fermentation

first crystal protein　virulence factor　sucking insect　pathogenic process

Ⅲ Translate the following Chinese phrases into English.

杀虫晶体　锉吸式昆虫　蛋白质内含物　微生物农药

害虫管理　非靶标昆虫　水溶性蛋白质　中肠组织

Ⅳ Choose the best answer for each of the following questions according to the text.

1. Which developmental stage of insects is effective of Bt?

A. egg　B. larva　C. pupa　D. adult

2. Which of the following does not belong to the baculoviruses?

A. *Helicoverpa zea* nuclear polyhedrosis virus (NPV)

B. *Cydia pomonella* granulosis virus (GV)

C. *Spodoptera exigua* NPV

D. *Bacillus thuringiensis* (Bt)

3. Which of the following does not belong to the entomopathogenic fungi?

A. *Bacillus subtilis*　B. *Verticillium lecanii*

C. *Metarhizium anisopliae*　D. *Beauveria bassiana*

4. Which of the following does not belong to microbial pesticides?

A. Bt　B. sabadilla

C. *Beauveria bassiana*　D. NPV

5. Usually, the use of *Bacillus subtilis* is much more in ________ environment.

A. the soil-rhizoplane　B. foliar treatment

C. spray　D. injection

Ⅴ Translate the following sentences into Chinese.

1. Bt is almost exclusively active against larval stages of different insect orders and kills the insect by disruption of the midgut tissue followed by septicemia caused probably not only by Bt but probably also by other bacterial species.

2. Baculoviruses are ideal, because as far as is known, they are safe for nontarget insects, humans, and the environment. Baculoviruses might be the only effective biocontrol agents available for controlling insect species and they provide an avenue for overcoming specific problems, such as resistance.

3. Fungi infect a broader range of insects than do other microorganisms, and infections of lepidopterans (moths and butterflies), homopterans (aphids and scale insects), hymenopterans (bees and wasps), coleopterans (beetles), and dipterans (flies and mosquitoes) are quite common.

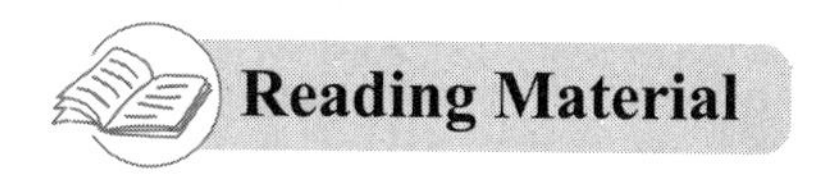

Reading Material

Bacillus thuringiensis

Bacillus thuringiensis (Bt) has become the leading biopesticide since the beginning of the 1970s, due to the lethality of the toxin to insects. It has attracted industry to use it worldwide as an effective weapon against agricultural pests and insect vectors of human diseases. Originally, *B. thuringiensis* was considered an entomopathogen. Within the last four decades the complexity and diversity of *B. thuringiensis* as an insecticidal microbe have been elucidated.

The first report on the crystalline parasporal body in the bacterium that might be associated with the insecticidal activity appeared by 1953. Angus demonstrated that this crystal contains an alkalinesoluble toxin for insects. *B. thuringiensis* produces a *b*-exotoxin known also as the fly-toxin, thermostable toxin, or thuringiensin, but this toxin was not approved for use in agriculture because its toxicity was not limited to insect pests. The *d*-endotoxin showed the most promising characteristics of an insect-specific bioinsecticide. By the end of the 1950s, the toxicity of the spore-crystal complex was classified into "insect types" — an early indication that the spore contributes to the insecticidal power of *B. thuringiensis*.

In the 1960s, initial efforts were made to quantify the spore-crystal mixtures by means of standardized bioassays. The introduction of commercial *B. thuringiensis* products by the agrochemical industry and the growing knowledge of the insecticidal toxin led to an urgent need to replace the spore count that had labelled the commercial products with a standardized value of insecticidal power. The idea of using a standard microbe for potency determination of the bioinsecticide product was accepted at the 1966 Colloquium of Insect Pathology and Microbial Control at Wageningen, Netherlands. The E-61 formulation from the Institute Pasteur, Paris, France, with an assigned potency of 1000 IU/mg, was first adopted as the standard, but was later replaced in 1971 by the HD-1-S-1971 with an assigned potency of 18,000 IU/mg. As from 1980 the HD-1-S-80 standard with a potency of 16,000 IU/mg has been available from international *B. thuringiensis* cultures. The standardized bioassay procedures proposed by Dulmage et al. were reviewed and further standardized for several insects species. The "activity ratio" in which a microbe powder is assayed against two insect species in parallel is useful for measuring differences between the tested microbe powder and the standard. Their potencies may be identical in one insect but different in another insect species. This ratio for any two insect species can be compared between laboratories, provided that the microbe powder is the same.

So far, however, the possibility of using new international standardization of Bt has not been accepted after discussing this issue in a recent Society for Invertebrate Pathology (SIP) meeting, mainly because: (i) insect strains and qualities differ between laboratories, and therefore mortalities are not comparable; and (ii) each of the Bt manufacturers has developed its own bioassay procedures for labeling the Bt products and changing the protocols will not be feasible for commercial considerations. Moreover, in the last two decades, additional bioassay protocols had to be designed in view of the following new developments and discoveries: (i) identification of new Cry proteins and their use in genetically engineered products; (ii) development of new conventional formulations; (iii) an accumulating evidence of insect resistance to d-endotoxin and Cry proteins; and (iv) analytical assays for the d-endotoxin, Cry proteins and insecticidal crystal protein (ICP) genes, as complementary information on the activity of Bt proteins in the microbe isolates.

Selected from: Navon A, Ascher K R S. Bioassays of Entomopathogenic Microbes and Nematodes. UK: CABI Publishing, 2000: 1-2.

Words and Expressions

lethality /lɪ'θæləti/ *n*. 致命性，毁坏性
vector /'vektə(r)/ *n*. [数] 向量，矢量，带菌者；*vt*. 无线电导引
entomopathogen 昆虫病原体
crystalline /'krɪstəlaɪn/ *adj*. 水晶的，结晶的
alkaline soluble 碱性溶液
exotoxin /'eksə(ʊ)tɒksɪn/ *n*. [基医] 外毒素，外泌毒
thermostable /θɜːməʊ'steɪb(ə)l/ *adj*. [化] 耐热的，热稳定的
spore-crystal 芽孢晶体

PART Ⅲ

FORMULATIONS

Unit 5 Liquid Formulations

After a pesticide is manufactured in its relatively pure form—the technical grade material, whether herbicide, insecticide, fungicide, or other classification—the next step is formulation. It is processed into a usable form for direct application or for dilution followed by application. The formulation is the final physical condition in which the pesticide is sold for use.

Formulation is the processing of a pesticidal compound by any method that will improve its properties of storage, handling, application, effectiveness, or safety. The term formulation is usually reserved for commercial preparation prior to actual use and does not include the final dilution in application equipment.

The real test for a pesticide is acceptance by the user. And, to be accepted for use by the homeowner, the grower, or commercial applicator, a pesticide must be effective, safe; easy to apply, and relatively economic although householders commonly pay 10 to 30 times the price that growers pay for a given weight of a particular pesticide. Price depends to a great extent on the formulation; for instance. The most expensive form of insecticide is the pressurized aerosol.

Pesticides, then, are formulated into many usable forms for satisfactory storage, for effective application, for safety to the applicator and the environment, for ease of application with readily available equipment, and for economy. These goals are not always simply accomplished, due to the chemical and physical characteristics of the pesticide. Fox example, some materials in their "raw" or technical condition are liquids, others solids; some are stable to air and sunlight, whereas others are not; some are volatile, others not; some are water soluble, some oil soluble, and others may be insoluble in either water or oil. These characteristics pose problems to the formulator, since the final formulated product must meet the standards of acceptability by the user.

Emulsifiable Concentrates

Spray formulations are prepared for insecticides, herbicides, miticides, fungicides, algicides, growth regulators, defoliants, and desiccants. Consequently, more than 75 percent of all pesticides are applied as sprays. The bulk of these are currently applied as water emulsions made from emulsifiable concentrates (EC).

Emulsifiable concentrates, synonymous with emulsible concentrates, are concentrated oil solutions of the technical grade material with enough emulsifier added to make the concentrate mix readily with water for spraying. The emulsifier is a detergent like material that makes possible the suspension of microscopically small oil droplets in water to form an emulsion.

When an emulsifiable concentrate is added to water, the emulsifier causes the oil to disperse immediately and uniformly throughout the water, if agitated, giving it an opaque or milky appearance. There are a few rare formulations of invert emulsions, which are water-in-oil suspensions, and are opaque in the concentrated form, resembling salad dressing or face cream. Invert emulsions are employed almost exclusively as herbicide formulations. The thickened sprays result in very little drift and can be applied in sensitive situations.

Emulsifiable concentrates, if properly formulated, should remain suspended without further agitation for several days after dilution with water. A pesticide concentrate that has been held over from last year can be easily tested for its emulsifiable quality by adding 1 ounce to 1 quart of water and allowing the mixture to stand after shaking. The material should remain uniformly suspended for at least 24h with no precipitate. If a precipitate does form, the same condition may occur in the spray tank, resulting in clogged nozzles and uneven application. In the home, this can be remedied by adding 2 tablespoons of liquid dishwashing detergent to each pint of concentrate and mixing thoroughly. In an agricultural or other circumstance, where several gallons of costly pesticide are involved, additional emulsifier can be obtained from the formulator. This should be added to the concentrate at the rate of 0. 2 to 0. 5 pound for each gallon of outdated material. The bulks of pesticides available to the homeowner are formulated as emulsifiable concentrates and generally have a shelf of about 3 years.

Every gallon of emulsifiable concentrate contains from 4 to 7 pounds of petroleum solvent, usually one of the more expensive aromatic solvents such as xylene. Within the past three years petroleum solvents have increased in price 300 percent or more, increasing significantly the costs of formulation and consequently the cost of the formulated product to the foreseeable future, and formulators are searching diligently for other, more economical formulations for their products.

The innovation in emulsifiable concentrates is the transparent emulsion concentrates (TEC). These are the mixture of the pesticide and the emulsifier with little or no hydrocarbon solvent. In these formulations the emulsifier or surfactant serves as the solvent, replacing the aromatic solvents, yet yielding a product that can be readily diluted with water.

Water-miscible Liquids

Water-miscible liquids are mixable in water. The technical grade material may be initially water miscible, or it may be alcohol miscible and formulated with an alcohol to become water miscible. These formulations resemble the emulsifiable concentrates in viscosity and color, but do not become milky when diluted with water. Water-miscible liquids are labeled as water-soluble concentrate (WSC), liquid (L), soluble concentrate (SC), or solution (S).

Oil Solutions

In their commonest form, oil solutions are the ready-to-use household and garden insecticide sprays sold in an array of bottles, cans, and plastic containers, all usually equipped with a handy spray atomizer. Not to be confused with aerosols. These sprays are intended to be used directly on pests or places they frequent oil solutions may be used as roadside weed sprays. For marshes and standing pools to control mosquito larvae, in fogging machines for mosquito and fly abatement programs, or for household insect sprays purchased in supermarkets. Commercially they may be sold as oil concentrates of the pesticide to be diluted with kerosene or diesel fuel before application or in the dilute, ready-to-use form. In either case the compound is dissolved in oil and is applied as an oil spray.

Flowable or Sprayable Suspensions

There are several types of flowable (F) or sprayable (S) suspensions. Physically they all are thick and creamy, and they vary in appearance from tan to white. The original flowable was an ingenious solution to a formulation problem. Some pesticides are soluble in neither oil nor water. But are soluble in one of the exotic solvents, which makes the formulation quite expensive and may price it out of the marketing competition. To handle the problem, the technical material is wet-milled with a clay diluent and water, leaving the pesticide-diluent mixture finely ground but wet. This "pudding" mixes well with water and can be sprayed but has the same tank-settling characteristic as the wettable powders.

An example of a second kind of innovative flowable is the blending of the finely ground insecticide carbaryl with molasses. This formulation reduces to some extent the pesticide's drift off target during aerial application, increases its adherence to foliage and thus reduces removal by rain, and increases the mortality of moths that are attracted to feed on the molasses.

A third flowable is made by mixing an emulsifiable concentrate, containing a very high percentage of a water-stable toxicant and a thickener, with two to four volumes of water. The result is a thick, concentrated normal emulsion, which is then diluted for use with the appropriate volume of water just before using. This formulation responds to the need for an emulsifiable concentrate made with the least solvent possible to avoid foliage burn in citrus groves. These formulations are occasionally identified as spray concentrates.

A fourth innovation is the flowable microencapsulated formulation. The insecticide is by a special process in small, permeable spheres of a polymer or plastic, 15 to 50 μm in diameter. These spheres are then mixed with wetting agents, thickeners, and water to give the desired concentration of insecticide in the flowable.

Ultralow-volume Concentrates

Ultralow-volume concentrates (ULV) are available only for commercial use in the control of public health, agricultural, and forest pests. They are usually the technical product in its original liquid form or, if solid, the original product dissolved in a minimum of solvent. They are usually applied without further dilution, by special aerial or ground spray equipment that limits the volume from 0.6 liter (L) to a maximum of 4.7 L per hectare (ha), as an extremely fine spray. The ULV formulations are used where good results can be obtained while economizing through the elimination of the normally high spray volumes, varying from 11 to 38 L/ha. This technique has proved extremely useful where insect control is desired over vast areas.

Fogging Concentrates

Fogging concentrates are the formulations sold strictly for public health use in the control of nuisance or disease vectors, such as flies and mosquitoes, and to pest control operators. Fogging machines generate droplets whose diameters are usually less than 10 μm but greater than 1 μm. They are of two types. The thermal fogging device utilizes a flash heating of the oil solvent to produce a visible vapor or smoke. The ambient fogger atomizes a tiny jet of liquid in a venturi tube through which passes an ultrahigh-velocity air stream. The materials used in fogging machines depend on the type of fogger. Thermal devices use oil only, whereas ambient generators use water, emulsions, or oils. In dwellings and food establishments the oil most commonly used is deodorized kerosene.

Aerosols

To produce an aerosol, the active ingredients must be soluble in the volatile, petroleum solvent in its pressurized condition. The pressure is provided by a propellant gas. When the petroleum solvent is atomized, it evaporates rapidly, leaving the microsized droplets of toxicant suspended in air. These products are available for homeowners as well as for commercial pest control operators.

Caution: Aerosols commonly produce droplets well below 10 μm in diameter. Such droplets are respirable, that is, they are absorbed by alveolar tissue in the lungs rather than impinging on the bronchioles, as do larger droplets. Consequently, aerosols of all varieties should be handled with discretion, and the user should inhale as little as possible.

Selected from: Ware G W. Pesticides: Theory and Application. W. H. Freeman and company, 1978: 21-27.

Words and Expressions

formulation /ˌfɔːmjuˈleɪʃn/ *n*. 用公式表示，明确地表达，剂型
emulsifiable concentrate ［化］乳油
growth regulator ［植］［生化］生长调节剂

defoliant /diː'fəʊliənt/ *n*.（美）［农药］落叶剂，［农药］脱叶剂
desiccant /'desɪk(ə)nt/ *adj*. 去湿的，使干燥的；*n*.［助剂］干燥剂
emulsion /ɪ'mʌlʃn/ *n*.［药］乳剂，［物化］乳状液，感光乳剂
emulsifier /ɪ'mʌlsɪfaɪə(r)/ *n*. 乳化剂，黏合剂
invert emulsion 逆乳状液，反相乳液
nozzle /'nɒzl/ *n*. 喷嘴，管口，鼻
dishwashing detergent 餐具洗涤剂，洗洁精
xylene /'zaɪliːn/ *n*.［有化］二甲苯
hydrocarbon solvent 烃类溶剂，碳氢化合物溶剂
surfactant /sɜː'fæktənt/ *n*. 表面活性剂；*adj*. 表面活性剂的
aromatic solven 芳烃溶剂
water-miscible liquid 水溶性液剂
viscosity /vɪ'skɒsəti/ *n*. 黏性，黏度
spray atomizer 喷雾器
oil solution 油剂
carbaryl /'kɑːbəraɪl/ *n*.［农药］甲萘威（接触性杀毒剂）
molasses /mə'læsɪz/ *n*. 糖蜜，糖浆
drift off 漂移
aerial application 空中喷药，飞机喷施
mortality /mɔː'tæləti/ *n*. 死亡数，死亡率，必死性，必死的命运
toxicant /'tɒksɪk(ə)nt/ *n*. 有毒物，［毒物］毒药；adj. 有毒的
thickener /'θɪkənə(r)/ *n*.［助剂］增稠剂，浓缩机
microencapsulated formulation 微胶囊剂
wetting agent 润湿剂
ultralow-volume concentrate 超低容量剂型
fogging concentrate 烟雾剂
thermal fogging device 热雾装置，热雾机
ambient fogger 室温雾化器，环境温度雾化器
aerosol /'eərəsɒl/ *n*.［化］气溶胶，气雾剂，烟雾剂

Notes

[1] technical grade：工业级。用于工业生产的原料，杂质相对较多。化学纯、分析纯、色谱纯，是指试剂的纯度级别。化学纯试剂是一般化学试验用的，有较少的杂质；分析纯试剂是做分析测定用的，杂质更少；色谱纯试剂是进行色谱分析时使用的标准试剂。

[2] ounce：盎司（此处指常衡盎司），重量单位。1 盎司＝28.3495g，1 盎司＝16 打兰（dram），16 盎司＝1 磅（pound）。

[3] quart：夸脱，容量单位，主要在英国、美国及爱尔兰使用。1 英制夸脱＝8 及耳＝2 品脱＝1/4 加仑＝1/32 蒲式耳。

Exercises

Ⅰ Answer the following questions according to the text.

1. What should be considered while formulating many different usable forms of pesticides?

2. How many kinds of liquid formulations of pesticides are mentioned in the text? What are they? Please give a short summary of their advantages and disadvantages.

3. How can a pesticide in concentrate be tested?

4. What does TEC mean? Why is TEC regarded as the most innovation in emulsifiable concentrates?

5. Why was the original flowable an ingenious solution to a formulation problem?

6. Why are aerosols effective only against residual flying and crawling insects?

7. What should be paid great attention to while applying aerosols?

Ⅱ Translate the following English phrases into Chinese.

water-in-oil	disease vector	spray formulation	liquid formulation
inert diluent	technical grade	physical condition	fogging concentrate
brand name	water soluble	plastic container	handy spray atomizer

Ⅲ Translate the following Chinese phrases into English.

油剂	悬浮剂	可溶性液剂	经济阈值	烟雾机
乳油	微乳剂	喷洒悬浮剂	物理状态	超低容量液剂

Ⅵ Choose the best answer for each of the following questions according to the text.

1. The most expensive form of insecticide is ________.

   A. emulsifiable concentrates　　B. water-miscible liquids

   C. wettable powders　　D. the pressurized aerosol

2. The shelf of the most pesticides formulated as EC available to homeowners is ________.

   A. about one year　　B. about two years

   C. about three years　　D. about four years

3. Ultralow-volume concentrates are available only for commercial use in the control of ________.

   A. public health　　B. agricultural pests

   C. forest pests　　D. all of the above

4. Which of the following statement is not true according to the text?

   A. Fogging concentrates are the formulations sold strictly for public health use in the control of nuisance and disease vectors.

   B. In dwelling and food establishments, the most commonly used is deodorized kerosene.

   C. Formulation is the final physical condition in which the pesticides are sold for use.

   D. Home and garden pesticides are only water miscibles.

5. Why should the user inhale droplets of aerosols as little as possible?

   A. Because they are respirable.

   B. Because they are too small to be inhaled.

C. Because they are toxic and harmful to the user's health.

D. Because they are very expensive.

Ⅴ **Translate the following short paragraphs into Chinese.**

1. Formulation is the processing of a pesticidal compound by any method that will improve its properties of storage, handling, application, effectiveness, or safety.

2. Emulsifiable concentrates, synonymous with emulsible concentrates, are concentrated oil solutions of the technical grade material with enough emulsifier added to make the concentrate mix readily with water for spraying.

3. Ultralow-volume concentrates (ULV) are available only for commercial use in the control of public health, agricultural, and forest pests. They are usually the technical product in its original liquid form or, if solid, the original product dissolved in a minimum of solvent.

4. Fogging concentrates are the formulations sold strictly for public health use in the control of nuisance or disease vectors, such as flies and mosquitoes.

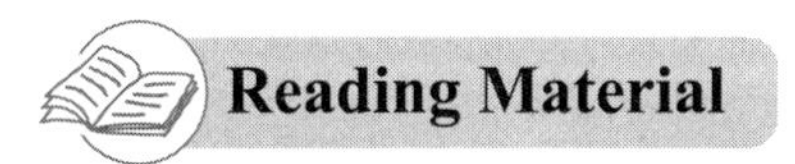

Advantages and Disadvantages of Liquid Formulations

Liquid formulations are generally mixed with water, but in some instances labels may permit the use of crop oil, diesel fuel, kerosene, or some other light oil as a carrier.

Emulsifiable Concentrates (EC)

An emulsifiable concentrate formulation usually contains a liquid active ingredient, one or more petroleumbased solvents, and an agent that allows the formulation to be mixed with water to form an emulsion. They are used against agricultural, ornamental and turf, forestry, structural, food processing, livestock, and public health pests. They are adaptable to many types of application equipment, from small, portable sprayers to hydraulic sprayers, lowvolume ground sprayers, mist blowers, and low-volume aircraft sprayers.

Advantages:

- Relatively easy to handle, transport and store.
- Little agitation required—will not settle out or separate when equipment is running.
- Not abrasive.
- Will not plug screens or nozzles.
- Little visible residue on treated surfaces.

Disadvantages:

- High a. i. concentration makes it easy to overdose or underdose through mixing or calibration errors.
- May cause damage to desirable plants (phototoxicity).
- Easily absorbed through skin of humans or animals.

• Solvents may cause rubber or plastic hoses, gaskets, and pump parts and surfaces to deteriorate.

• May cause pitting or discoloration of painted finishes.

• Flammable—should be used and stored away from heat or open flame.

• May be corrosive.

Ultra low-volume (ULV)

These concentrates may approach 100 percent active ingredient. They are designed to be used as is or to be diluted with only small quantities of a specified carrier and are used at rates of no more than 1/2 gallon per acre. These special purpose formulations are used mostly in outdoor applications, such as in agricultural, forestry, ornamental, and mosquito control programs.

Advantages:

• Relatively easy to handle, transport, and store.

• Remain in solution; little agitation required.

• Not abrasive to equipment.

• Will not plug screens and nozzles.

• Leave little visible residue on treated surfaces.

Disadvantages:

• Difficult to keep pesticide on target—high drift hazard.

• Specialized equipment required.

• Easily absorbed through skin of humans or animals.

• Solvents may cause rubber or plastic hoses, gaskets, and pump parts and surfaces to deteriorate.

• Calibration and application must be done very carefully because of the high concentration of active ingredient.

Aerosols (A)

These formulations contain one or more active ingredients and a solvent. Most aerosols contain a low percentage of active ingredients. There are two types of aerosol formulations—the ready-to-use type commonly available in pressurized sealed containers, and those products used in electrical or gasolinepowered aerosol generators that release the formulation as a "moke" or "fog".

These formulations are usually small, self-contained units that release the pesticide when the nozzle valve is triggered. The pesticide is driven through a fine opening by an inert gas under pressure, creating fine droplets. These products are used in greenhouses, in small areas inside buildings, or in localized outdoor areas. Commercial models, which hold 5 to 10 pounds of pesticide, are usually refillable.

Advantages:

• Ready to use.

• Portable.

- Easily stored.
- Convenient way to buy a small amount of a pesticide.
- Retain potency over fairly long time.

Disadvantages:

- Practical for only very limited uses.
- Risk of inhalation injury.
- Hazardous if punctured, overheated, or used near an open flame.
- Difficult to confine to target site or pest.

Formulations for Smoke or Fog Generators

These aerosol formulations are not under pressure. They are used in machines that break the liquid formulation into a fine mist or fog (aerosol) using a rapidly whirling disk or heated surface. These formulations are used mainly for insect control in structures such as greenhouses and warehouses and for mosquito and biting fly control outdoors.

Advantages:

- Easy way to fill entire enclosed space with pesticide.

Disadvantages:

- Highly specialized use and equipment.
- Difficult to confine to target site or pest.
- May require respiratory protect ion to prevent risk of inhalation injury.

Selected from: Randall C, Hock W, Crow E, *et al*. National Pesticide Applicator Certification Core Manual. The National Association of State Departments of Agriculture Research Foundation, 2012: 52-78.

Words and Expressions

crop oil　作物油，植物油
diesel fuel　柴油，柴油机燃料
kerosene　/'kerəsiːn/ *n*. 煤油，火油
hydraulic sprayer　液压喷雾器
aircraft sprayer　飞机喷雾器
overdose　/'əuvədəus/ *n*. 药量过多，(有害物) 过量；*v*. 服药过量
underdose　/'ʌndəˌdos/ *n*. (药物等的) 不足剂时；*vt*. 给……用药剂量不足

Unit 6 Solid Formulations

Solid formulations can be divided into two types: ready-to-use and concentrates that must be mixed with water to be applied as a spray.

Dusts

Historically, dusts have been the simplest formulations of pesticides to manufacture and the easiest to apply. Examples of the undiluted toxic agent are sulfur dusts, sodium fluoride. An example of the toxic agent with active diluent is the garden insecticides having sulfur dust as its carrier or diluent. A toxic agent with an inert diluent is the most common type of dust formulation in use today. Insecticide-fungicide combinations are applied in this manner, the carrier being an inert clay, such as pyrophyllite. In this instance, particles small enough to pass through a 60-mesh screen are considered dusts. Mesh is a unit that refers to the number of grids per inch through which they will pass.

Despite their ease in handling, formulation, and application, dusts are the least effective and, ultimately, the least economical of the pesticide formulations. The reason is that dusts have a very poor rate of instance, an aerial application of a standard dust formulation of pesticide will result in 10 percent to 40 percent of the material reaching the crop. The remainder drifts upward and downwind. Psychologically, dusts are annoying to the nongrower who sees great clouds of dust resulting from an aerial application, in contrast to the grower who believes he or she is receiving a thorough application for the very same reason.

Wettable Powders

Wettable powders (WP) are essentially concentrated dusts containing a wetting agent to facilitate the mixing of the powder with water before spraying. The technical material is added to the inert diluent, in this cases a finely ground talc or clay, in addition to a wetting agent or surfactant and mixed thoroughly in a ball mill. Without the wetting agent, the powder would float when added to water, and the two would be almost impossible to mix. Because wettable powders usually contain from 50 to 75 percent clay or talc, they sink rather quickly to the bottom of spray tanks unless the spray mix is agitated constantly.

Wettable powders are one of the most widely used pesticide formulations. They can be used for most pest problems and in most types of spray equipment where agitation is possible. Wettable powders have excellent residual activity. Because of their physical properties, most of the pesticide remains on the surface of treated porous materials such as concrete, plaster, and untreated wood. In such cases, only the water penetrates the material.

Advantages:

- Easy to store, transport, and handle.

- Less likely than ECs and other petroleum-based pesticides to cause unwanted harm to treated plants, animals, and surfaces.
- Easily measured and mixed.
- Less skin and eye absorption than ECs and other liquid formulations.

Disadvantages:

- Inhalation hazard to applicator while measuring and mixing the concentrated powder.
- Require good and constant agitation (usually mechanical) in the spray tank and quickly settle out if the agitator is turned off.
- Abrasive to many pumps and nozzles, causing them to wear out quickly.
- Difficult to mix in very hard, alkaline water.
- Often clog nozzles and screens.
- Residues may be visible on treated surfaces.

Soluble Powders

Water-soluble powders (SP) is a finely ground water-soluble solid and may contain a small amount of wetting agent to assist its solution in water. It is simply added to the spray tank, where it dissolves immediately. Unlike the wettable powders and flowables, these formulations do not require constant agitation; they are true solutions and form no precipitate. Because of their sometimes dusty quality, soluble powders may be packaged in convenient, water-soluble bags to be dropped into the spray tank.

Tracking Powders

Special dusts known as tracking powders are used for rodent and insect monitoring and control. For rodent control, the tracking powder consists of finely ground dust combined with a stomach poison. Rodents walk through the dust, pick it up on their feet and fur, and ingest it when they clean themselves.

Tracking powders are useful when bait acceptance is poor because of an abundant, readily available food supply. Non-toxic powders, such as talc or flour, often are used to monitor and track the activity of rodents in buildings.

Granulars

Granulars (G) overcome the disadvantages of dusts in their handling characteristics. The granules are small pellets formed from various inert clays and sprayed with a solution of the toxicant to give the desired content. After the solvent has evaporated, the granules are packaged for use. Granular materials range in size from 20 to 80 mesh. Only insecticides and a few herbicides are formulated as granules. They range from 2 to 25 percent active ingredient and are used almost exclusively in agriculture, although systemic insecticides as granules can be purchased for lawn and ornamentals. Granular materials may be applied at virtually any time of day, since they can be applied aerially in winds up to 20 mph without problems of drift, an impossible task with sprays or dusts.

Granular pesticides are most often used to apply chemicals to the soil to control weeds, nematodes, and insects living in the soil, or for absorption into plants through the roots. Granular formulations are sometimes applied by airplane or helicopter to minimize drift or to penetrate dense vegetation.

Once applied, granules release the active ingredient slowly. Some granules require soil moisture to release the active ingredient. Granular formulations also are used to control larval mosquitoes and other aquatic pests. Granules are used in agricultural, structural, ornamental, turf, aquatic, right-of-way, and public health (biting insect) pest control operations.

Advantages:

- Ready to use—no mixing.
- Drift hazard is low, and particles settle quickly.
- Little hazard to applicator—no spray, little dust.
- Weight carries the formulation through foliage to soil or water target.
- Simple application equipment needed, such as seeders or fertilizer spreaders.
- May break down more slowly than WPs or ECs because of a slow-release coating.

Disadvantages:

- Often difficult to calibrate equipment and apply uniformly.
- Will not stick to foliage or other uneven surfaces.
- May need to be incorporated into soil or planting medium.
- May need moisture to activate pesticide.
- May be hazardous to non-target species, especially waterfowl and other birds that mistakenly feed on the seedlike granules.
- May not be effective under drought conditions because the active ingredient is not released in sufficient quantity to control the pest.

Water-dispersible Granules

Water-dispersible granules (WDG), also known as dry flowables, are like wettable powders except instead of being dustlike, they are formulated as small, easily measured granules. Water-dispersible granules must be mixed with water to be applied. Once in water, the granules break apart into fine particles similar to wettable powders. The formulation requires constant agitation to keep them suspended in water. The percentage of active ingredient is high, often as much as 90 percent by weight.

Water-dispersible granules share many of the same advantages and disadvantages of wettable powders except: They are more easily measured and mixed. Because of low dust, they cause less inhalation hazard to the applicator during handling.

Baits

A bait formulation is an active ingredient mixed with food or another attractive substance. The bait either attracts the pests or is placed where the pests will find it. Pests are killed by eating the bait that contains the pesticide. The amount of active ingredient in most bait formulations is quite low, usually less than 5 percent.

Baits are used inside buildings to control ants, roaches, flies, other insects, and rodent control. Outdoors they sometimes are used to control snails, slugs, and insects such as ants and termites. Their main use is for control of vertebrate pests such as rodents, other mammals, and birds. Spot application—the placing of the bait in selected places accessible only to the target species—permits the use of very small quantities of oftentimes highly toxic materials in a totally safe manner, with no environmental disruption.

Advantages:

- Ready to use.
- Entire area need not be covered because pest goes to bait.
- Control pests that move in and out of an area.

Disadvantages:

- Can be attractive to children and pets.
- May kill domestic animals and non-target wildlife outdoors.
- Pest may prefer the crop or other food to the bait.
- Dead vertebrate pests may cause odor problem.
- Other animals may be poisoned as a result of feeding on the poisoned pests.
- If baits are not removed when the pesticide becomes ineffective, they may serve as a food supply for the target pest or other pests.

Injectable Baits

Pastes and gels are mainly used in the pest control industry for ants and cockroaches. Insecticides formulated as pastes and gels are now the primary formulations used in cockroach control.

They are designed to be injected or placed as either a bead or dot inside small cracks and crevices of building elements where insects tend to hide or travel. Two basic types of tools are used to apply pastes and gels—syringes and bait guns. The applicator forces the bait out of the tip of the device by applying pressure to a plunger or trigger.

Advantages:

- They are odorless, produce no vapors, have low human toxicity, and last for long periods.
- Applicator exposure is minimal.
- Hidden placements minimize human and pet exposure.
- Very accurate in their placement and dosage.
- Easily placed in insect harborage for maximum effectiveness.

Disadvantages:

- Can become contaminated from exposure to other pesticides and cleaning products.
- When exposed to high temperatures, gels can run and drip.
- May stain porous surfaces.
- Repeated applications can cause an unsightly buildup of bait.

Selected from: Randall C, Hock W, Crow E, *et al*. National Pesticide Applicator Certification Core Manual. The National Association of State Departments of Agriculture Research Foundation, 2012: 52-78.

Words and Expressions

sodium fluoride ［无化］氟化钠

pyrophyllite /ˌpaɪərəʊˈfɪlaɪt/ *n*.［矿物］叶蜡石

talc /tælk/ *n*. 滑石粉，爽身粉，滑石

plaster /ˈplɑːstə(r)/ *n*. 石膏，灰泥，膏药

concrete /ˈkɒŋkriːt/ *adj*. 混凝土的，实在的，具体的，有形的；*n*. 具体物，凝结物

slugs /slʌg/ *n*.［无脊椎］蛞蝓，懒汉，子弹（slug 的复数）

Notes

［1］mesh screen：目筛。目前国际上通用的是泰勒标准筛，所谓的多少目是指在每英寸（=2.54cm）的长度上有多少筛孔，如果有 100 个孔，就是 100 目筛，孔数越多，孔眼也就越小。但由于制作材料不同，比如有不锈钢筛、尼龙筛、铜筛等，它们的粗细不同，所以同是 100 目筛，大小实际上也有区别。

［2］mph：mile per hour 的缩写，速度计量单位，表示英里/时。俗称"迈"，1 迈＝1.609344km/h。

Exercises

Ⅰ Answer the following questions according to the text.

1. Why are dusts the least effective and the least economical of the pesticide formulations?
2. What are the advantages of using micro granules?
3. Why are many of insecticides sold for garden use in the form of wettable powders?
4. What are the advantages and disadvantage of granules and baits?

Ⅱ Translate the following English phrases into Chinese.

60-mesh screen	toxic agent	wettable powder	water-soluble powder
active diluents	granular	residual activity	water-dispersible granule
stomach poison	hand sprayer	residual effect	soil application

Ⅲ Translate the following Chinese phrases into English.

固体剂型	硫黄粉	追踪粉	惰性稀释剂	空中喷药	内吸性杀虫剂
多孔材料	球磨机	诱饵枪	缓释杀虫剂	微胶囊剂	局部施用

Ⅳ Choose the best answer for each of the following questions according to the text.

1. What are the advantages of dusts according to the text?

   A. their ease in handling　　B. their ease in formulation

   C. their ease in application　　D. all of the above

2. Compared with dusts, what is the advantage of granular pesticides?

   A. their handling characteristics　　B. their price

   C. their application　　D. their effect

3. Which are the compositions of the wettable powders?

A. wetting agent
B. inert dilution
C. surfactant
D. all of the above

4. Which pests are controlled using baits?
A. ants and roaches
B. flies and rodent
C. soil insect
D. all of the above

5. Which of the following statement is not true according to the text?
A. Spot application of baits permits the use very large quantities of highly toxic materials in a totally safe manner, without no environmental disruption.
B. Insecticidal adhesive tapes work as contact insecticides against arawling insects.
C. Depending on the fumigant, the treated soils may require covering with plastic sheets foe several days to retain the volatile chemical, allowing it to exert its maximum effect.
D. Ozone is by now a deadly issue, so far as aerosols are concerned.

Ⅴ Translate the following short paragraphs into Chinese.

1. dusts have been the simplest formulations of pesticides to manufacture and the easiest to apply. Examples of the undiluted toxic agent are sulfur dusts, sodium fluoride.

2. Wettable powders (WP) are essentially concentrated dusts containing a wetting agent to facilitate the mixing of the powder with water before spraying. The technical material is added to the inert diluent, in this cases a finely ground talc or clay, in addition to a wetting agent or surfactant and mixed thoroughly in a ball mill.

3. The granules are small pellets formed from various inert clays and sprayed with a solution of the toxicant to give the desired content. After the solvent has evaporated, the granules are packaged for use.

4. A bait formulation is an active ingredient mixed with food or another attractive substance. The bait either attracts the pests or is placed where the pests will find it. Pests are killed by eating the bait that contains the pesticide.

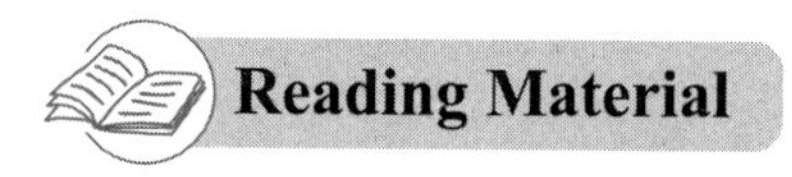

Adjuvants

Adjuvants are chemicals that do not possess pesticidal activity. Adjuvants are either premixed in the pesticide formulation or added to the spray tank to improve mixing or application or to enhance pesticidal performance. They are used extensively in products designed for foliar applications. Adjuvants can be used to customize the formulation to specific needs and compensate for local conditions.

Adjuvants are designed to perform specific functions, including wetting, spreading, sticking, reducing evaporation, reducing volatilization, buffering, emulsifying, dispersing, reducing spray drift, and reducing foaming. No single adjuvant can perform all these functions, but compatible adjuvants often can be combined to perform multiple functions simultaneously.

Surfactants

Surfactants, also called wetting agents and spreaders, physically alter the surface tension of a spray droplet. For a pesticide to perform its function properly, a spray droplet must be able to wet the foliage and spread out evenly over a leaf. Surfactants enlarge the area of pesticide coverage, thereby increasing the pest's exposure to the chemical. Surfactants are particularly important when applying a pesticide to waxy or hairy leaves. Without proper wetting and spreading, spray droplets often run off or fail to cover leaf surfaces adequately. Too much surfactant, however, can cause excessive runoff and reduce pesticide efficacy.

Surfactants are classified by the way they ionize or split apart into electrically charged atoms or molecules called ions. A surfactant with a negative charge is anionic. One with a positive charge is cationic, and one with no electrical charge is nonionic. Pesticidal activity in the presence of a nonionic surfactant can be quite different from activity in the presence of a cationic or anionic surfactant. Selecting the wrong surfactant can reduce the efficacy of a pesticide product and injure the target plant.

Anionic surfactants are most effective when used with contact pesticides (i. e., pesticides that control the pest by direct contact rather than being absorbed systemically). Cationic surfactants should never be used as stand-alone surfactants because they usually are phytotoxic. Nonionic surfactants, often used with systemic pesticides, help pesticide sprays penetrate plant cuticles.

Stickers

A sticker is an adjuvant that increases the adhesion of solid particles to target surfaces. These adjuvants can decrease the amount of pesticide that washes off during irrigation or rain. Stickers also can reduce evaporation of the pesticide, and some slow down the degradation of pesticides by sunlight. Many adjuvants are formulated as spreader-stickers to make a general-purpose product.

Extenders

Some adjuvant manufacturers have named their products "extenders". Extenders' function like stickers by retaining pesticides longer on the target area, slowing evaporation, and inhibiting degradation by sunlight.

Plant Penetrants

These adjuvants have a molecular configuration that enhances penetration of some pesticides into plants. An adjuvant of this type may increase penetration of a pesticide on one species of plant but not another. Enhanced penetration increases the activity of some pesticides.

Compatibility Agents

Pesticides are commonly combined with liquid fertilizers or other pesticides. Certain combinations can be physically or chemically incompatible, which causes clumps and uneven distribution in the tank. Occasionally the incompatible mixture plugs the pump and distribution lines resulting in expensive cleanup and repairs. A compatibility agent may eliminate these problems.

After adding the desired pesticides and the compatibility adjuvant to the jar, shake the mixture and then check for clumping, separation, thickening, and heat release. Any one of these signs indicates an incompatibility problem.

Buffers or pH Modifiers

Most pesticide solutions or suspensions are stable between pH 5.5 and pH 7.0 (slightly acidic to neutral). Above pH 7.0 (alkaline or basic), the pesticide may be subject to degradation. Once a pesticide solution becomes alkaline, the risk exists that the pesticide degrades. Buffers and acidifiers are adjuvants that acidify and stabilize the water in the spray tank. Buffers must be added to the tank mix water first. The water must be neutralized or slightly acidified prior to adding pesticides and adjuvants.

Drift Control Additives

Drift is a function of droplet size. Small, fine drops with diameters of 100 microns or less tend to drift away from targeted areas. Drift control additives, also known as deposition aids, improve on-target placement of the pesticide spray by increasing the average droplet size. Drift reduction can be very important near sensitive sites and may well be worth the small reduction in efficacy that may result from the change in droplet size.

Defoaming Agents

Some pest icide formulations create foam or a frothy "head" in spray tanks. This is often the result of both the type of surfactant used in the formulation and the type of spray tank agitation system. The foam usually can be reduced or eliminated by adding a small amount of a defoaming agent.

Thickeners

As the name suggests, thickeners increase the viscosity (thickness) of spray mixtures. These adjuvants are used to control drift or slow evaporation after the spray has been deposited on the target area. Slowing evaporation is important when using systemic pesticides because they can penetrate the plant cuticle only as long as they remain in solution.

Selected from: Randall C, Hock W, Crow E, *et al*. National Pesticide Applicator Certification Core Manual. The National Association of State Departments of Agriculture Research Foundation, 2012: 52-78.

Words and Expressions

adjuvant /'ædʒʊvənt/ *adj*. 辅助的; *n*. 助剂，辅助物

surfactant /sɜːˈfæktənt/ *n*. 表面活性剂；*adj*. 表面活性剂的

anionic /ˌænaɪˈɒnɪk/ *adj*. 阴离子的，带负电荷的

cationic /ˌkætaɪˈɒnɪk/ *adj*. 阳离子的

nonionic /ˌnɔnaiˈɔnik/ *adj*. 在溶液中不分解成离子的；*n*. 非离子物质

sticker /ˈstɪkər/ *n*. 黏结剂，张贴物，坚持不懈的人，尖刀，难题

extender /ɪkˈstendə/ *n*. [助剂] 填充剂，延伸部分

defoam /diːˈfəʊm/ *v*. 去除……的泡沫，阻止……上形成泡沫

thickener /ˈθɪkənər/ *n*. [助剂] 增稠剂，浓缩机

viscosity /vɪˈskɒsəti/ *n*. 黏性，黏度

PART Ⅳ
NANOPESTICIDE

Unit 7 Nanopesticide

The term "nano" in "nanotechnology" has been originated from the Greek word meaning "dwarf". Accurately, the word "nano" means one-billionth part of a metre (i. e., 10^{-9}). The term nanotechnology is usually used for materials having size range between 1 and 100 nm. It is also obvious that these materials should exhibit different properties from bulk (i. e., micrometric and larger) materials due to their size. Usually these differences in properties include chemical reactivity, physical strength, magnetism, electrical conductance, and optical effects.

The term "nanobiotechnology" is a multidisciplinary combination of nanotechnology, biotechnology, material science, chemical processing, system engineering, nanocrystals, and nanobiomaterials. Researchers in biology and chemistry field are actively engaged in the synthesis of organic, inorganic, hybrid, and metal nanomaterials including different kinds of nanoparticles having unusual optical, physical, biological properties. Due to these unusual properties, nanoparticles have tremendous applications in various fields including electronic, nanomedicine, pharmaceuticals, biomaterial engineering, and agriculture.

Nanotechnology is one of the emerging fields and it has the prospective to change the current scenario of the agricultural and food industry with the help of newly developed methods. These includes the treatment of plant diseases, improving the ability of plants to absorb nutrients, nanobased kits mediated rapid detection of pathogens, and so on. Currently developed nanobiosensors and other smart delivery systems have enormous application in agriculture field. This will help the agricultural industry to contest against different crop pathogens. Nanostructured catalysts will be developed in near future and it will increase the efficacy of commercially available pesticides and insecticides.

Agricultural production also constrained by a number of abiotic and biotic factors like insect pests, diseases, and weeds cause substantial damage to agricultural production. Literature suggests that pests cause 5%～10% in wheat, 25% in rice, 35% in oilseeds, 30% in pulses, 20% in sugar cane, and 50% loss in cotton. These also infest the food and other stored products. These products include storage structures and packages, bins, causing huge amount of loss to the stored food and also deterioration of food quality. The crop loss was estimated to be US $2000 billion per year.

Nanotechnologies have already discovered new and improved way to deliver pesticides through encapsulation and controlled release procedures. Pesticides can be applied more easily and safely in combination with the use of nanoemulsions. Nanotechnologies have a huge potential and offer a great opportunity to develop new products against pests. Nanotechnology improves and enhances its performance and acceptability by increasing effectiveness, safety, patient adherence, as well as ultimately reducing health care costs. Nanoscale devices are capable to detect and treat nutrient deficiency, an infection, or other health problem, long before symptoms were visibly evident.

Nanotechnology principally deals with the introduction of novel nanomaterials that can lead to revolutionary new structures and devices with the help of extremely sophisticated biological tools. Nowadays pesticides are typically applied through drench or spray application to plants or soil as either as "preventative" treatment, or after a disease establishes. In this circumstance, nanotechnology can provide great chance to develop new pesticides. Actually nanotechnology improves the performance and acceptability of traditional or novel pesticides by increasing its effectiveness, targeted action, safety, disease adherence, along with eventually reducing the management costs. Along with the advancement of nanotechnology, nanoscale devices are being proposed that have the potentials to sense and treat an infection at a very beginning stage, actually long before symptoms were apparent at the macroscale. This type of treatment using nanotechnology is possible if the affected area is targeted with a greater awareness of the hazards, which is associated with the use of synthetic organic insecticides and for that reason, exploration of appropriate alternative products for pest control becomes an urgent need. Therefore, nanoparticles are thought to be needed to reduce the number of unnecessary problems associated with agriculture. Development of targeted pesticide delivery will improve efficacy of the pesticide through reductions in dosing intervals, and diminishing its environmental toxicities. We can use nanoparticles for the preparation of new generation of pesticides, insect repellents, and insecticides.

Well-engineered nanoparticles can serve as "magic sphere" which can accommodate pesticides, herbicides, chemicals, or genetic materials, and can be the target to precise parts of plant to release their content in a targeted manner. Porous hollow silica nanoparticles (PHSNs) or mesoporous silica nanoparticles (MSNs) have engrossed the attention of nanotechnologists due to their prospective applications. These relatively inert nanomaterials have features like high surface area, variable pore volume, and stability under specific physicochemical conditions. Also these pore networks are tightly ordered and extremely homogeneous in size. It has been found that MSNP/DNA complexes enhanced receptor-mediated endocytosis efficiency through mannose receptors. These results give an indication of the use of MSNP both as a potential gene carrier to antigen presenting cells and pesticides carrier in the future. These pesticides can be directed very precisely to the target cells of target pest species adopting the similar principle as aforementioned of gene delivery. These could prove themselves extremely effective in sustainable agro-ecosystem. Pesticide validamycin loaded PHSNs can be considered for efficient delivery system for its controlled release. This type of controlled release behavior of the hollow silica nanoparticles provides the possibility of employing it as a promising carrier for pesticide controlled delivery in agricúlture because immediate, as well as prolonged release of pesticide is needed for plants.

To understand the action on nanosilica, the cuticular chemistry of insects must be known. Insects use a variety of cuticular lipids, which help them to maintain their cuticular barrier and thus stop desiccation-related death. When the nanosilica is applied on plant surface, they got rapidly absorbed by the cuticular lipids of insect coat due to their hydrophobic nature. This causes death of the insect by physical means like destroying the hydrophobic coat. It can be said that modified surface charged hydrophobic nanosilica (3～5 nm) could be successfully implemented to manage a variety of ectoparasites of animals and agricultural insect pests. Besides silica, polyethylene-glycol (PEG) coated nanoparticles are useful in pest management. The insecticidal activity of these nanoparticles loaded with garlic essential oil are well documented against the adult of stored products pest *Tribolium castaneum* with an efficacy up to 80% against control. This high efficacy value is probably due to the unique feature of the nanoparticles, which ensure slow but persistent release of the active components from these nanoparticles. Amorphous nanosilica, found in various natural sources (like epidermis of vegetables, burnt pretreated rice hulls, the shell wall of phytoplankton, straw at thermoelectric plants, and volcanic soil, etc.) shows promising potential as a biopesticide. The silica nanoparticles undergo physisorption by the cuticular lipids and were reported to cause the death of insects purely applying physical means of disrupting protective barrier. Neither photosynthesis nor respiration, were altered by the application of nanoparticles on the leaf and stem surface in several groups of crop and horticultural plants.

World Health Organization (WHO) also considers the use of amorphous silica as a safe nanobiopesticide for humans. A recent report suggests that silica nanoparticles caused 100% mortality in rice weevil (*Sitophilus oryzae*). Furthermore, silica nanoparticles (3～5 nm), modified by hydrophobic charge on the surface, were successfully utilized to regulate a range of agricultural insect pests and veterinary importance animal ectoparasites. This particle has also been successfully applied as a thin film on seeds to reduce the fungal growth and boost cereal germination.

Nanocides are pesticides encapsulated at nanoscale. Packaging of nanoscale active materials inside a tiny "envelope " or "shell" could be considered as a more sophisticated encapsulation approach for the formulation of nanoscale pesticides. Nanocapsules are usually composed of thin external layers, considered as shells, having huge space inside. At present, the methods of both controlled release and encapsulation are playing a major role in controlling the use of various pesticides, herbicides, as well as fertilizers in agriculture worldwide. Nanoencapsulation is the method of encapsulation of the so-called nanomaterial and is probably the most promising technology used to protect host plants against insect pests. Nanocapsules are generally composed of polymers and this technology helps to formulate nanocide containing the pesticides inside the shell. In this preparation, the chemical is protected by the shell and remains safe from the damage induced by the external agents. Besides, it improves the solubility of the active material and helps to enhance the penetration power through the target tissues. By adjusting the external factors (e. g. , pH), the chemicals inside the shells can be released through the controlled opening which makes these capsules more attractive toward the balanced unloading of the materials as they reach target places of a particularly defined conditions.

Selected from: Grumezescu A M. New Pesticides and Soil Sensors-Nanotechnology in the Agri-Food Industry, Volume 10. Academic Press, Elsevier Inc, 2017: 47-60.

Words and Expressions

nano /ˈnænəu/ *n.* 纳米
nanotechnology /ˌnænəutekˈnɔlədʒi/ *n.* 纳米技术
dwarf /dwɔːf/ *n.* 侏儒，矮子
micrometric /ˌmaikrəuˈmetrik/ *adj.* 测微的，测微术的
multidisciplinary /ˌmʌltidisəˈplinəri/ *adj.* (涉及) 多门学科的，多专业的
nanoparticle /ˈnænəupɑːtikl/ *n.* 纳米粒子，纳米颗粒，毫微粒
tremendous /trəˈmendəs/ *adj.* 极大的，巨大的，惊人的，极好的
emerging /iˈməːdʒiŋ/ *adj.* 新兴的
scenario /səˈnɑːriəu/ *n.* 方案，情节，剧本，设想
nanobased kit 纳米试剂盒
enormous /iˈnɔːməs/ *adj.* 庞大的，巨大的
constrain /kənˈstrein/ *v.* 限制，限定，约束
substantial /səbˈstænʃl/ *adj.* 大量的，实质的
sugar cane 甘蔗
infest /inˈfest/ *vt.* 骚扰，感染，扰乱，寄生于
encapsulation /inˌkæpsjuˈleiʃn/ *n.* 封装，包装
nanoemulsion /ˌnænəuiˈmʌlʃn/ *n.* 纳米乳液，纳米乳剂
nanoscale /ˈnænəuskeil/ *n.* 纳米级，纳米尺度
deficiency /diˈfiʃnsi/ *n.* 缺点，缺陷，缺乏
sophisticated /səˈfistikeitid/ *adj.* 复杂的，精细的，精致的
drench /drentʃ/ *n.* 浸泡，浸液
spray /sprei/ *n.* 喷雾
preventative /priˈventətiv/ *adj.* 预防性的，防止的
symptom /ˈsimptəm/ *n.* [临床] 症状，征兆
macroscale /ˈmækrəskeil/ *n.* 宏观尺度，大规模
dosing interval *n.* [医] 给药间隔
diminishing /diˈminiʃiŋ/ *v.* 减少，递减，衰减
repellent /riˈpelənt/ *n.* 驱虫剂
engross /inˈgrəus/ *vt.* 吸引，独占，使全神贯注
homogeneous /ˌhɔməˈdʒiːniəs/ *adj.* 均匀的，[数] 齐次的，同类的，同质的

endocytosis /ˌendəusaiˈtəusis/ *n*.［细胞］内吞作用

mannose /ˈmænəuz; -s/ *n*.［生化］甘露糖

antigen presenting cell ［组织］抗原递呈细胞

species /ˈspiːʃiːz/ *n*.［生物］物种，种类

cuticular /kjuˈtikjələ/ *adj*.［昆］表皮的，角质层的

desiccation /ˌdesiˈkeiʃn/ *n*. 干燥

hydrophobic /ˌhaidrəˈfəubik/ *adj*. 疏水的，憎水的

ectoparasite /ˌektəuˈpærəsait/ *n*. 体表寄生虫

garlic /ˈgɑːlik/ *n*. 大蒜，蒜头

essential oil ［林］［化工］香精油，精油

amorphous /əˈmɔːfə/ *adj*. 无定形的，无组织的，［物］非晶形的

epidermis /ˌepiˈdəːmisː/ *n*. 上皮，表皮

rice hull ［粮食］稻壳，稻谷壳，粗糠

phytoplankton /ˈfaitəuˌplæŋ(k)t(ə)n/ *n*.［植］浮游植物（群落）

straw /strɔː/ *n*. 稻草，秸秆

thermoelectric /ˌθəːməuiˈlektrik/ *adj*.［电］热电的

physisorption /ˌfiziˈsɔːpʃən/ *n*. 物理吸附

photosynthesis /ˌfəutəuˈsinθəsis/ *n*. 光合作用

respiration /ˌrespəˈreiʃn/ *n*. 呼吸，呼吸作用

horticultural /ˌhɔːtiˈkʌltʃərəl/ *adj*. 园艺的

veterinary /ˈvetnri; ˈvetrənəri/ *adj*. 兽医的

cereal /ˈsiəriəl/ *n*. 谷类植物，谷类，谷物

germination /ˌdʒəːmiˈneiʃn/ *n*. 发芽，萌发

capsule /ˈkæpsjuːl/ *n*.［药］胶囊

Notes

［1］porous hollow silica nanoparticles（PHSNs）：多孔空心二氧化硅纳米颗粒。

［2］mesoporous silica nanoparticles（MSNs）：介孔二氧化硅纳米颗粒。

［3］validamycin：井冈霉素。井冈霉素是一种放线菌产生的抗生素类低毒杀菌剂，由中国工程院院士沈寅初于20世纪70年代研制。

［4］polyethylene-glycol（PEG）：聚乙二醇。聚乙二醇是分子量在200～8000或者8000以上的乙二醇高聚物的总称。

［5］*Tribolium castaneum*：赤拟谷盗。赤拟谷盗也称面粉虫，是鞘翅目拟步甲科拟谷盗属中常见的贮藏物害虫。

［6］World Health Organization（WHO）：世界卫生组织。简称世卫组织，是联合国系统内卫生问题的指导和协调机构，总部设置在瑞士日内瓦。它负责拟定全球卫生研究议程，制定规范和标准，向各国提供技术支持以及监测和评估卫生趋势。世卫组织只有主权国家才能参加，是国际上最大的政府间卫生组织。

［7］rice weevil（*Sitophilus oryzae*）：米象。米象俗称蛘子，属鞘翅目象虫科，是贮藏谷物的主要害虫。

Exercises

Ⅰ Answer the following questions according to the text.

1. Where does the term "nano" in "nanotechnology" originate from? And how is it defined?

2. Why nanotechnology can provide great chance to develop new pesticides?

3. Why the nanoparticles MSNP can be considered both as a potential gene carrier to antigen presenting cells and pesticides carrier in the future?

4. What are nanocides? At present, what are the possible approaches to formulate it?

Ⅱ Translate the following English phrases into Chinese.

prolonged release	optical effect	pesticides carrier	electrical conductance
controlled release	surface area	physical strength	targeted pesticide delivery
hydrophobic coat	pore volume	chemical reactivity	sustainable agro-ecosystem

Ⅲ Translate the following Chinese phrases into English.

纳米生物农药	纳米粒子	纳米乳液	纳米胶囊	纳米生物传感器
纳米二氧化硅	纳米材料	生物因素	寄主植物	角质层脂质

Ⅳ Do the following statements agree with the information given in the text (True, False or Not Given)?

1. Nanotechnology is a frontier field that can lead to revolutionary in the agricultural and food industry with the help of newly developed methods.

2. Nanostructured catalysts have increased the efficacy of commercially available pesticides and insecticides.

3. It has been found that MSNP/DNA complexes enhanced receptor-mediated endocytosis efficiency through glucose receptors.

4. Nanosilica has an excellent hydrophobicity that can get rapidly absorbed by the cuticular lipids of insect coat.

5. Nanoemulsions are usually composed of thin external layers, considered as shells, having huge space inside.

Ⅴ Translate the following short passages into Chinese.

1. The term "nanobiotechnology" is a multidisciplinary combination of nanotechnology, biotechnology, material science, chemical processing, system engineering, nanocrystals, and nanobiomaterials.

2. Well-engineered nanoparticles can serve as "magic sphere" which can accommodate pesticides, herbicides, chemicals, or genetic materials, and can be the target to precise parts of plant to release their content in a targeted manner.

3. Nowaday's pesticides are typically applied through spray application to plants or soil as either as "preventative" treatment, or after a disease establishes. In this circumstance, nanotechnology can provide great chance to develop new pesticides. Actually nanotechnology improves the performance and acceptability of traditional or novel pesticides by increasing its effectiveness, targeted action, safety, disease adherence, along with eventually reducing the management costs.

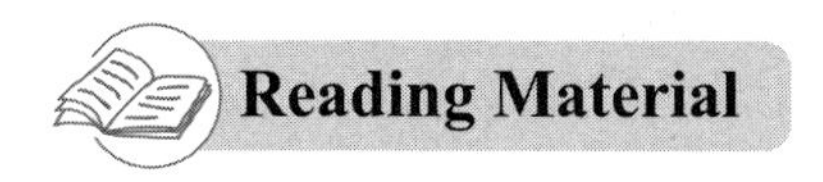

Nanoparticles

Pesticides, such as insecticides, acaricides, rodenticides, nematicides, molluscicides, fungicides, and herbicides are commonly used in agriculture to improve crop yield and efficiency and in general protect plants from damaging influences, such as weeds, plant diseases, or insects. It was found that some of used pesticides are toxic for environment; moreover, some of them are hazardous or even toxic for human and other animals. Since the development of effective and safe pesticides is difficult and expensive, new technologies have been tested. Nanoscale science and nanotechnology have been demonstrated to have a great potential in providing novel and improved solutions, and thus the number of contributions increases exponentially.

Nanotechnology is regarded as one of the key technologies of the 21st century. Nanosize materials change their physical and chemical properties; a variety of industrial, pharmaceutical, and medical products have been improved and innovated in such a way. Nanotechnology application in agriculture and the food sector is relatively recent compared to its use in pharmacy. The use of nanotechnologies can significantly contribute to sustainable intensification of agricultural production, because they can facilitate not only protection of plants against pesticides but also monitoring of plant growth, secure rising of global food production, guarantee enhanced food quality, and minimize the waste. Food and agricultural production are among the most important fields of nanotechnology application.

According to the adopted recommendation on the definition of a nanomaterial by the European Commission, nanomaterial means "a natural, incidental, or manufactured material containing particles, in an unbound state or as an aggregate or as an agglomerate and where, for 50% or more of the particles in the number size distribution, one or more external dimensions is in the size range 1～100 nm. In specific cases and where warranted by concerns for the environment, health, safety, or competitiveness the number size distribution threshold of 50% may be replaced by a threshold between 1% and 50%. By derogation from the aforementioned, fullerenes, graphene flakes, and single wall carbon nanotubes with one or more external dimensions below 1 nm should be considered as nanomaterials".

In general, it can be stated that the ability of nanoparticles (NPs) "to permeate anywhere" connects primarily with their particle size and shape. Three categories of NPs were defined based on particle size: (i) ultrafine particles, $<$100 nm in diameter (largely consisting of primary combustion products); (ii) accumulation-mode particles between 100 nm and 2.5 μm in diameter; and (iii) coarse-mode particles $>$2.5 μm in diameter.

Based on aforementioned particle size definitions, a concept of a classification system was suggested that classify NPs not only according to their particle sizes but also according to biodegradability (ability of a compound to be degrade in organism/environment) into four classes: (i) size $>$100 nm and biodegradable; (ii) size $>$100 nm and nonbiodegradable;

(iii) size ＜100 nm and biodegradable; (iv) size ＜100 nm and nonbiodegradable.

NPs are characterized by large surface area that is able to bind, absorb and transport compounds. Besides their extremely small particle size (that causes easy permeability through any biomembrane), they are used for carrying compounds.

Therefore, nanosystems consist of two basic components: an active ingredient and a nanocarrier that stabilizes the active ingredient in the nanoform. Nanoformulations usually consist of several surfactants, polymers, or inorganic (e. g. , metal). NPs in the nanometer size range and therefore cannot be considered as a single entity. Current NP platforms can be classified into three major categories including: (i) inorganic-based (solid) NPs (nonbiodegradable), for example, nanocrystals, shells, quantum dots based on gold, silver, copper, iron, various semiconductors, ceramic (silica-based NPs), and carbon nanotubes or fullerenes; (ii) organic-based NPs (frequently biodegradable), for example, liposomes, solid lipid NPs, polymeric NPs, micelles, and capsules; and (iii) hybrid NPs (combination of inorganic and organic components). Most frequently organic-based NPs consist of water-soluble biodegradable biocompatible polymers, such as polylactide homopolymers, poly (lactide-co-glycolide), poly (propylene glycol), poly (ε-caprolactone), poly (amidoamine), polyacrylamide, and polysorbate 80-coated poly (butylcyanoacrylate), or from natural compounds, such as chitosan, lignin, phospholipids, lecithin, starch, cellulose, and alginates.

The formulations based on biodegradable organic-based matrices allow designing controlled release nanocarriers. The surface of prepared nanocarriers can be modified by various other molecules or specific compounds, and by this approach aqueous solubility (lipophilicity or hydrophilicity) can be modified, and targeted biodistribution can be ensured.

Selected from: Grumezescu A M. New Pesticides and Soil Sensors-Nanotechnology in the Agri-Food Industry, Volume 10. Academic Press, Elsevier Inc. , 2017: 81-84.

Words and Expressions

intensification /inˌtensifiˈkeiʃn/ *n*. 强化，集约化

European Commission 欧盟委员会

aggregate /ˈæɡriɡət/ *n*. 集合体

agglomerate /əˈɡlɔməreit/ *n*. 团块，凝聚物

dimension /daiˈmenʃn; diˈmenʃn/ *n*. [数] 尺寸

threshold /ˈθreʃhəuld/ *n*. 临界值，阈值，极限

derogation /ˌderəˈɡeiʃn/ *n*. 毁损，减损

ultrafine particles 超微颗粒，超细颗粒物

accumulation-mode particles 积聚模颗粒

coarse-mode particles 粗模态颗粒，粗颗粒物

biodegradable /ˌbaiəudiˈɡreidəbl/ *adj*. 可生物降解的

biomembrane /ˌbaiəuˈmembrein/ *n*. [分子生物] 生物膜

nanoformulation /ˈnænəuˌfɔːmjuˈleiʃn/ *n*. 纳米剂型，纳米制剂

surfactant /səːˈfæktənt/ *n*. 表面活性剂

quantum dot 量子点

semiconductor /ˌsemikənˈdʌktə(r)/ *n*.［电子］［物］半导体

ceramic /səˈræmik/ *n*. 陶瓷，陶瓷制品

carbon nanotube 碳纳米管

fullerene /ˈfuləriːn/ *n*. 富勒烯，球壳状碳分子

liposome /ˈlipəsəum/ *n*.［生化］脂质体

polymeric /ˌpɔliˈmerik/ *adj*. 聚合的，聚合体的，聚合物的

micelle /miˈsel:maiˈsel/ *n*.［分子生物］胶束；［生物］微团

polylactide homopolymer 聚丙交酯均聚物

lactide-co-glycolide 丙交酯-乙交酯共聚物

propylene glycol 丙烯乙二醇

ε-caprolactone ε-己内酯

polyacrylamide /ˌpɔliəˈkriləmaid/ *n*.［高分子］聚丙烯酰胺

butylcyanoacrylate /ˈbjuːtailˌsaiənəuˈækrileit/ *n*. 氰基丙烯酸正丁酯

lignin /ˈlignin/ *n*.［木］木质素

phospholipid /ˌfɔsfəˈlipid/ *n*.［生化］磷脂

lecithin /ˈlesiθin/ *n*.［生化］卵磷脂，蛋黄素

starch /stɑːtʃ/ *n*. 淀粉

cellulose /ˈseljuləus/ *n*. 纤维素

alginates /ˈældʒəˌnet/ *n*. 海藻酸盐

organic-based matrices 有机基质

lipophilicity /lipəfiˈlisiti/ *n*.［化学］亲油性，亲脂性

hydrophilicity /ˈhaidrəfilisiti/ *n*.［化学］亲水性，亲和性

Unit 8 Nanoformulation

Nanoformulations, similarly to other pesticide or drug formulations, are designed to increase the apparent solubility of poorly soluble active ingredients and to release the active ingredient in a slow/targeted manner and/or to protect the active ingredient against premature degradation.

Nanopesticide formulations can be classified according to several criteria. For instance, they can be classified according to their intended purpose as: (i) formulations that increase the solubility of poorly-water soluble pesticides; (ii) formulations that slow down the release rate of the encapsulated pesticide; (iii) formulations able to achieve targeted delivery and to provide protection against premature degradation.

Herein nanoformulations will be divided into different categories based on the chemical nature of the nanocarrier, such as organic polymer-based formulations, lipid-based formulations, nanosized metals and metal oxides, clay based nanomaterials, layered double hydroxides, silica nanoparticles, and so on. Furthermore, depending on the particular manufacturing method nanoparticles falling into any of the aforementioned classes can have various structures and morphologies with different properties like nanocapsules, nanospheres, micelles, nanoliposomes, nanoemulsions, nanosuspensions, and so on.

Nanocapsules

Nanocapsules are vesicular or reservoir type structures comprising an inner central cavity surrounded by a polymer coating or membrane. The internal cavity, which confines the active ingredients (AIs) may be either hydrophilic or hydrophobic. In the former case, the structure is also referred to as a polymersome, the aqueous cavity being surrounded by a bilayer of an amphiphilic di- or triblock (ABA) copolymer coating, or by a monolayer membrane of an amphiphilic triblock copolymer (ABA), where A and B stand for the hydrophilic, and the hydrophobic segments, respectively. Unlike nanocapsules which can entrap only hydrophobic AIs, polymersomes can encapsulate both hydrophilic AIs which are solubilized in the aqueous inner cavity and hydrophobic AIs which are confined within the amphiphilic membrane through hydrophobic interactions with the hydrophobic B blocks.

Polymeric nanocapsules can be fabricated from preformed polymers or alternatively during the polymerization of suitable monomers. Both methods have been used to prepare pesticide loaded nanocapsules.

1. Preparation Nanocapsules Starting with Monomers

Methyl methacrylate (MMA)-styrene (St) copolymer nanocapsules containing the lansiumamide B (LB) have been prepared by Yin. LB isolated from kernels of *Clausena lansium* was dissolved together with the two monomers in a little amount of petroleum ether and chloroform. Next, this solution was emulsified by ultrasonication in an aqueous solution of the anionic emulsifier sodium dodecyl sulfate and the cosurfactant, *n*-amyl alcohol. Then, the radical polymerization initiator azoisobutyronitrile was added to the clarify emulsion and the reaction mixture was heated in 70℃ water bath for 3 h under magnetic stirring. Eventually, a translucent aqueous suspension containing LB-loaded poly (MMA-co-St) nanocapsules was obtained.

2. Preparation Nanocapsules from Preformed Polymers

There are several methods to prepare pesticide-loaded nanocapsules from preformed polymers, such as nanoprecipitation, emulsion-solvent diffusion, emulsion-solvent evaporation, layer by layer self-assembly, ionic gelation, polyelectrolyte complexation, and melt-dispersion techniques.

Biodegradable polyesters and polysaccharides have been mostly used. Using an oil-in-water (O/W) emulsion-solvent evaporation technique, Pereira has prepared poly (ε-caprolactone) nanocapsules containing the herbicide atrazine. In brief, atrazine was dissolved in the water miscible organic solvent acetone, while the polymer was dissolved in the water immiscible organic solvent methylene chloride and myritol was added into the latter solution. The organic phase was produced by mixing and sonicating the above two solutions for 1 min. Then, the aqueous phase containing polyvinyl alcohol as a hydrophilic surfactant was added and the O/W emulsion was sonicated for 8 min. Eventually, organic solvents were removed in a rotary evaporator. The resulted formulation contained 1 mg/mL of atrazine.

Micelles

Amphiphilic block copolymers self-assemble in water to form colloidal particles with a core-shell morphology called micelles. The self-assembly process is driven by the hydrophobic interactions developed on the formation of the hydrophobic micellar core which is surrounded by a hydrophilic shell or corona. However, aggregation occurs only after the concentration of the individual amphipathic molecules in solution exceeds a certain threshold concentration named the critical micellar concentration (CMC) and it is marked by an abrupt change in many physicochemical properties of the aqueous solution of the amphiphile. So, micelles are dynamic structures in the sense that if the amphiphile concentrations drops below CMC, then disassembly takes place.

There is no clean cut division between micelles and nanospheres prepared from amphiphilic block copolymers. That means that as the length of the hydrophobic block increases, the central hydrophobic core of the micelles becomes more dense resembling nanospheres.

Nanoliposomes

Nanoliposomes are vesicular structures consisting of a phospolipid bilayer enclosing an internal aqueous cavity at the nanoscale level. When placed in water amphiphilic phospholipids undergo a self-organization process forming a bilayer membrane. During this process, the hydrophobic tails of phospholipid molecules approach each other becoming tightly packed in a parallel alignment within the bilayer and squeezing out water molecules from this region, while the polar heads of the phospholipid molecules are positioned at the margins of the supramolecular aggregate being in contact with the surrounding polar water molecules. With the aid of an energy input, these flat membranes can be converted into nanoparticulate delivery systems with a bilayer vesicular morphology.

The methods to produce liposomes can be divided in two main categories: mechanical and nonmechanical. Mechanical methods include sonication/ultrasonication, high pressure homogenization, extrusion, microfluidization, colloid mill, and others. Nonmechanical methods include reversed-phase evaporation and depletion of mixed detergent-lipid micelles. The fluidity of the lipid bilayer can be modulated by incorporation into the phospholipid membrane of cholesterol which prevents crystallization of the acyl chains of phospholipids and enhances the shape stability of nanoliposomes. However, long-term storage of these nanocarriers is problematic because of the physical and chemical instabilities of liposomes in aqueous dispersions.

Nanoemulsions

Nanoemulsions are emulsions with droplet size in the nanometer range. They are kinetically stable but not thermodynamically stable. So, one can say that nanoemulsions are metastable systems and their stability depends on the method of preparation. There are two main types of methods to prepare nanoemulsions: high-energy emulsification methods which make use of high-shear stirring, high-pressure homogenizers, and ultrasonic generators, and low-energy emulsification methods which take advantage of the energy stored in the system to promote the formation of small droplets.

Nanoemulsion is a complex systems entailing of water, surfactant and oil phase. The resulting colloidal solution is kinetically stable and optically isotropic. It has a droplet size, which is in 20 ~ 200 nm range. Presently, due to wide range of nanoparticle sizes nanoemulsions are widely used in different branches of applied science including agriculture.

Nanosuspensions

Nanosuspensions, also called nanodispersions are formed by dispersion of crystalline or amorphous nanoparticles of the AI in liquid media. For example: the preparation of an aqueous suspension of pyridalyl containing alginate (ALG) nanocapsules. Saini synthesized pyridalyl-loaded ALG nanocapsules by an ionotropic pregelation method. Sodium ALG was dissolved in distilled water at 50~60℃, and the hydrophilic surfactant Tween 80 was added to this solution within 15 min under continuous stirring. Subsequent dropwise addition of an ethanolic solution of pyridalyl at room temperature afforded an emulsion, which was further sonicated for 15 min. Next, a solution (2 mg/mL) of $CaCl_2$ was dropped into the latter emulsion over a course of 60 min. The stirring was continued for another 60 min and the resulting nanocapsule suspension was allowed to stand overnight. Then, the solvent was evaporated at 45~50℃ under reduced pressure for 30 min eventually yielding an aqueous suspension of pyridalyl nanocapsules.

In controlled release formulations (CRFs) for pesticide delivery most frequently nanospheres, nanocapsules, hydrophilic nanogels and micelles are used. The benefit of such formulations is that the application of smaller amount of active compound per area is sufficient, as long as the formulation may provide an optimal concentration delivery for the target pesticide for longer times. Since reapplications are not necessary, they decrease the costs and reduce phytotoxicity and environmental damage to other untargeted organisms and even the crops themselves.

Nanoencapsulation can enhance the dispersion of hydrophobic pesticides in aqueous media and allows a controlled release of the active ingredient. Such smart delivery systems are characterized by high selectivity, without hindering the bioactive agents to target the pathogen.

Moreover, insoluble compounds in preparations of NP pesticides can be relatively readily dispersed in solution, the interaction with the target insect could be more effective, and the risk of environment contamination is reduced. It is important for CRFs that they must remain inactive until the active compound is released.

The benefits of polymer encapsulated formulations in comparison with nonencapsulated commercial formulations are extended activity, reduced evaporative, degradation losses and leaching, and these formulations are especially advantageous when a half-life of a potent pesticide is too short.

Selected from: Grumezescu A M. New Pesticides and Soil Sensors-Nanotechnology in the Agri-Food Industry, Volume 10. Academic Press, Elsevier Inc, 2017: 95-184.

Words and Expressions

premature /'premətʃər/ *adj*. 过早的，提前的
degradation /ˌdegrə'deiʃn/ *n*. 降解，退化
criteria /krai'tiəriə/ *n*. 标准，条件（criterion 的复数）
morphology /mɔː'fɔlədʒi/ *n*. 形貌，形态
nanosphere /'nænəusfiə/ *n*. 纳米微球
nanosuspension /ˌnænəusə'spenʃn/ *n*. 纳米悬浮剂
vesicular /vi'sikjulə/ *adj*. 囊的，水泡的，有泡的
reservoir /'rezəvwɑː(r)/ *n*. 水库，蓄水池
bilayer /'baileiə/ *n*. 双层，双分子层（膜）
amphiphilic /ˌæmfi'filik/ *adj*. 两亲的，两亲性质的
triblock /'traiblɔk/ *n*. 三嵌段
copolymer /kəu'pɔlimə/ *n*.［高分子］共聚物
monolayer /'mɔnəleiə/ *n*. 单层，单分子层（膜）
fabricate /'fæbrikeit/ *vt*. 制造，装配
encapsulate /in'kæpsjuleit/ *vt*. 压缩，将……装入胶囊，将……封进内部
polymerization /ˌpɔlimərai'zeiʃn/ *n*.［高分子］聚合（作用）

monomer /'mɔnəmə/ *n.* 单体，单元结构
styrene /'stairi:n/ *n.* [有化] 苯乙烯
kernel /'kə:nl/ *n.* 核，籽粒，种子
petroleum ether 石油醚
chloroform /'klɔrəfɔ:m/ *n.* 氯仿，三氯甲烷
emulsify /i'mʌlsifai/ *v.* 乳化
ultrasonication /ˌʌltrəˌsɔni'keiʃən/ *n.* 超声破碎，超声处理
anionic emulsifier 阴离子乳化剂
n-amyl alcohol [有化] 正戊醇
radical polymerization initiator 自由基聚合引发剂
water bath 水浴
translucent /trænz'lu:snt/ *adj.* 透明的，半透明的
diffusion /di'fju:ʒn/ *n.* 扩散
evaporation /iˌvæpə'reiʃn/ *n.* 蒸发
polyelectrolyte /ˌpɔlii'lektrəlait/ *n.* [物化] 聚合电解质，聚电解质
dispersion /di'spə:ʃn/ *n.* 分散
polyester /ˌpɔli'estər/ *n.* 聚酯
polysaccharide /ˌpɔli'sækəraid/ *n.* [有化] 多糖，多聚糖
oil-in-water 水包油
atrazine /'ætrəzi:n/ *n.* [农药] 莠去津
miscible /'misəbl/ *adj.* [化学] 易混合的，可溶混的，能混溶的
immiscible /i'misəbl/ *adj.* 互不相溶的
myritol 桃金娘油，香桃木油
polyvinyl alcohol [高分子] 聚乙烯醇
colloidal /kə'lɔidəl/ *adj.* 胶体的，胶质的，胶状的
micellar /mi'selə/ *adj.* 胶束的，微胞的
corona /kə'rəunə/ *n.* 冠状物
abrupt /ə'brʌpt/ *adj.* 突然的，意外的
dense /dens/ *adj.* 稠密的，浓厚的
resemble /ri'zembl/ *vt.* 类似，像
squeeze /skwi:z/ *v.* 挤，压榨
supramolecular /ˌsu:prəmə'lekjulə/ *adj.* 超分子的
homogenization /hɔˌmɔdʒənai'zeiʃən/ *n.* 均质化，均匀化，均化（作用）
extrusion /ik'stru:ʒn/ *n.* 挤出，喷出
microfluidization /maikrəuflu:idai'zeiʃn/ *n.* [医] 微流化
colloid mill [化工] 胶体磨
depletion /di'pli:ʃn/ *n.* 消耗，损耗
detergent /di'tə:dʒənt/ *n.* 清洁剂，去垢剂
fluidity /flu'idəti/ *n.* [流] 流动性，流质，易变性
modulate /'mɔdjəleit/ *v.* 调节，调制，调整
cholesterol /kə'lestərɔl/ *n.* [生化] 胆固醇

problematic /ˌprɔbləˈmætik/ *adj*. 问题的，有疑问的，不确定的

droplet /ˈdrɔplət/ *n*. 小滴，微滴

metastable /ˌmetəˈsteibəl/ *adj*. [物] [化学] 亚稳的，相对稳定的

shear /ʃiə(r)/ *n*. 剪切

homogenizer /hɔˈmɔdʒənaizə/ *n*. 均质机

emulsification /iˌmʌlsifiˈkeiʃən/ *n*. [化学] 乳化，乳化作用

isotropic /ˌaisəˈtrɔpik/ *adj*. [物] [数] 各向同性的，等方性的

ionotropic /aiˌɔnəˈtrɔpik [化学] *adj*. 离子移变的

phytotoxicity /ˌfaitətɔkˈsisiti/ *n*. 危害植物的毒性，药害

hinder /ˈhində(r)/ *v*. 阻碍，妨碍，干扰

leaching /ˈliːtʃiŋ/ *n*. [矿业] 沥滤，浸出

Notes

[1] layered double hydroxides：层状双氢氧化物，英文缩写 LDHs，是一种典型的无机层状材料，具化学组分可调控性以及相应的物理化学性能可调控性，被广泛应用于吸附、催化、离子交换以及纳米复合材料等领域。

[2] methyl methacrylate：甲基丙烯酸甲酯，简称甲甲酯，是一种重要的化工原料，易燃，有强刺激性气味，中等毒性。

[3] lansiumamide B：黄皮新肉桂酰胺 B。从植物黄皮种子中提取分离得到的一种酰胺类生物碱，具杀虫、抑菌等生物活性。

[4] *Clausena lansium*：黄皮。黄皮主要分布于热带、亚热带地区，我国主要分布于长江以南各省。黄皮植物除具有药用价值外，其提取物还具有优异的杀虫、抑菌及除草等生物活性。

[5] sodium dodecyl sulfate：十二烷基硫酸钠，英文缩写 SDS，可用作阴离子表面活化剂、乳化剂及发泡剂等。

[6] nanoprecipitation：纳米沉淀法。一种制备纳米颗粒的方法，主要通过控制溶质与非溶剂的混合（产生局部过饱和）来制备纳米颗粒。

[7] pyridalyl：啶虫丙醚。一种新型高效、低毒杀虫剂，由日本住友化学公司研发，主要用于防治鳞翅目害虫幼虫。

[8] layer by layer self-assembly：层层自组装。20 世纪 90 年代快速发展起来的一种简易、多功能的表面修饰方法。它利用逐层交替沉积的方法，借助各层分子间的弱相互作用（如静电引力、氢键、配位键等），使层与层自发缔合形成结构完整、性能稳定、具有某种特定功能的分子聚集体或超分子结构。

Exercises

Ⅰ Answer the following questions according to the text.

1. What are the advantages of nanoformulations? How are they classified?

2. What are the characteristics of nanocapsules? What methods can be used to prepare pesticide loaded nanocapsules?

3. What are the characteristics of nanoliposomes? Do they have any? limitations?

4. Which nanoformulations are most frequently used in controlled release formulations for pesticide delivery? And why?

Ⅱ Translate the following English phrases into Chinese.

melt-dispersion techniques	emulsion-solvent evaporation	ionic gelation
emulsion-solvent diffusion	polyelectrolyte complexation	high-shear
critical micellar concentration	amphiphilic triblock copolymer	anionic emulsifier

Ⅲ Translate the following Chinese phrases into English.

纳米载体	纳米微球	纳米脂质体	纳米分散剂
纳米凝胶	纳米包封	纳米悬浮剂	纳米乳剂

Ⅳ Do the following statements agree with the information given in the text (True, False or Not Given)?

1. Nanopesticide formulations are mainly classified by their intended purpose or the chemical nature of the nanocarrier.

2. Polymersomes can encapsulate both hydrophilic AIs and hydrophobic AIs.

3. Nanoemulsions are emulsions with droplet size in the nanometer range, which have good thermodynamic stability.

4. Saini synthesized pyridalyl-loaded ALG nanocapsules by an iononic gelation method.

5. It is a challenge that CRFs must remain inactive until the active compound is released.

Ⅴ Translate the following short passages into Chinese.

1. Nanoformulations, similarly to other pesticide or drug formulations, are designed to increase the solubility of poorly soluble active ingredients and to release the active ingredient in a slow manner or to protect the active ingredient against premature degradation.

2. Nanocapsules are vesicular or reservoir type structures comprising an inner central cavity surrounded by a polymer coating or membrane. The internal cavity, which confines the active ingredients (AIs) may be either hydrophilic or hydrophobic.

3. During this process, the hydrophobic tails of phospholipid molecules approach each other becoming tightly packed in a parallel alignment within the bilayer and squeezing out water molecules from this region, while the polar heads of the phospholipid molecules are positioned at the margins of the supramolecular aggregate being in contact with the surrounding polar water molecules.

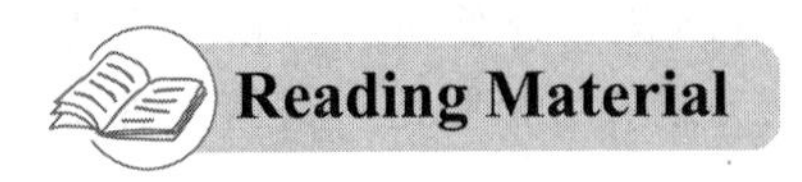

Advantages of Nanoemulsions and Nanodispersions

While the main objective of preparing colloids of chemical pesticides is to improve water solubility, and hence delivery of the active ingredients, recent research has shown that reformulation of chemical pesticides as a colloid can also influence other important properties of the pesticide.

Bioavailability

In many cases, a significant increase in bioavailability is observed, due to a combination of greater surface area for exposure (smaller particle sizes in addition to enhanced wetting and spreading of the pesticide on treated surfaces) and enhanced penetration through waxy plant and insect cuticles. For example, studies by Anjali on neem oil nanoemulsions showed that the median lethal concentration (LC_{50}) against southern house mosquitoes (*Culex quinquefasciatus*) was highly dependent on the emulsion droplet size, with smaller droplets being substantially more effective than larger droplets. This was hypothesized to be due to an increase in the surface area of the droplets, which increases the rate of accumulation by the larvae of the insecticidal components of the oil.

The herbicide glyophosate has also been shown to have improved, efficacy when it is formulated as a colloid. Lim prepared glyphosate nanoemulsions (200 nm diameter) with various proportions of fatty acid methyl esters, organo-silicones, and alkyl glucosides, and compared their efficacy with commercially available formulation Roundup against a variety of plant pests. These nanoemulsions yielded improved efficacy against cowsfoot grass (*Eleusine indica*), and had a similar efficacy against creeping foxglove (*Asystasia gangetica*), slender buttonweed (*Diodia ocimifolia*), and buffalo grass (*Paspalum conjugatum*), despite having significantly lower spray deposition on the leaf surface. This increased efficacy was ascribed to enhance penetration of the pesticide active ingredients through the plants cuticles. Nanodispersions of permethrin were also found to be more effective than permethrin microparticles. that permethrin nanodispersions had greater efficacy against yellow fever mosquito larvae, with a 24h LC_{50} of 6.3 μg/L compared with 20 μg/L for microparticles.

Enhanced Stability

Another area of research concerns the stability of the pesticides prepared as colloids. Many chemical pesticides are known to be susceptible to degradation, which can occur through a variety of mechanisms. One of the most important degradation processes of a pesticide in water is abiotichydrolysis, where the pesticide reacts with H_3O^+ and OH^- species in water, converting it into a molecule with greatly reduced biological activity. Incorporation of chemical pesticides into an O/W nanoemulsion may restrict degradation by hydrolysis, as the pesticide experiences a different surrounding environment from the bulk water phase. For example, Song et al. prepared O/W nanoemulsions of triazophos, and compared the decomposition rates with a coarse emulsion of triazophos (over a 48 hour period) under a range of temperatures and pH conditions. They reported that decomposition rates were lower in the nanoemulsions over the temperature range 25～45℃, and over a pH range of 5～9. This was ascribed to limited diffusion of the H_3O^+ and OH^- species from the aqueous phase into the nonpolar micelle cores.

Controlled Release

Previous studies showed that the release kinetics of pesticide active ingredients from colloidal formulations, can be used to slow the release process, resulting in more sustained exposure and longer term efficacy. for example, Yang et al. investigated the long term insecticidal activity of lipid nanodispersions (< 240 nm diameter) loaded with garlic essential oil against adult red flour beetle (*Tribolium castaneum*). They found that the nanodispersion efficacy against adult *T. castaneum* remained over 80% after 5 months, which was hypothesised to be due to the slow and persistent release of the essential oils from the nanoparticles.

Selected from: Grumezescu A M. New Pesticides and Soil Sensors-Nanotechnology in the Agri-Food Industry, Volume 10. Academic Press, Elsevier Inc., 2017: 206-210.

Words and Expressions

bioavailability /ˌbaiəuəˌveilə'biləti/ *n*. 生物利用率，生物利用度，生物药效率
penetration /ˌpenə'treiʃn/ *n*. 渗透，穿透，穿透力
median lethal concentration [农药] 致死中浓度
southern house mosquito 南方库蚊
hypothesize /hai'pɔθəsaiz/ *v*. 假设，假定
glyphosate /'glaifəseit/ *n*. [农药] 草甘膦
organo-silicone 有机硅
glucoside /'glu:kəsaid/ *n*. [生化] 葡萄糖苷
cowsfoot grass 牛筋草
creeping foxglove 赤道樱草
slender buttonweed 细叶纽扣草
buffalo grass 野牛草
spray deposition 雾化沉积
permethrin /pə:'meθrin/ *n*. [农药] 二氯苯醚菊酯
abiotichydrolysis /ˌeibai'ɔtikhai'drɔlisis/ *n*. 非生物水解
triazophos /triæzəu'fəus/ *n*. [农药] 三唑磷
decomposition /ˌdi:ˌkɔmpə'ziʃn/ *n*. 分解，腐烂，变质

PART V
APPLICATION OF PESTICIDES

Unit 9 Application Technologies

The optimum use of pesticides requires not only correct timing, but also efficient transfer of active ingredients to those areas within a crop where the pests, weeds or diseases are located.

Simple changes in spray application can result in dramatic changes to the distribution of pesticide within a crop providing significant changes in biological efficacy. However, the optimization of application parameters is often neglected because the effects of changes are often complex and little understood by growers. However, in the face of environmental and economic pressures, field pesticide doses are being reduced. This will inevitably increase the pressure on growers to maximize the efficiency of their spray targeting, since the effects of poor application will be more evident.

Arable crops are normally considered as being plane or two-dimensional spray targets. Pesticide label recommendations for arable crops usually define doses and application methods, with little regard for the development of the crop. Limited application specifications such as "apply 1 L/ha in 200 L/ha of water using a boom sprayer" are ubiquitous.

However, from the beginnings of spray research, it was understood that the quantity of pesticide reaching the canopy of arable crops could be influenced by application method. Courshee recorded that, although only 20% of the volume sprayed by a typical application was deposited on the crop canopy, it could be influenced by crop growth and the volume applied. With the higher volume rates applied at that time, runoff spraying was very common, and losses to the ground underneath the crop were very high. Courshee suggested that, proportionately, spray deposition on the crop increased with decreasing spray volume.

Clearly, through the growing season, crops change considerably in their size and structure. A cereal crop, which may have provided little ground cover and flat spray targets early in the season, rapidly develops into a dense canopy with significant structure as it matures. To ignore this structure is to neglect the nature of the crop and miss an opportunity for optimization. The situation is somewhat different in row crops such as potatoes and beet. Row crops have been more easily identified as being three-dimensional targets, and for many years, growers have accepted the need to direct their sprays towards the intended target and to account for growing foliage.

Bush crops such as vines, and top fruit, provide linear vertical targets for spray and require particular consideration. Specialist equipment has been developed and there is awareness amongst growers of the effect of foliage density. However, since many bush-crop sprayers use air-jets to distribute the spray, improved targeting requires a practical knowledge of the interaction of air-jets and foliage. Full-size trees, such as are found in traditional orchards, provide a complex and three-dimensional target. Spray deposition often relies on air assistance, and the influence of the structure of the target is great. Thus, with applications to trees, the prospect for improving the targeting sprays and the likelihood of poor application is great.

The aim of an optimised pesticide application must be to maximize the proportion of the active ingredient of the spray that deposits on the target site. The target site must be that area within the crop where the pesticide is most biologically active. It can be seen that the "target" not only varies with the pest but also with the pesticide. Since different pesticides have different modes of action, the choice of spray target will also vary with the pesticide. An insect pest may be controlled with a direct-acting contact material and the target will be the pest itself. A highly translocated pesticide that acts as an insect stomach poison will require a more indirect route and require another choice of target. Weeds may be controlled by a pre-emergence application of a residual herbicide to bare soil, or may be controlled by a post-emergence spray by direct contact with the growing plant. In each case, consideration must be given to both the pest and the chemical. For an optimum application, this should then affect the choice and operation of the sprayer.

Sprays for crop protection are generated by various means, but most sprays used in agriculture are generated using hydraulic pressure. Flat-fan and hollow-cone nozzles are the most common, but deflector or anvil nozzles are used. Hydraulic flat-fan nozzles are available in a range of sizes, producing sprays varying from Very Fine to Coarse. Typical drop-size distributions are invariably wide, and drop size distributions varying from 10 to 1000 μm diameter are not uncommon.

Although wide drop-size distributions can sometimes be an advantage the large numbers of small drops produced in hydraulic nozzle sprays can result in spray drift and inadequate targeting. Rotary atomizers can reduce the breadth of drop-size distributions and provide a more targeted size distribution. They have not been widely adopted in broad acre ground crops, because their spray volumes and drop trajectories often cause control difficulties, but they have found some acceptance in orchard sprayers.

Advanced nozzle designs, such as twin-fluid, pre-orifice and air-inclusion nozzles, can also be used. Most are designed to reduce spray drift. Atomisation in twin-fluid nozzles occurs because the interaction of air and liquid. Different spray qualities can be produced by changing both liquid and air pressures. Low spray volume rates (75～150 L/ha) and high work rates (ha/hour) are possible.

Drops in a typical hydraulic nozzle are formed in a liquid sheet that travels at 15～25 m/s. Following sheet disintegration, drops move in an air-jet caused by the interaction of the spray plume and the surrounding air. Close to the nozzle, all drops move at the same speed, but, as the air-jet decays, fine drops with their greater drag to mass ratio become detrained. They can then become influenced by atmospheric air movements and cause spray drift. Lower spray volumes usually require smaller orifice nozzles that, in turn, produce finer sprays and increase the potential for spray drift.

Many modern sprayers are equipped with additional features designed to decrease spray-drift and improve deposition. Features include air assistance, where airflow, created by a fan placed on the sprayer, creates an air curtain behind the spray. The air is usually ducted via a flexible sleeve mounted on the boom. Fine drops leaving the nozzles are entrained in the air curtain and projected towards the crop.

Large drops, with their greater kinetic energy, can cause problems as they impact on crops. Drops greater than 200 μm diameter have the potential to cause spray runoff and contamination of the soil. With the lowering of spray volume rates, and the use of more fine nozzles, this problem has reduced in importance in recent years, but the trend towards the use of coarse sprays for drift reduction may reintroduce the problem in the near future.

As crops develop from seeds to maturity, their development can be described by the growth stages that can be distinguished during the growing season, and there are several established scales in use. Crop growth stages are defined by the number of developed leaves, the number of branches, or by a combination of stem elongation, number of developed leaves and reproduction organs. Decimal crop growth stage codes have been defined to describe the development of potatoes and cereals.

During the growing season, as the crop develops, leaf mass increases until fruit-setting or tuber-filling, when leaf mass normally starts to decline. At this point, translocation occurs from leaf to reproduction organs—the product to be harvested. This means that the amount of leaf mass, and therefore the area to be covered by the pesticide, alters during the growing season. These changes also have an effect on where, for example, disease occurs. In dense crop canopies, fungi and spore survival is higher than in an open-crop canopy. Since the leaf area of an arable crop can cover five times the soil surface area (leaf area index, LAI=5), the spray volume may be required to cover not just one hectare per hectare of land, but five hectares per hectare of land. Because the target sites of diseases are not often known, the spray may be required to cover the whole of the plant tissue, on both sides of the leaves, requiring as much as 10 ha coverage per 1 ha of land. Yet, despite these large areas of plant tissue, shielding of the soil by leaves is never complete. Drops can penetrate through dense crop canopies and reach the soil. Losses to soil depend on spray quality and the application system, in relation to crop type and its development.

Any pesticide must reach its site of intended biological action to be useful. This is ultimately the target for the pesticide, because if it does not reach this site it cannot work. The range of potential pest species, compounded by variations in the possible sites of pesticide action among these species, creates an almost overwhelming complexity that could prevent any meaningful general description of the location of target sites. However, sites of biological action are predominantly located within the pest organism, which will require the pesticide to be ingested or further absorbed. The pest surface is often regarded as a biological target site, either when the surface is the site of biological action or when the pesticide is absorbed through the surface to act internally.

To describe the main characteristics of biological target sites and alternative targeting pathways, it is useful to discuss groups of pests that have similar characteristics. Perhaps the most obvious distinction lies between pests that present either mobile or static targets for pesticides. The life stage of the species to consider is that which is vulnerable to pesticide action. Mobility creates both challenges and opportunities for targeting. Although moving targets may be harder to intercept if their surface itself is the target, this motion can be helpful should a sufficient dose of pesticide need to be accumulated by the target pest. Pesticide must reach static pests directly or indirectly, depending upon the availability of supplementary environmental forces that can redistribute residues toward their target. There are some classic scenarios that illustrate these distinctions. Insecticides may be placed as baits to be orally ingested by mobile targets. Herbicides and fungicides that act upon germinating seeds or spores may be placed upon the surface where the weeds or fungi will try to develop. Translocated pesticides may be applied to a suitable matrix within the translocation system responsible for moving the pesticide towards its targets. Foliar-applied pesticides need to be deposited upon the foliage in question, which, for nonselective herbicides, demands a selective application technique to exploit differences in foliar morphology.

The form and distribution pattern required of deposits must facilitate the pesticide reaching its intended biological target, although an additional consideration is the need to allow for protection of non-target species, possibly using physical means of selectivity. In general, mobile pests will require less evenly distributed deposits, because they can intercept them, especially if some means for adding an attractant to the bait is available. Prophylactic deposits have to be very well distributed in order not to leave gaps where control will not be effective. Static pests may require an adequate dose to be deposited on their surface, but the exact location of that deposit may be more or less critical, depending on the way that the pesticide acts or is absorbed.

Agricultural pesticide application practice covers treatments ranging across the whole crop growth cycle, from protecting seeds pre- and post-planting, through safeguarding growing plants, to protecting the yield both before and after harvest. Products may be applied in the widest possible variety of physical forms. Those applied directly as formulated may involve solids, liquids and gases, while those applied following preparation of use, as diluted mixtures, generally involve liquids. The propensity for the applied formulation or mixture to be dispersed, in order to achieve appropriate distribution, will vary with the physical form that is applied. The intrinsic mobility of solids is less than that of liquids, which are, in turn, less mobile than gases. It should be noted that more intrinsically mobile forms offer a greater capability of reaching challenging targets, but are correspondingly more difficult to manage in terms of hazard containment.

The scope to alter both the efficiency and safety of application through control of the fate of applied material has been explored in a classification of application equipment by hazard. Application techniques as well as equipment are required to be taken into account for such a classification, which covers all four possible distinct styles of application i. e. *direct*, *liquid*, *solid* and *space*. The "direct" style of application, which is typified by the need to bring treated items into intimate contact with pesticide, is exemplified by admixture processes, e. g. seed treatments. Direct applications take place under highly controllable conditions and often within relatively confined environments, which can potentially mitigate the exposure risks from applied material. The "liquid" style of application is typified by the use of a liquid vehicle to carry the product, usually as a spray, to cover treated surfaces with deposits varying in their size and frequency. The "solid" style of application also achieves distribution of product in more or less finely divided forms, in a similar way. In the liquid and solid application styles the application equipment and technique determine the fate and behavior of the applied product. The "space" treatment style of application is self-explanatory—utilizing the gaseous or vapor phase, or particles of neutral buoyancy, to transport applied product throughout an enclosed space.

Selected from: Wilson M F. Optimising Pesticide Use. England: John Wiley & Sons Ltd, 2003: 23-29.

Words and Expressions

canopy /'kænəpi/ *n*. 天篷，遮篷
translocate /'trænsləkeɪt/ *v*. 改变……的位置
preemergence /ˌpriːɪ'mɜːdʒəns/ *adj*. (植物种子) 出土前的，出土前施用的
postemergence /ˌpəusti'mɜːdʒəns/ *adj*. (农作物) 出苗后至成熟前的
hydraulic /haɪ'drɒlɪk/ *adj*. 水力的，水压的
flat-fan nozzle 扇形雾锥喷嘴
hollow-cone 空心锥，空心圆锥体
deflector /dɪ'flɛktə/ *n*. 变流装置，偏针仪
rotary atomizer 旋转雾化器
anvil /'ænvɪl/ *n*. [解] 砧骨
trajectory /trə'dʒektəri/ *n*. [物] (射线的) 轨道，轨线
twin-fluid 双流体 (喷嘴)
orifice /'ɒrɪfɪs/ *n*. 孔，口
atomization /ˌætəumaɪ'zeɪʃən/ *n*. 雾化，[分化] 原子化
disintegration /dɪsˌɪntɪ'greɪʃn/ *n*. 分裂，瓦解，(人格的) 崩溃，蜕变，衰变
detrain /ˌdiː'treɪn/ *v*. (使) 下火车
duct /dʌkt/ *n*. 管，输送管，排泄管；*vt*. 通过管道输送
elongation /ˌiːlɒŋ'geɪʃn/ *n*. 延长
ingest /ɪn'dʒest/ *vt*. 摄取，咽下，吸收
vulnerable /'vʌlnərəbl/ *adj*. 易受攻击的，易受……的攻击
intercept /ˌɪntə'sept/ *vt*. 中途阻止，截取

supplementary /ˌsʌplɪˈmentri/ *n*. 增补者，增补物；*adj*. 辅助的
germinate /ˈdʒɜːmɪneɪt/ *vt*. 使发芽，使生长；*vi*. 发芽，生长
morphology /mɔːˈfɒlədʒi/ *n*. [生物] 形态学，形态论
prophylactic /ˌprɒfəˈlæktɪk/ *adj*. 预防疾病的；*n*. 预防药，避孕药
propensity /prəˈpensəti/ *n*. 倾向
dispersed /dɪˈspɜːst/ *adj*. 被驱散的，被分散的，散布的
admixture /ədˈmɪkstʃər/ *n*. 混合，混合物
buoyancy /ˈbɔɪənsi/ *n*. 浮性，浮力，轻快

Notes

[1] leaf area index：叶面积指数，指单位土地上作物的全部叶面积（仅一面）与土地面积之比，是衡量群体结构的一个重要指标。系数过高影响作物通风透光；过低不能充分利用日光。

[2] …producing sprays varying from Very Fine to Coarse：产生的雾滴大小范围是从细小到粗大。

[3] twin-fluid nozzles：双流体雾化喷嘴。此种喷嘴结构简单，流量小，能提供最细密的液滴雾化效果（1～20μm），所以用途广泛。

Exercises

Ⅰ Answer the following questions according to the text.

1. How does the feature of air assistance of some modern sprayers work to decrease spray-drift and improve deposition?

2. Why may the spray volume be required cover five hectares per hectare of land in dense crop canopies?

3. Give examples of some classic scenarios that illustrate the distinctions lies between pests that present either mobile or static targets for pesticides.

4. Please briefly state the characteristics of the four application styles：direct，liquid，solid and space.

Ⅱ Translate the following English phrases into Chinese.

row crop	spray volume	runoff spraying	intrinsic mobility
spray drift	rotary atomizer	spray deposition	hydraulic pressure
liquid sheet	intended target	spray application	drop-size distribution

Ⅲ Translate the following Chinese phrases into English.

细雾滴	移动靶标	苗后喷药	叶面积指数	选择性施用技术
粗雾滴	喷洒流失	静态靶标	苗前施用	非选择性除草剂

Ⅳ Choose the best answer for each of the following questions according to the text.

1. The followings are factors that need be taken into account to achieve the optimization of arable pesticide application except ________.

A. size and structure changes of crops   B. crop growth
C. spray volume   D. a common application specifications

2. Which of the following statement is true about tree and bush crop application?
A. Vines and top fruit present linear flat targets for spray.
B. Full-size trees usually provide a complex and two-dimensional target.
C. A practical knowledge of the interaction of air-jets and foliage is necessary for an improved targeting because many bush-crop sprayers use air-jets to distribute the spray.
D. Growers haven't been aware of the effect of foliage density yet.

3. Which of the following nozzle designs can easily lead to spray drift and inadequate targeting?
A. pre-orifice nozzles   B. twin-fluid nozzles
C. air-inclusion nozzles   D. hydraulic nozzles

4. Which of the following statements is wrong about rotary atomizers?
A. Rotary atomizers can reduce the breadth of drop-size distributions and provide a more targeted size distribution.
B. Rotary atomizers have been widely adopted in broad acre ground crops.
C. The spray volumes and drop trajectories of rotary atomizers often cause control difficulties.
D. Rotary atomizers have been accepted in some orchard sprayers.

5. Drops greater than ________ diameter have the possibility to cause spray runoff and pollution of the soil.
A. 300 μm   B. 250 μm   C. 200 μm   D. 150 μm

6. The growth stages of potatoes are defined by ________.
A. stem elongation   B. the number of branches
C. reproduction organs   D. the number of developed leaves

7. With regard to dense crop canopies, which of the following statement is not true?
A. Dense crop canopies can prevent drop penetration and reaching the soil.
B. In dense crop canopies, there is a higher survival of fungi and spore than in an open-crop canopy.
C. The spray volume may be require to cover five hectares per hectare of land since the leaf of an arable crop can cover five times the soil surface area.
D. In spite of dense crop canopies, complete protection of the soil by leaves is impossible.

8. Which physical form of products applied will provide a greater capability of reaching challenging targets?
A. solid   B. liquid   C. gas   D. None of the above

9. What style application has the need to involve treated items closely exposed to pesticides?
A. the " direct" style of application   B. the " liquid" style of application
C. the " solid" style of application   D. the " space" style of application

V **Translate the following short passages into Chinese.**

1. The optimum use of pesticides requires not only correct timing, but also efficient transfer of active ingredients to those areas within a crop where the pests, weeds or diseases are located.

2. Simple changes in spray application can result in dramatic changes to the distribution of pesticide within a crop providing significant changes in biological efficacy.

3. Many modern sprayers are equipped with additional features designed to decrease spray-drift and improve deposition. Features include air assistance, where airflow, created by a fan placed on the sprayer, creates an air curtain behind the spray.

4. The aim of an optimised pesticide application must be to maximize the proportion of the active ingredient of the spray that deposits on the target site. The target site must be that area within the crop where the pesticide is most biologically active. It can be seen that the "target" not only varies with the pest but also with the pesticide. Since different pesticides have different modes of action, the choice of spray target will also vary with the pesticide.

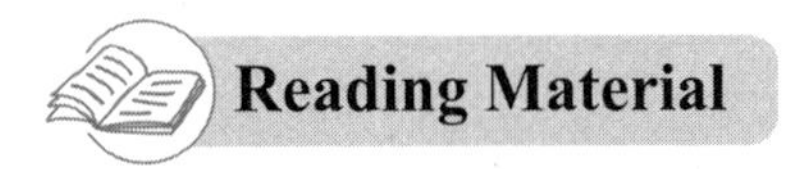

Minimizing Particle Drift

The misapplication of crop protectant products is a major concern in the application industry. One form of misapplication is spray drift. Although drift cannot be completely eliminated, the use of proper equipment and application techniques will maintain drift deposits within acceptable limits. The initial recommendation for drift control is to read the pesticide label. Instructions are given to ensure the safe and effective use of pesticides with minimal risk to the environment. Chemical company surveys indicate that a large percentage of drift complaints involve application procedures not specified on the label.

There are two ways that chemicals move downwind to cause damage: vapor drift and particle drift. Vapor drift is associated with the volatilization of pesticide molecules and then movement off-target. Particle drift is the off-target movement of spray particles formed during or after the application. The amount of particle drift depends mainly on the number of small "driftable" particles produced by the nozzle. Although excellent coverage can be achieved with extremely small droplets, decreased deposition and increased drift potential limit the minimum size that will provide effective pest control.

Several equipment and application factors greatly affect the amount of spray drift that occurs: the type of nozzle and orientation, pressure, boom height, and spray volume. The ability to reduce drift is no better than the weakest component in the spraying procedure. As previously mentioned, the potential for drift must be considered when selecting a nozzle type. Of the many types of nozzles available for applying pesticides, a few, especially those using the newer technology, are specifically designed for reducing drift by reducing the amount of small driftable spray particles in the spray pattern. Higher pressures and nozzles with lower flow rates will also lead to more drift by producing finer spray droplets. Changing pressure alone will also change the flow rate per nozzle and the overall application rate. Spray height is also an important factor in reducing drift losses. Mounting the boom closer to the ground can reduce drift. Nozzle spacing and spray angle determine the correct spray height for each nozzle type. Wide-angle nozzles can be placed closer to the ground than nozzles producing narrow spray angles. On the other hand, older style wide-angle nozzles also produce smaller droplets. When this occurs, the advantages of lower boom height are negated to some extent. However, the newer technology wide-angle drift reduction nozzles have actually been designed to reduce the number of small droplets and will assist in the reduction of drift at lower heights.

The use of larger nozzles is another means of minimizing drift. Increasing the spray volume by using higher capacity spray tips results in larger droplets that are less likely to move off-target. The only effective means of reducing drift by increasing spray volume is to increase the nozzle size.

Although not directly an equipment factor, one of the best tools available for minimizing drift damage is the use of drift control additives in the spray solution to increase the spray droplet size. Tests indicate that in some cases downwind drift deposits are reduced by 50%～80% with the use of drift control additives.

Drift control additives make up a specific class of chemical adjuvants and should not be confused with products such as surfactants, wetting agents, spreaders, and stickers. Drift control additives are formulated to produce a droplet size spectrum with fewer small droplets.

A number of drift control additives are commercially available, but they must be mixed and applied according to label directions in order to be effective. Some products are recommended for use at a rate of 2～8 oz per 100 gal of spray solution. Increased rates may further reduce drift but may also cause nozzle distribution patterns to be nonuniform. Drift control additives will vary in cost depending on the rate and formulation but are comparatively inexpensive for the amount of control provided. It is wise to test these products in each spray system to ensure that they are working properly before adapting this practice. Not all products work equally well for all systems. They do not eliminate drift, however, and common sense must still remain the primary factor in reducing drift damage.

Selected from: Wheeler W B. Pesticides in Agriculture and the Environment. New York: Marcel Dekker, Inc., 2002: 281-284.

Words and Expressions

downwind /ˌdaunˈwɪnd/ *adj*. 顺风的； *adv*. 顺风
boom height 动臂高度，喷杆作业高度
sticker /ˈstɪkər/ *n*. 黏附剂

Unit 10 Pesticide Equipments

Because it is essential to protect our environment during the use of pesticides, marked improvements in application technologies have been developed. Variable rate applications, prescription rates of crop protection products, direct injection, closed handling systems, onboard dry and liquid application systems, control systems, spot sprayers, shielded sprayers, air assist systems, new nozzle designs, and tank-rinsing devices are examples of technological changes that have affected the pesticide application industry.

Basic Application Systems

Better application equipment and new techniques that allow for smaller dosages of pesticides and reduced drift have become increasingly important in minimizing harmful effects of pesticides on applicators and the environment. Changes in the application equipment places increased responsibility on those who apply pesticides to be knowledgeable about the equipment being used. It is not essential to know about all types of application equipment, but a very good understanding of application equipment in general will be beneficial to the readers.

The types of sprayers used to apply pesticide products include hand-operated sprayers, low-pressure powered sprayers, high-capacity powered sprayers, airplane sprayers, and special sprayers for selective application of pesticides. Devices for granular application are also used for a variety of pesticides, either by broadcast application or by row or band application for covering wide swaths or narrow strips over the crop row.

1. Hand-Operated Sprayers

Hand-operated sprayers, such as compressed air and knapsack sprayers, are designed for spot treatment and for areas unsuitable for larger units. They are relatively inexpensive, simple to operate, maneuverable, and easy to clean and store. Compressed air or carbon dioxide is used to apply pressure to the supply tank and force the spray liquid through a nozzle.

2. Hand-Held Spray Guns

Spray guns range from those that can produce a low flow rate with a wide-cone spray pattern or a flooding or showerhead nozzle pattern to those that can produce a high flow rate with a solid narrow-stream spray pattern. Spray guns with showerhead nozzles are commonly used to make commercial lawn applications. Four factors are critical for delivering the correct rate uniformly over the application area when using a showerhead type of nozzle: (1) The exact pressure must be monitored; (2) a proper spraying speed must be maintained; (3) a uniform motion technique must be used; and (4) a constant nozzle height and angle with reference to the ground must be maintained. When the spray gun is used, one should be aware of the difficulty in obtaining a uniform spray.

3. Low-Pressure Field Sprayers with Booms

Low-pressure sprayers equipped with spray booms are more commonly used than any other kind of application equipment. Tractor-mounted, pull-type, and selfpropelled sprayers are available in many models, sizes, and prices. Application volumes can vary from 5 to over 100 gallons per acre.

Drop and Rotary Spreaders

Drop and rotary spreaders are available for applying granular pest control products. Drop spreaders are usually more precise and deliver a more uniform pattern than rotary spreaders. Because the granules drop straight down, there is also less chemical drift. Some drop spreaders will not handle larger granules, however, and ground clearance can be a problem. Moreover, because the edges of a drop-spreader pattern are well defined, any steering error will cause missed or doubled strips. Drop spreaders also usually require more effort to push than rotary spreaders.

Every drop or rotary spreader should be calibrated for proper delivery rate with each product and operator because of variability in the product, the operator's walking speed, and environmental conditions. The easiest method for checking the delivery rate of a spreader is to spread a weighed amount of product on a measured area and then weigh the product remaining in the speader to determine the rate actually delivered.

Electrostatic Spray

An electrostatic charge is now being used commercially to aid in the transfer and attachment of the spray particle to the target. Electrostatic spray systems are commercially available for both aerial and ground applications. With the ground application system, the process uses the principle of contact charging the liquid solution before it reaches the nozzles. The electric charge produced by the Energized Spray Process (ESP) system creates a high intensity electrostatic field that helps propel the spray droplets toward the target at a high velocity. Contact charging differs from earlier electrostatic systems that used induction charging of the spray solution at the nozzle. Contact charging adds 40,000 V to the liquid spray solution in a charging chamber and then distributes the solution in the charged state to the boom and nozzles. The electrostatic spray process shows promise of increasing coverage to both the upper and lower sides of the target leaves.

Hoods and Spray Shields

The use of mechanical shielded booms on sprayers offers applicators and growers another potential method to reduce drift. Several design options exist with this technology. Shielded booms are designed to protect the spray from the wind as it leaves the nozzle and travels to the target. A very important concern with shielded booms is the design. Improperly designed shields can result in more drift because negative pressures may build up inside the hood and force pesticide sprays out of the hood and into the environment, resulting in drift. Research has shown the potential for reduced drift when hoods are used rather than unshielded booms. Most of the studies reported that the drift potential is very closely related to the droplet size spectrum. The smaller the droplets are sprayed, the less potential there is for a dramatic reduction in drift with the hoods. Research has also shown that hoods do not perform as well in higher wind speeds as in weaker winds. It is a common belief that full boom shields provide little potential for drift reduction in row crops although their use for cereal grains or on fallow ground may result in reduced drift. Shielded boom sprayers have not been universally adopted throughout the spray industry. However, because of the uniformity of the target area, shielded booms are becoming popular in turf applications. The use of individual row hoods, another variation of hooded spraying, in row crop settings is gaining popularity. Hoods of this type are designed to shield certain plants while spraying nonselective herbicides between the rows. Research shows that such systems may allow growers greater flexibility with their weed control program while reducing chemical costs, improving chemical efficiency, and reducing drift.

A hood spray system that incorporates optical sensors inside the individual row hoods is being developed commercially. This system uses a beam of light to detect weeds under the hoods and between the crop rows. When the sensor detects the weed, a spray nozzle is activated to spray the detected weed. With this system, the hood can protect sensitive plants in the rows from the nonselective spray materials. It is difficult for these sensors to distinguish weeds from the growing crop; thus the hood performs two critical functions in this system: It provides a protected area in which to sense the weeds and then shields the sensitive crop from the emitted spray. An additional benefit with this technology is the increased potential for using reduced amounts of herbicides. This translates to reduced crop protection costs and could also result in less drift. This technology also provides an excellent opportunity for use in site-specific crop protectant applications in the future.

Injection Systems

Efficient and safe use of inputs has always been the goal of applicators. Direct injection is an important technological development that can be used to help the application industry reduce the problems associated with chemical application. Direct injection is a technology that may possibly have the greatest effect on the method of applying pesticides. With direct injection, the spray tank contains only water or the carrier. Prior to exiting the nozzle, chemical formulations or specially blended materials are injected directly into the spray lines that are applying the carrier as the sprayer travels through the field. The type of mixing that occurs depends on whether the injection occurs before or after the carrier spray pump. The type of metering pump used distinguishes the types of injection systems. The systems currently on the market use either piston or cam metering pumps to inject the chemical into the carrier. Either the chemical is injected into an in-line mixer prior to spraying or a series of peristaltic pumps meter the chemical and inject it on the inlet side of the carrier spray pump. The early direct injection systems had several limitations. These included a lag time for the chemical to reach the nozzles, improper mixing of the chemical before spraying, and inability of the units to distribute wettable powder formulations.

Many of the early problems with this technology have been resolved. Improved metering pump systems have reduced chemical lag time. The use of inline mixers has resulted in more uniform mixing. The addition of agitation to mix wettable powders allows the use of a wide variety of formulations. Systems also exist that allow for the injection of dry formulations. Direct injection technology is becoming more prominent in the agricultural application industry. Control of injection with computers makes this technology well suited to adjusting rates on-the-go and for prescription applications. Rates can be accurately controlled to take advantage of site-specific needs that require precise application. On-line printers are available to produce a permanent record of chemical use and job location. Either the injection systems are included in the electronic controlling device or they can be added on as a module to existing control devices.

Another driving force behind much of the newly developed application technology is the development of sensors and the application of controllers. Spray controllers are being integrated into spray monitor systems. Electronic devices to control application rates have been widely used for years. Controllers are designed to automatically compensate for changes in speed and application rates on-the-go. Some are computer-based and work well with new application techniques such as direct injection and variable rate application. Computers and controllers work together to place pesticide in the precise desired position at the prescribed amount. The applicator's ability to precisely place pesticides is an important environmental factor.

The acceptance of direct injection technology has been spurred by environmental concerns, concern for operator safety, regulations, and the development of new products that are effective at very low rates of application. Direct injection eliminates the need to tank-mix chemicals; thus pesticide compatibility problems are eliminated. Cleanup of equipment is minimal, and with no leftover solutions, disposal of rinsates is not a major concern. If the chemicals are in returnable containers and are handled in a closed system, the potential for operator exposure is greatly reduced. Because of the added precision and the ability to spot-spray only where the pesticides are needed with the direct injection process, a substantial savings to the producer is realized and the environmental impact is reduced. Success or failure in the pesticide application industry rests on how well we manage and reduce the negative impacts on the environment.

Handling Systems

A major emphasis for chemical companies and equipment manufacturers has been to develop new and innovative ways to make the handling of chemicals more convenient and to reduce exposure for the people who use pesticide products. Bulk-handling and mini-bulk-handling systems are available to store, transport, and handle liquid and granular pesticides. The closed systems associated with bulk tanks reduce operator contact with the chemicals and eliminate potential spillage, and, with the returnable 250 ~ 300 gal containers, container disposal is eliminated. Closed handling systems are also being developed to store, transport, and transfer dry granular pesticides. For example, pneumatic handling systems are used to transfer granular herbicides from bulk storage at the fertilizer plant into tendering vehicles that will deliver the product by air to the applicator units in the field.

Technological improvements in the application industry have occurred at a very rapid rate in recent years. As scientists continue to focus on the precision farming of tomorrow, the equipment industry will work to improve and develop the new equipment needed to achieve the goal of more effective application.

Selected from: Wheeler W B. Pesticides in Agriculture and the Environment. New York: Marcel Dekker, Inc., 2002: 271-291.

Words and Expressions

spot sprayer 点式喷雾器
shielded sprayer 屏蔽式喷雾机
air assist system 空气辅助系统
tank rinsing devices 槽冲洗装置
hand-operated sprayer 手动喷雾器
swath /swɒθ/ *n*. 收割的刈痕，细长的列
knapsack sprayer 喷雾器，[植保] 背负式喷雾器，背负式喷粉机
knapsack /'næpsæk/ *n*. 背包
maneuverable /mə'nuːvərəbl/ *adj*. 有机动性的，容易操作的，可调动的
wide cone 宽锥
showerhead nozzle 淋喷头喷嘴，莲蓬头喷嘴
pulsation /pʌl'seɪʃn/ *n*. 脉动，搏动，震动，悸动
boom /buːm/ *n*. 繁荣，吊杆，(起重机) 臂；*v*. 发隆隆声，激增
steer /stɪər/ *v*. 驾驶，掌舵
calibrate /'kælɪbreɪt/ *v*. 校准
electrostatic /ɪˌlektrəʊ'stætɪk/ *adj*. 静电的，静电学的
Energized Spray Process 电喷雾工艺
hood /hʊd/ *n*. 头巾，车篷，引擎罩；*v*. 用头巾包裹，给……加罩
fallow /'fæləʊ/ *n*. 休耕地；*adj*. [农] 休耕的，未孕的；*vt*. 使 (土地) 休闲，休整
cam /kæm/ *n*. 凸轮
peristaltic /ˌpɛrɪ'stɔltɪk/ *adj*. 蠕动的
lag /læg/ *n*. 落后，迟延，防护套；*adj*. 最后的；*vi*. 缓缓而行，滞后
module /'mɒdjuːl/ *n*. [计] 模块，组件，模数
spur /spɜː(r)/ *n*. 刺激物，刺激；*v*. 鞭策，刺激
compatibility /kəmˌpætə'bɪləti/ *n*. [计] 兼容性
rinse /rɪns/ *v*. 嗽口，(用清水) 刷；*n*. 嗽洗，漂洗
spillage /'spɪlɪdʒ/ *n*. 溢出，溢出量
pneumatic /njuː'mætɪk/ *adj*. 装满空气的，有气胎的，风力的；*n*. 气胎
tender /'tendər/ *adj*. 嫩的，温柔的，软弱的

Notes

[1] electrostatic spray：静电喷雾技术，通过高压静电发生装置，使雾滴带电喷施的方法。药液在植株叶片表面的沉积量显著增加，可将农药有效利用率提高到90%。

[2] precision farming 或 precision agriculture：精准农业。它是当今世界农业发展的新潮流，是由信息技术支持的根据空间变异，定位、定时、定量地实施一整套现代化农事操作技术与管理的系统。

Exercises

Ⅰ Answer the following questions according to the text.

1. What are the four factors that are essential for delivering the correct rate uniformly over the application area when using a showerhead type of nozzle?

2. What are advantages and disadvantages, compared with rotary spreaders, of drop spreaders for applying granular pest control products?

3. Why is an accurate design of shielded booms a very important concern?

4. What are the limitations of early direct injection and how are these early problems resolved? What are benefits of direct injection?

5. which do application equipments have types?

Ⅱ Translate the following English phrases into Chinese.

optical sensor	air assist system	application technology
hand-operated sprayer	direct injection	airplane sprayer
closed handling system	hand-held spray gun	manual sprayer
knapsack sprayer	application equipment	low-pressure powered sprayer

Ⅲ Translate the following Chinese phrases into English.

静电喷雾	点喷雾器	精准农业	散装传递系统	颗粒状除草剂
高负载动力喷雾器		压缩空气喷雾器	保护性喷雾器	液体及颗粒剂型

Ⅳ Choose the best answer for each of the following questions according to the text.

1. Which sprayers are aimed at spot treatment and areas that are not suitable for large unit?

A. low-pressure powered sprayers　　B. high-capacity powered sprayers

C. hand-operated sprayers　　D. airplane sprayers

2. Among the application equipment mentioned in this article, which one is the most commonly used?

A. hand-held spray guns

B. low-pressure sprayers equipped with spray booms

C. drop spreaders

D. rotary spreaders

3. Which of the following statement is not true about drop spreaders and rotary spreaders?

A. Drop spreaders are usually more precise and deliver a more uniform pattern than rotary spreaders.

B. There is not any chemical drift by applying drop spreaders.

C. Drop spreaders usually require more effort to push.

D. For drop spreaders, any steering error will cause missed or doubled strips.

4. Which of the following application system provides a decided benefit with fungicide and insecticide?

A. closed handling systems B. onboard dry and liquid application system

C. electrostatic spray systems D. air assist systems

5. Which of the following statement is not true about drift induction for some applications?

A. There is less pesticide drift when using drop and rotary spreaders.

B. Direct injection can effectively reduce drift.

C. The use of individual row hoods in row crop setting may reduce drift.

D. It has been proved that the electrostatic spray can result in drift induction.

6. What of the following is not the benefit of the hood spray system?

A. The hood provides a protected area in which to sense the weeds and then shields the sensitive crop from the emitted spray.

B. This technology increases the potential for using reduced amounts of herbicides, thus reduces crop protection costs and could also result in less drift.

C. This spray system shows promise of increasing coverage to both the upper and lower sides of the target leaves.

D. This technology also provides an excellent opportunity for use in site-specific crop protectant applications in the future.

7. According to the text, which method of applying pesticide may possibly have greatest effect?

A. direct injection B. electrostatic spray

C. manual spray D. using mechanical shielded booms on sprayers

8. The benefits of closed handling system associated with bulk tanks are the following except ________.

A. eliminate container disposal

B. reduce operator's exposure to chemicals

C. avoid potential spillage

D. make the pesticides more effective in controlling pest

Ⅴ Translate the following short passages into Chinese.

1. Better application equipment and new techniques that allow for smaller dosages of pesticides and reduced drift have become increasingly important in minimizing harmful effects of pesticides on applicators and the environment.

2. Hand-operated sprayers, such as compressed air and knapsack sprayers, are designed for spot treatment and for areas unsuitable for larger units.

3. Electrostatic spray systems are commercially available for both aerial and ground applications. With the ground application system, the process uses the principle of contact

charging the liquid solution before it reaches the nozzles.

4. Pesticides will continue to play a significant role in helping farmers provide an abundant and safe food supply for people throughout the world. The application industry will continue to change to make the use of pesticides as safe as possible. Technological improvements in the application industry have occurred at a very rapid rate in recent years. As scientists continue to focus on the precision farming of tomorrow, the equipment industry will work to improve and develop the new equipment needed to achieve the goal of more effective application.

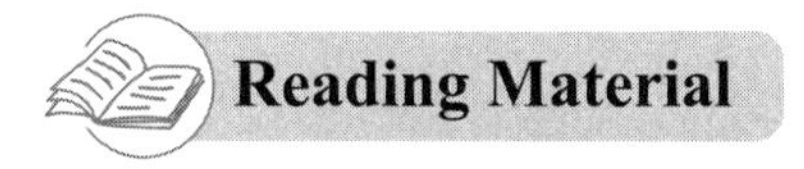

Sprayer Components

All low-pressure sprayers have several basic components, including a pump, a tank, agitation devices, flow-control assemblies, strainers, hoses and fittings, booms, nozzles, and, typically, electronic or computerized components to help improve the accuracy of the application process. A brief description of each of these components follows.

Pumps

The pump is the "heart" of the sprayer. Sprayer pumps are used to create the hydraulic pressure required to deliver the spray solution to the nozzles and then atomize it into droplets. The most common types of pumps available for applying pesticides are roller, centrifugal, diaphragm, and piston pumps. For low-pressure sprayers the centrifugal and roller pumps are the most common, but the diaphragm pump is becoming more popular. Either a diaphragm or piston pump is commonly used where higher pressures are needed to move spray product through long lengths of hose such as in turf or roadside applications.

Tanks

The spray tank should have adequate capacity for the job. Tanks should also be clean, corrosion-resistant, easy to fill, and suitably shaped for mounting and effective agitation. The openings on the tank should be suitable for pump and agitator connections. Tanks that are not transparent should have a sight gauge or other external means of determining the fluid level. Sight gauges should have shutoff valves to permit closing in case of failure. The primary opening of the tank should be filled with a cover that can be secured to avoid spills and splashes. It also should be large enough to facilitate cleaning of the tank. A drain should be located at the bottom so that the tank can be completely emptied.

Tanks are commonly constructed of stainless steel, polyethylene, and fiberglass. The materials used will influence the cost of the tank, its durability, and its resistance to corrosion.

Agitation Devices

Agitation requirements depend largely on the formulation of the chemical being applied. Soluble liquids and powders do not require special agitation once they are in solution, but emulsions, wettable powders, and liquid and dry flowable formulations will usually separate if they are not agitated continuously. Separation causes the concentration of the pesticide spray to vary greatly as the tank empties. Improper agitation may also result in plugging of the parts of the spray distribution system. For these and other reasons, thorough agitation is essential. Hydraulic jet agitation is the most common method used with low-pressure sprayers. Jet agitation is simple and effective. A small portion of the spray solution is circulated from the pump output back to the tank, discharging it under pressure through holes in a pipe or through special agitator nozzles.

Flow Control Assemblies

Roller pumps, diaphragm pumps, and piston pumps usually have a flow control assembly consisting of a bypass-type pressure regulator or relief valve, a control valve, a pressure gauge, and a boom shutoff valve. Bypass pressure relief valves usually have a spring-loaded ball, disk, or diaphragm that opens with increasing pressure so that excess flow is bypassed back to the tank, thus preventing damage to the pump and other components when the boom is shut off. When the control valve in the agitation line and the bypass relief valve in the bypass line are adjusted properly, the spraying pressure will be regulated.

Because the output of a centrifugal pump can be reduced to zero without damaging the pump, a pressure relief valve and separate bypass line are not needed. The spray pressure can be controlled with simple gate or globe valves. It is preferable, however, to use special throttling valves designed to accurately control the spraying pressure. Electrically controlled throttling valves are becoming popular for remote pressure control.

Because nozzles are designed to operate within certain pressure limits, a pressure gauge must be included in every sprayer system. The pressure gauge must be used for calibrating and while operating in the field. Select a gauge that is suitable for the pressure range that you will be using.

A quick-acting boom cutoff or control valve allows the sprayer boom to be shut off while the pump and the agitation system continue to operate. Electric solenoid valves, which eliminate inconvenient hoses and plumbing, are also available.

Strainers

Three types of strainers are commonly used on low-pressure sprayers: tank filler strainers, line strainers, and nozzle strainers. The strainer size numbers (20 mesh, 50 mesh, etc.) indicate the number of openings per inch. Strainers with high mesh numbers have smaller openings than strainers with low mesh numbers. Coarse-basket strainers are placed in the tank filler opening to prevent twigs, leaves, and other debris from entering the tank as it is being filled. A 16 or 20 mesh tank filler strainer will retain lumps of wettable powder until they are broken up, helping to provide uniform tank mixing.

A suction line strainer is used between the tank and a roller pump to prevent rust, scale, or other material from damaging the pump. A 40 or 50 mesh strainer is recommended. A suction line strainer is not usually needed to protect a centrifugal pump, except against large pieces of foreign material.

The inlet of a centrifugal pump must not be restricted. If a strainer is used, it should have an effective straining area several times larger than the area of the suction line. It should also be no smaller than 20 mesh and should be cleaned frequently. A line strainer (usually 50 mesh) should be located on the pressure side of the pump to protect the spray nozzles and agitation nozzles.

Small-capacity nozzles must have a strainer of the proper size to stop any particle that might plug the nozzle orifice. Nozzle strainers vary in size depending on the size of the nozzle tip used, but they are commonly 50 or 100 mesh.

Hoses and Fittings

All hoses and fittings should be of a suitable quality and strength to handle the chemicals at the selected operating pressure. A good hose is flexible and durable and resistant to sunlight, oil, and chemicals. It should also be able to hold up under the rigors of normal use, such as twisting and vibration. Two widely used materials that are chemically resistant are ethylene vinyl acetate (EVA) and ethylene propylene diene monomer (EPDM). A special reinforced hose must be used for suction lines to prevent their collapse.

Sometimes the pressure greatly exceeds the average operating pressures. These peak pressures usually occur as the spray boom is shut off. For this reason, the sprayer hoses and fittings must always be in good condition to prevent a possible rupture that could cause spills or cause the operator to be sprayed with the chemical.

As liquid is forced through the spray system, the pressure drops due to the friction between the liquid and the inside surface of the hoses, pipes, valves, and fittings. The pressure drop is especially high when a large volume of liquid is forced through a small-diameter hose or pipe.

To minimize pressure drop, spray lines and suction hoses must be the proper size for the system. The suction hoses should be airtight, noncollapsible, as short as possible, and as large as the opening on the intake side of the pump. A collapsed hose can restrict flow and "starve" a pump, decreasing the flow as well as causing damage to the pump or the pump seals.

Other lines, especially those between the pressure gauge and the nozzles, should be as straight as possible with a minimum of restrictions and fittings. The proper size for these lines varies with the size and capacity of the sprayer. A high fluid velocity should be maintained throughout the system. If the lines are too large, the velocity will be low and the pesticide may settle out from the suspension and clog the system. If the lines are too small, an excessive drop in pressure will occur.

Booms

The boom on the sprayer provides a place to attach the nozzles in order to obtain a uniform distribution of the pesticide across the application target. Boom length and height will vary depending on the type of application. Boom stability is important in achieving uniform spray application. The boom should be relatively rigid in all directions. It should not swing back and forth or up and down. The boom should be constructed to permit folding for transport. The boom height should be adjustable.

Nozzles

The spray nozzle is the final part of the distribution system. The selection of the correct type and size is essential for each application. The nozzle determines the amount of spray applied to an area, the uniformity of the application, the coverage of the sprayed surface, and the amount of drift. One can minimize the drift problem by selecting nozzles that give the largest droplet size while providing adequate coverage at the intended application volume and pressure. Although nozzles have been developed for practically every kind of spray application, only a few types are commonly used in pesticide applications. An emphasis on nozzle design over the past few years has resulted in a vast improvement in spray quality.

Selected from: Wheeler W B. Pesticides in Agriculture and the Environment. New York: Marcel Dekker, Inc., 2002: 273-278.

Words and Expressions

strainer /'streɪnər/ *n*. 滤网，松紧扣，过滤器
roller /'rəulər/ *n*. 滚筒，辊子
fiberglass /'faɪbɚ,glæs/ *n*. 玻璃纤维，玻璃丝
durability /,djuərə'bɪləti/ *n*. 持久性，耐久力
plug /plʌg/ *n*. 插头，塞子；*v*. 塞住
throttle /'θrɒtl/ *n*. 节流阀，[车辆] 风门，喉咙；*vt*. 压制，扼杀，使……窒息
solenoid /'sɒlənɔɪd/ *n*. [电] 螺线管
debris /'debriː/ *n*. 碎片，残骸
suction /'sʌkʃn/ *n*. 吸入，抽气，抽气机，抽水泵，吸引
orifice /'ɒrɪfɪs/ *n*. 孔，口
friction /'frɪkʃn/ *n*. 摩擦，摩擦力
airtight /'eətaɪt/ *adj*. 密封的，无懈可击的

PART VI
ALLELOPATHY

Unit 11 Allelopathy

Serendipity is an apparent aptitude to make fortunate discoveries accidentally and make unexpected, good things happen. Serendipity may be available in weed science if the presence of allelopathic interactions can be confirmed and used to control weeds. Organisms from microbes to mammals find food, seek mates, ward off predators, and defend themselves against disease via chemical interactions. Allelopathic interactions are chemical, and discovery of the cause and mechanism of these interactions may yield a treasure of biological and chemical approaches to control weeds. At least 25% of human medicinal products originated in the natural world or are synthetic derivatives of naturally occurring substances. Many natural interactions are chemical interactions and some of them could influence the course of weed science.

Interference is the term assigned to adverse effects that plants exert on each other's growth. Competition is part of interference and occurs because of depletion or unavailability of one or more limiting resources. Allelopathy, another form of interference, occurs when one plant, through its living or decaying tissue, interferes with growth of another plant via a chemical inhibitor. Allelopathy comes from the Greek allelo (= each other), which is similar to the Greek allelon (=one another). The second root is the Greek patho or pathos that means suffering, disease, or intense feeling. Allelopathy is therefore the influence, usually detrimental (the pathos), of one plant on another, by toxic chemical substances from living plant parts, through their release when a plant dies, or their production from decaying tissue.

There is a subset of allelochemicals known as kairomones that have favorable adaptive value to organisms receiving them. A natural kairomone from water hyacinth is a powerful insect attractant for a weevil (*Neochetina eichhorniae*) and the water hyacinth mite (*Orthogalumna terebrantis*). The kairomone is liberated when water hyacinth is injured by surface wounding or by the herbicide 2, 4-D. The kairomone enhances control of water hyacinth by attracting large numbers of weevils and mites to the area of the plant's wound. Thus the kairomone has favorable value to the insects but not to the water hyacinth. Control of water hyacinth is enhanced when insect damage is combined with herbicide stress.

For weed management purposes, allelopathy is considered a strategy of control. Corn cockle and ryegrass seeds fail to germinate in the presence of beet seeds. If tobacco seeds germinate and grow for 6 days in petri dishes and then an extract of soil, incubated for 21 days with timothy residue is added, the root tips of tobacco blacken within 1 h while radicle elongation is unaffected. If an extract of soil, incubated with rye residue is added, the symptoms are reversed. Residues of timothy, maize, rye, and tobacco all reduce the respiration rate of tobacco seedlings.

Kooper, a Dutch ecologist, observed the large agricultural plain of Pasuruan on the island of Java, Indonesia, where sugarcane, rice, and maize grew. After harvest, the fallowed fields developed a dense cover of weeds. Kooper observed that the postharvest floristic composition of each community was stable year after year. He found that floristic composition was determined at the earliest stages of seed germination, not by plant survival or a struggle for existence, but by differential seed germination. He showed that seeds of other species were present but could not germinate unless removed from their environment. Kooper concluded that previous vegetation established a soil chemical equilibrium (an allelopathic phenomenon), determined which seeds could germinate, and subsequently, which plants dominated.

The word allelopathy was first used by Molisch (1937), an Austrian botanist. an Austrian botanist. He included toxicity exerted by microorganisms and higher plants and that use has continued. The phenomenon, however, had been observed much earlier by several scientists. A classic example of allelopathy is found in the black walnut forests of Central Asia. Few other plants survive under the forest plant canopy because of the presence of juglone, a quinone root toxin derived from black walnut trees. The effect of juglone couldn't be reproduced in the greenhouse because some plant metabolites, including phenolics, require ultraviolet light for their biosynthesis.

Another classic study is the work by Muller and Muller in California who observed that California chaparral often occurred near, but not intermixed with, California sagebrush. Neither species grew in the zones of contact between the respective communities; other species grew between the communities. They found volatile terpenes, particularly camphor (a monoterpene ketone) and cineole (a terpene ether) produced by the chaparral, were responsible for the no-contact zones. They concluded that plants, in this case the chaparral, are fundamentally leaky systems. Other studies are described by Rice and Thompson. Zhang et al. found that volatile allelochemicals released from leaf tissue of crofton weed affected seedling growth of upland (non-paddy) rice. β-tyrosine, an isomer of the common amino acid tyrosine, has been shown to contribute to the allelopathic potential (a defense function) of rice.

Research in China showed that the allelopathic potential of the invasive species, creeping daisy, was increased by acid rain. Simulated acid rain increased total carbon and nitrogen content, available phosphorus, decreased soil pH, and accelerated creeping daisy's litter decomposition, thereby enhancing the allelopathic potential of the weed's litter.

One plant does not consciously set out to affect another, rather the effect occurs as a normal, perhaps serendipitous, ecological interaction with evolutionary implications. Allelopathic species have been selected by evolutionary pressure, because they can outcompete neighbors through energy-expensive biochemical processes that produce allelochemicals. The energy expense is not a waste of resources because no species evolves successfully by wasting resources. Exploration of the phenomena will lead to better understanding of plant evolutionary strategies and, possibly, provide clues for herbicide synthesis and development.

Allelopathy has also been explored with a number of crops, and there have been attempts to find crop cultivars with a competitive allelopathic edge. Residues of several crops have phytotoxic activity on other plants. This effect is often incorrectly attributed to an allelochemical. All plants produce chemicals that are weakly phytotoxic.

Laboratory studies have often demonstrated allelopathy, but the evidence produced should not be regarded as conclusive of the existence of allelopathy in the environment until it is confirmed by field studies. Field studies are essential to obtain ecologically relevant data. Clues from laboratory studies are not sufficient without field confirmation. For example, Norsworthy demonstrated the allelopathic potential of aqueous extracts of wild radish in controlled environment studies. The evidence indicated that aqueous extracts of wild radish or incorporated wild radish residues suppressed seed germination, radicle growth, seedling emergence, and seedling growth of "certain crops and weeds", but subsequent field confirmation is essential to establish the reality of allelopathy as an ecological phenomenon.

Plants produce a myriad of metabolites of no known utility to their growth and development. They are often referred to as secondary plant metabolites and are defined as compounds having no known essential physiological function. The idea that these compounds may injure other forms of life is not without a logical base. However, proof is questionable because most allelochemical effects occur through soil, a complex chemical matrix. Conclusive studies require extraction and isolation of the active agent from soil. Any allelopathic chemical may be chemically altered prior to or during extraction. That which is extracted, isolated, and studied may not be what the plant produced.

Secondary plant metabolites, also known as natural products, are regarded by many as "a vast repository of materials and compounds with evolved biological activity, including phytotoxicity". It is proposed that some of these compounds may be useful directly as herbicides or as templates for herbicide development.

Allelochemicals vary from simple molecules such as ammonia to the quinoneejuglone, the terpenes camphor and cineole, complex conjugated flavonoids such as phlorizin, or the heterocyclic alkaloid caffeine. Putnam lists several chemical groups from which allelopathic agents come: organic acids and aldehydes, aromatic acids, simple unsaturated lactones, coumarins, quinones, flavonoids, tannins, alkaloids, terpenoids and steroids, a few miscellaneous compounds such as long-chain fatty acids, alcohols, polypeptides, nucleosides, and some unknown compounds. The diversity suggests several mechanisms of action, a multiplicity of effects, and is one reason for the slow emergence of a theoretical framework. The chemistry of allelopathy is as complex as synthetic herbicide chemistry but it is a chemistry of discovery as opposed to one of synthesis.

There is little doubt that allelopathy occurs in plant communities; but there are questions about how important it is in nature and if it can be exploited in cropped fields. It has been reported for many crop and weed species, but proof of its importance in nature is lacking. Proof will require something similar to the application of Koch's postulates (1912) proposed for plant pathology in 1883 and amended by Smith (1905). The analogous postulates applied to allelopathy are:

1. Observe, describe, and quantify the degree of interference in a natural community.

2. Isolate, characterize, and synthesize the suspected toxin produced by the suspected allelopathic plant.

3. Reproduce the symptoms by application of the toxin at appropriate rates and times in nature.

4. Monitor release, movement and uptake, and show they are sufficient to cause the observed effects.

These four steps describe difficult, expensive, complex scientific research. Rigorous proof has rarely been applied to any ecological interaction, but such proof is vital if allelopathic research is to move from description to causation. Duke (2015) added the necessity of determining whether the compounds identified are produced in sufficient quantity in time and space in soil to exert an allelopathic effect. He goes on to assert that most articles purporting to deal with allelopathy are not designed to prove it. They are designed "to discover phytotoxins that might have utility as herbicides or herbicide leads".

In short, it is insufficient to make an observation and suspect a toxin. It is insufficient to demonstrate the toxin is produced by one plant. Specific cause and effect must be demonstrated through chemical and plant studies. It may not be necessary to prove that plant X is the source of allelochemical Y. If an allelochemical, effective as a natural herbicide, can be isolated and identified, then, in theory, it might be useful without absolute proof of its plant origin or physiological mode of action.

Selected from: Zimdahl R L. Allelopathy. Fundamentals of Weed Science, 2018: 253-270.

Words and Expressions

allelopathy /əli'lɑpəθi/ *n*.［植］化感作用，植化相克，相互影响
serendipity /ˌserən'dɪpəti/ *n*. 意外发现
allelopathic /ˌælɪlə'pæθɪk/ *adj*. 对抗疗法的，异株克生的
depletion /dɪ'pliːʃn/ *n*. 消耗，损耗，放血
detrimental /ˌdetrɪ'mentl/ *adj*. 不利的，有害的；*n*. 有害的人（或物）
allelochemicals /æleləʊ'kemɪklz/ *n*. 化感化合物
kairomone /'kairəməun/ *n*.［植保］利它素，［生理］种间激素
water hyacinth ［植］水葫芦，凤眼蓝
weevil /'wiːvl/ *n*.［昆］象鼻虫

Neochetina eichhorniae 水葫芦象甲，凤眼兰象
mite /maɪt/ *n*.小虫，螨，小动物，微粒
Orthogalumna terebrantis 叶螨
corn cockle 麦仙翁；瞿麦
ryegrass /'raɪgræs/ *n*.［肥料］黑麦草
timothy /'tɪməθɪ/ *n*.［植］梯牧草
radicle /'rædɪkəl/ *n*.幼根，［植］胚根，（神经，血管等的）根，基，原子团
rye /raɪ/ *n*.黑麦，吉卜赛绅士；*adj*.用黑麦制成的
postharvest /'pəust'hɑːvist/ *adj*.用于采收后的，收割期后的
floristic composition ［植］植物区系组成，植物种类成分
canopy /'kænəpi/ *n*.天篷，华盖，遮篷，苍穹；*vt*.用天蓬遮盖，遮盖
juglone /dʒuː'gləʊn/ 胡桃醌；胡桃酮
quinone /'kwɪnəʊn;kwɪ'nəʊn/ *n*.醌（等于 chinone）
phenolics /fi'nɔliks/ 酚类物质，酚醛树脂
chaparral /ˌʃæpə'ræl/ *n*.丛林，茂密的树丛
sagebrush /'seɪdʒbrʌʃ/ *n*.［植］蒿属植物
terpene /'tɜːpiːn/ *n*.萜（烃），萜烯
camphor /'kæmfər/ *n*.莰酮，樟脑
monoterpene /ˌmɔnə'təːpiːn/ *n*.［有化］单萜
ketone /'kiːtəʊn/ *n*.酮
cineole /'sɪnɪəʊl/ *n*.［林］桉树脑
ether /'iːθər/ *n*.乙醚
crofton weed 紫茎泽兰
tyrosine /'taɪrəsiːn/ *n*.［生化］酪氨酸
isomer /'aɪsəmər/ *n*.［化学］同分异构体
invasive species 入侵物种
daisy /'deɪzi/ *n*.雏菊；菊科植物，极好的东西；*adj*.极好的，上等的
serendipitous /ˌserən'dɪpətəs/ *adj*.偶然发现的
radicle growth 胚根生长
seedling emergence 出苗，幼苗出土
seedling growth 幼苗生长
repository /rɪ'pɒzətri/ *n*.贮藏室，仓库，知识库，智囊团
ammonia /ə'məʊniə/ *n*.［无化］氨
flavonoid /'fleɪvənɔɪd/ *n*.黄酮类，［有化］类黄酮
phlorizin /'flɒrɪzɪn/ *n*.根皮苷，果树根皮精
heterocyclic /ˌhetərə'saɪklɪk/ *adj*.杂环的，不同环式的
alkaloid /'ælkəlɔɪd/ *n*.［有化］生物碱，植物碱基
caffeine /'kæfiːn/ *n*.［有化］［药］咖啡因
aldehyde /'ældɪhaɪd/ *n*.醛；乙醛
aromatic acid ［有化］芳香酸
unsaturated lactone 不饱和内酯

coumarin /ˈkuːmərɪn/ *n*. [有化] 香豆素

tannin /ˈtænɪn/ *n*. 丹宁酸；鞣酸

terpenoid /təːˈpinɔɪd/ *n*. [有化] 萜类化合物，[有化] 类萜

steroid /ˈsterɔɪd:ˈstɪərɔɪd/ *n*. 类固醇，[有化] 甾族化合物

miscellaneous /ˌmɪsəˈleɪniəs/ *adj*. 混杂的，多方面的，多才多艺的

long-chain fatty acid 长链脂肪酸

Notes

allelopathy：化感作用，异株克生。指植物产生的次生代谢产物（化感化合物）在植物生长过程中，通过信息抑制其他植物的生长发育并加以排除的现象。

Exercises

Ⅰ Answer the following questions according to the text.

1. What is the difference and connection between interference and competition?
2. What is allelopathy? and what is function of allelopathy?
3. Give an example of the role of kairomones.
4. Allelopathic phenomena are widespread in nature. Please give examples.
5. Why can't be used as evidence for its presence in the environment when allelopathy are found in the laboratory?
6. There are many allelochemicals in natural products，Please give examples.

Ⅱ Translate the following English phrases into Chinese.

chemical equilibrium   ecological interaction   weed sciences   plant communities
floristic composition   seedling emergence   synthetic derivatives   allelopathic effect

Ⅲ Translate the following Chinese phrases into English.

杂草管理   种子萌发   幼苗生长   入侵物种   酸雨
化感现象   胚根生长   不饱和内酯   不利影响

Ⅳ Choose the best answer for each of the following questions according to the text.

1. Which of the following methods can control water hyacinth?

   A. weevil   B. water hyacinth mite
   C. 2,4-D   D. all of the above

2. Which seed can not germinate in the presence of beet seeds?

   A. ryegrass   B. rice   C. wheat   D. corn

3. The analogous postulates applied to allelopathy are ________.

   A. observe，describe，and quantify the degree of interference in a natural community.
   B. isolate，characterize，and synthesize the suspected toxin produced by the suspected allelopathic plant.
   C. reproduce the symptoms by application of the toxin at appropriate rates and times in nature.
   D. monitor release，movement and uptake，and show they are sufficient to cause the

observed effects.

E. all of the above

4. If we want to control tobacco seedlings, which of the following plants residues can be used for allelopathy?

A. timothy B. maize C. rye D. all of the above

5. The following description is correct ________.

A. Allelopathy is a common phenomenon in nature.

B. Allelochemicals must be organic compounds.

C. Allelopathy are demonstrated in the lab, which show that allelopathy exist in the enviroment.

D. Allelopathy postulates must be confirmed.

Ⅴ Translate the following short passages into Chinese.

1. Allelopathy occurs when one plant, through its living or decaying tissue, interferes with growth of another plant via a chemical inhibitor.

2. Interference is the term assigned to adverse effects that plants exert on each other's growth. Competition is part of interference and occurs because of depletion or unavailability of one or more limiting resources.

3. Allelochemicals vary from simple molecules such as ammonia to the quinone, the terpenes camphor and cineole, complex conjugated flavonoids such as phlorizin, or the heterocyclic alkaloid caffeine.

4. The analogous postulates applied to allelopathy are: (i) Observe, describe, and quantify the degree of interference in a natural community; (ii) Isolate, characterize, and synthesize the suspected toxin produced by the suspected allelopathic plant; (iii) Reproduce the symptoms by application of the toxin at appropriate rates and times in nature; (iv) Monitor release, movement and uptake, and show they are sufficient to cause the observed effects.

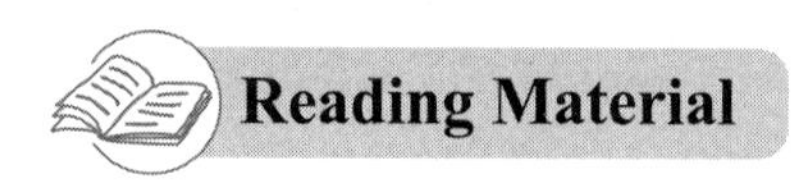

Allelochemicals

Concerned consequences related to commonly used herbicide in the environment have encouraged the farmers and agricultural producers to take advantage of technologies that manage weeds and be available to farmers. The use of allelopathy in various agricultural systems can be considered as one of these options. Allelopathy can be divided into two categories which include true allelopathy and functional allelopathy. Generally, direct release of allelochemical compounds in the environment, by donor species, can be considered as true allelopathy. However, the functional allelopathy is referred to the use of plants with allelopathic properties in the form of mulch or cover crops in agricultural fields, which the phytotoxic compounds produced and released after their decomposition through chemical and biochemical processes. Typically, the use of allelopathic cover crops involves the application of their residues as mulch at the surface of the field or spraying of their aqueous extracts similar to postemergence herbicides.

The use of allelopathic phenomenon as an ecofriendly strategy for weed management can be considered. Unlike the common herbicides, allelochemicals are of plant origin and do not cause residual effects and environmental pollution. Plants with allelopathic properties are a source of various allelochemical compounds that can be used to discover and produce new bioherbicides. Allelopathy refers to any positive or negative effects of a living organism on another living organism through direct or indirect release of allelochemicals into the environment. Actually, in plant allelopathic phenomenon, there is always a plant as a donor and a plant as a receptor of allelochemicals. In agro-ecosystems, the receptor and donor plants may be crops or weeds. Allelochemicals are referred to chemical substances, which are responsible for allelopathic effects. These compounds are synthesized in plants as secondary metabolites and play an important role in the relationships between plants and the surrounding environment.

Allelopathic Compounds

The allelochemicals produced by the plants are composed of various compounds such as alkaloids, cyanohydrins, lactones, organic acids, amino acids, fatty acids, flavonoids, tannins, and other compounds. Allelopathic compounds are synthesized in different plant tissues such as leaves, flowers, fruits, roots, rhizomes, stems, and pollen. The allelochemicals are released in the environment through evaporation, leaching, root leakage and plant decay.

Generally, the allelochemicals are environmentally friendly compounds and their presence in the environment is not a threat to human health and bio-ecosystems. Allelochemical compounds can affect processes such as germination, growth and development, nutrient uptake, cell division, membrane permeability, enzyme activity, and fatty acid metabolism. However, the concentration and efficiency of allelopathy depends on the age and plant tissue type. Nowadays, the allelopathic phenomenon is used to weed management and produce bioherbicides.

Potential Allelochemicals and Secondary Metabolites

Generally, the allelochemicals depending on their chemical similarities are classified in different groups, including alkaloids, flavonoids, phenols, tannins, cyanohydrins, amino acids, peptides, terpenes, ketones, cinnamic acids, benzoic acid, water-soluble organic acids, fatty acids, unsaturated lactones, quinones, polyacetylenes, coumarin, steroids, and benzoquinones. Some of these allelochemicals are commercially produced or in the process of commercialization by various companies. For example, the Cinch is produced from an allelochemical named Cineole and Callisto is commercially produced from the leptospermone. Another allelochemical that is successfully used as a natural herbicide for controlling weeds is sorgoleone, extracted from the sorghum plant. The phytotoxins derived from different microbial population have also been used as allelochemicals for weed management. Bialaphos is a natural synthesized product from *Streptomyces viridochromeogenes* and *Streptomyces hygroscopicus* that has been widely used to control weeds.

Although the physiological nature of allelochemicals has not yet been specifically addressed, it has been found that these compounds can interact the vital processes of plants, such as photosynthesis, cell division, enzymatic processes, and respiration. Understanding the biochemical and physiological effects of allelochemicals and identifying and purifying the chemical compounds responsible for phytotoxic effects on target plants is one of the most important factors in the development of new herbicides.

Allelopathic Interactions Between Crops and Weeds

Allelopathy refers to the interaction between living organisms and especially plant species in the environment that has occurred over centuries in agricultural ecosystems. Hence, allelopathy has an undeniable role in plant communications, including, crop-crop, crop-weed, and weed-weed interactions. Also, allelopathy plays a significant role in intra- and interspecific competition in the plant population. it reported that allelopathy phenomenon using cover crops has led to the successful control of different weeds including foxtail (*Setaria* spp.), morning glory (*Ipomoea* spp.), and yellow nutsedge (*Cyperus esculentus* L.) in field. The occurrence of allelopathy among various plant species may inhibit the growth and development of neighboring plant species and affect seed germination.

Although the protection of donor plant species from adverse biotic conditions can also be considered as one of the consequences of allelopathy. Today, in sustainable agricultural systems and especially organic farming and conservation agriculture, which minimize the use of agricultural pesticides, research on allelopathy and the use of allelochemicals for weed management are of great importance.

Methods to Study

During recent years, weed management researchers have been seeking to develop practical methods for using allelopathic phenomena to control weeds in agro-ecosystems. Accordingly, three major methods have been reported for the use of this biological phenomenon including (1) the use of a donor crop with allelopathic properties to suppress weeds, (2) the use of aqueous or hydro alcoholic extracts of allelopathic plants for postemergence sprays similar to herbicides, and (3) the use of living mulches or crop residues from plants with allelopathic properties in crop rotations. Plant allelopathic effects are also determined by various analytical methods such as high-performance liquid chromatography (HPLC), gas chromatography (GC), gas chromatography-mass spectrometry (GC-MS), and different bioassay methods in field or controlled conditions. Verdeguer et al. evaluated the allelopathic effects of different plants such as ambilateral (*Lantana camara*), red river gum (*Eucalyptus camaldulensis*), and wild rosemary (*Eriocephalus africanus*) through analytical methods (GC-MS) in order to their herbicidal activity on green amaranth (*Amaranthus hybridus*) and purslane (*Portulaca oleracea*). They reported that red river gum extracts were the most effective on inhibition of germination and seedling growth of both weeds.

Selected from: Mehdizadeh M, Mushtaq W. Biological Control of Weeds by Allelopathic Compounds from Different Plants: A Bioherbicide Approach. Natural Remedies for Pest, Disease and Weed Control.

Words and Expressions

allelochemical compound 化感化合物
phytotoxic compound 植物性毒素化合物
cover crop 覆盖作物，被护作物
ecofriendly 生态友好型，环保型
environmental pollution 环境污染
bioherbicide 生物除草剂
agro-ecosystem /ˌæɡrəu'iːkəsistəm/ *n.* 农业生态系统
alkaloid /'ælkəlɔɪd/ *n.* [有化] 生物碱；植物碱基
cyanohydrins /ˌsaɪənə'haɪdrɪn/ *n.* [有化] 氰醇，羟腈
lactone /'læktəʊn/ *n.* [有化] 内酯
nutrient uptake 营养吸收
cell division 细胞分裂
membrane permeability 膜通透性
phenol /'fiːnɒl/ *n.* 石碳酸，[有化] 苯酚
cinnamic acid [有化] 肉桂酸，苯丙烯酸
benzoic acid [有化] 苯甲酸
polyacetylene /ˌpɔliə'setiliːn/ *n.* 聚乙炔
benzoquinone /ˌbenzəu'kwinəun/ *n.* [有化] 苯醌
Callisto /kə'listəu/ *n.* 木卫四（木星最亮的四颗卫星之一，是距木星第六远的卫星）；卡利斯托（希腊神话中宙斯心爱的女神，被赫拉所憎恨并被她变为一只熊）
leptospermone 纤精酮
bialaphos 双丙氨膦，双丙氨膦钠，双丙氨磷
photosynthesis /ˌfəutəu'sɪnθəsɪs/ *n.* 光合作用
foxtail /'fɒksteɪl/ *n.* 狐尾，狐尾草
morning glory *n.* [园艺] 牵牛花（番薯属植物），昙花一现的人或物
yellow nutsedge 油莎草，铁荸荠，地栗
Cyperus esculentus L. 油莎草
seed germination 种子萌发
mulch /mʌltʃ/ *n.* 覆盖物，护盖物，护根；*vt.* 做护根，以护盖物覆盖
ambilateral /ˌæmbə'lætərəl/ *adj.* 双方面的，[解剖] 两侧的
Lantana camara 马缨丹
Eucalyptus camaldulensis 赤桉，小叶桉，桉木
wild rosemary (*Eriocephalus africanus*) 野迷迭香
green amaranth (*Amaranthus hybridus*) 绿穗苋
purslane /'pɜːslən/ (*Portulaca oleracea*) 马齿苋

Unit 12 Application of Allelopathy

Practical weed control can be achieved by using allelopathy. Importantly, such weed control will neither harm the environment nor increase weed management costs. Allelopathic weed control may be applied as a single strategy in certain cropping systems, such as organic farming. Further, it can be combined with other methods to achieve integrated weed management. Under allelopathic weed control, the allelopathic potential of crops is manipulated in such a way that the allelochemicals from these crops reduce weed competition. The living plants or their dead materials express the allelopathic activity through the exudation of allelochemicals. The processes for exudation of allelochemicals are: root exudation, leaching from dead or live plant tissues, and volatilisation from the aboveground plant parts. Several factors help the movement of allelochemicals to target species. Soil hyphae are important allelopathic transporters.

Allelopathic weed control can be implemented by growing allelopathic plants in close proximity to weeds which promote production of these chemicals; or by placing the allelopathic materials obtained from dead plants in close proximity to weeds. The decomposing plant material releases allelochemicals which are absorbed by the target weeds. The most important example for such cases includes the use of allelopathic plant residues for weed control. Allelopathic weed control can also be implemented by growing allelopathic plants in a field for a certain period of time, in order for their roots to exude allelochemicals. Crop rotation is the most important example for such allelopathic weed control.

Another way to control weeds through allelopathy includes obtaining allelochemicals in a liquid-solution by dipping the allelopathic chaff in water for a certain period of time. Several researchers have advocated using this way of weed control either alone or in combination with other methods of weed control. Recent research indicates that allelopathic plants not only suppress weeds but can have positive effects on the soil environment, that is, improved nutrient availability to crop plants through and enhanced soil microbial activities. Hence, the allelochemicals excreted from the microorganisms further helped to suppress crop weeds and diseases.

Allelopathic Cultivars

Crop cultivars demonstrating high productivity in farmers' fields are commercially acceptable. At the same time, the capability of crop cultivars to suppress weeds is being considered as a preferred criterion for cultivar selection in many parts of the world. The allelopathic potential of crop plants contributes to the weed suppressing ability of cultivars. Preferring weed-suppressive allelopathic cultivars over non-allelopathic cultivars can reduce weed infestation without incurring any extra cost, and would help to improve the efficacy of inputs and the method of weed control. A number of studies clearly elaborate the importance of sowing allelopathic cultivars in reducing weed pressure.

Although significantly influenced by environmental factors, the allelopathic potential of crop cultivars is a genetically controlled process. The allelopathic potential of crop cultivars against weeds can be increased through the breeding process. As a first step, the crop germplasm can be screened for its allelopathic potential. However, a weed suppressive allelopathic cultivar should also be high yielding. After selecting cultivars with desired traits, the genomic approaches can be applied for characterizing the relevant genes. These studies suggest that some of the crop cultivars possess allelopathic potential while others do not. The crop cultivars with allelopathic potential can be grown for inexpensive, easy and environment friendly weed control.

Allelopathic Intercrops

Compatible crops are grown together in order to harvest higher net yield and economic benefits. Further, growing crops in mixtures improves resource (land, water, nutrients and light) use efficiency. In addition to these benefits, intercropping can be used to suppress weeds for environment friendly and economical weed control. In particular, crops with allelopathic potential when intercropped with other crop plants help to reduce weed intensity, and hence improve crop productivity. For instance, intercropping maize and cowpea on alternate ridges helped reduce weed (*Echinochloa colona* L., *Portulaca oleracea* L., *Corchorus olitorius* L., and *Dactyloctenium aegyptium* L.) intensity by ~50% as well as improve land use efficiency. Therefore, intercropping allelopathic crops with the main crop can help to reduce weed intensity and improve yield gains.

Cover Crops

Cover crops are grown with the aim of maintaining the sustainability of an agro-ecosystem. Various objectives of growing cover crops include improving soil fertility and soil quality, and suppressing weeds and plant pathogens. Cover crops with allelopathic potential can suppress weeds. Several of the important cover crops include canola, rape seed, cereal rye, crimson clover, wheat, red clover, brown mustard, oats, cowpea, fodder radish, annual ryegrass, mustards, buckwheat, hairy vetch, and black mustard. Some of the cropping systems (e.g., organic cropping) heavily rely on cover cropping for weed management. The observations from farmers' fields and the results of experiments indicated that the release of allelochemicals from allelopathic cover crops and their physical effects were responsible for weed suppression in conservation organic farm fields.

Further, cover crops also possess several additional benefits other than weed management. For example, the cover crops also improved soil moisture retention, soil fertility, and crop productivity. Mixtures of cover crops have been found more effective in suppressing weeds compared to a single cover crop. Using more than one cover crop can produce higher quantities of diverse allelochemicals as well as higher biomass to suppress weeds more effectively.

Haramoto and Gallandt explored the role of brassica cover crops including white mustard and rape seed for weed suppression in agricultural systems. The authors argued that the brassica species exude allelochemicals, which are named glucosinolates. In natural environments, glucosinolates are decomposed into several compounds, most important of which are isothiocyanates. Isothiocyanates are biologically active and suppress the germination and growth of exposed plant species. The effects of brassica plants were more pronounced on germination of weed species than on their growth. The allelopathic effects of the brassica plants may be carried to the succeeding crops, which can be avoided through careful selection of the cover and the succeeding crops.

Cover crops are also useful in conservation tillage systems to effectively suppress weeds. For example, soybean could be successfully sown in a standing rye cover crop in a no-till soil. Planting soybean in a standing rye crop resulted in longlasting and effective weed control with no damage caused to the soybean crop. Similarly, rye and wheat cover crops helped to improve weed control in glyphosate-resistant cotton under a conservation till system. Biomass of weeds including *Eleusine indica* L., *Amaranthus palmeri* S., and *Ipomoea lacunose* L. was reduced by the cover crops, which helped to acquire season-long weed control. Moreover, cover crops can also reduce the weed seed bank in conservation till systems. For example, the cover crops hairy vetch and oat effectively reduced seed banks (30%~70%) of weeds, including *Datura stramonium* L., *Digitaria sanguinalis* L., *Amaranthus retroflexus* L., and *E. indica* in the upper soil layer.

Allelopathic Residues

In most cases, specific parts of crops are used for consumption, while the rest of plant portions are fed to animals, discarded, or incorporated in the soil as organic matter. For example, wheat, maize, and rice are salient grain crops whose grains are consumed as food while other plant parts are either fed to animals or left in the field. Similarly, cotton is the salient fibre crop of the world whose seed cotton is obtained for industrial uses while the rest of plant parts are either discarded or left in the field. Allelopathic plant residues left in the field either unintentionally or added manually express their activity to suppress weeds. For example, the plant residues of barley, rye, and triticale retained in a maize field were evaluated for their allelopathic effect against *E. crusgalli* and *Setaria verticillata* (L.) P. Beauv. in Greece. The allelopathic mulches decreased the emergence of *S. verticillata* (0%~67%) and *E. crusgalli* (27%~80%) compared with the nonmulched treatment. The maize plants received no harmful effect from the applied mulches. The grain yield of maize was increased by 45% in the plots applied with barley mulch compared with the ones with no mulch (non-treated control).

Similarly, the maize residues incorporated in an organically grown maize-broccoli rotation were found to reduce weed biomass in the following crop (broccoli) by 22%~47%. The incorporated mulch also helped to improve the soil nutrient status. Similarly, in another study, tomato seedlings were transplanted in mulch residues of three crops (oat, hairy vetch, and subterranean clover). The mulches were effective in suppressing weeds (35%~80%) in terms of density and biomass over the control treatment, while oat was the most effective among the mulches in suppressing weeds. However, oat also negatively affected tomato yield. Nevertheless, the highest increase in yield over the control resulted from hairy vetch. In a similar study, the residue mulch of oat and hairy vetch effectively reduced weed density (*A. retroflexus*, *Polygonum aviculare* L., *P. oleracea*, and *Chenopodium album*) in black pepper. Oat was more effective against weeds than hairy vetch; however, hairy vetch resulted in a higher increase in black pepper yield than oat. In conclusion, allelopathic plant residues can be applied as mulch to suppress weeds and improve grain yield. Enhanced moisture conservation and nutrient availability are additional benefits of using mulches for weed management.

Crop Rotation

Crop rotation is the sequence or arrangement of crops sown on a certain field. The objective of this sequence is to maintain soil productivity and sustainability. Certain changes in this crop sequence can reduce pest infestations. Crop rotation alone lowers weed infestation in crop fields while it enhances the effectiveness of weed control when combined with other methods. For example, crop rotation was among the major weed control strategies used by organic growers in New York, to suppress weeds. Crop rotation becomes more effective when no weed seeds from the neighboring land invade the field under rotation.

The allelochemicals added to the field from the previous allelopathic crop and the changed management practices together help to control weeds. Recent research has proven that allelopathic plants fill the soil with allelochemicals, which suppresses weeds in the following crop. The soil from allelopathic rice cultivars contained higher concentrations of allelochemicals, which suppressed the growth of *E. crusgalli*.

In conclusion, allelopathic crops express their allelopathic activity through exudation of allelochemicals. Growing allelopathic crop cultivars may become an important way to suppress weeds, especially when used under the umbrella of integrated weed management. Similarly, the use of allelopathic cover crops, allelopathic intercrops, the inclusion of allelopathic crops in rotation, and the use of allelopathic plant residues as mulches are important ways that can be practiced for economical, environment friendly weed management in agricultural systems. The allelopathic potential of crops is desired to be strengthened using conventional and modern plant breeding techniques.

Selected from: Jabran K, Mahajan G, Sardana V, *et al*. Allelopathy for weed control in agricultural systems. Crop Protection, 2015, 72: 57-65.

Words and Expressions

weed control 杂草防除

organic farming 有机农业，有机耕作

integrated weed management 杂草综合治理

root exudation　根系分泌物

hyphae　/'haifiː/ *n.* 菌丝

crop rotation　［农学］轮作

allelopathic cultivars　化感栽培种

crop cultivars　农作物品种

germplasm　/'dʒəːmplæzm/ *n.*［胚］［遗］种质

intercropping　/ˌɪntə'krɔpiŋ/ *n.* 间作，［农学］间混作；*v.*［农学］间作（intercrop 的 ing 形式）

Echinochloa colona L.　芒稷，光头稗

Portulaca oleracea L.　马齿苋

Corchorus olitorius L.　长果种黄麻，长蒴黄麻

Dactyloctenium aegyptium L.　龙爪茅

cover crops　遮盖作物

canola　/kə'nəulə/ *n.* 一种菜籽油

rape seed　油菜籽

crimson clover　绛车轴草，绛红三叶草

red clover　红三叶草

brown mustard　黑芥

oats　/əuts/ *n.* 燕麦，燕麦片（oat 的复数），燕麦粥

cowpea　/'kaupiː/ *n.* 豇豆

ryegrass　/'raɪgræs/ *n.*［肥料］黑麦草

buckwheat　/'bʌkwiːt/ *n.* 荞麦，荞麦粉，荞麦片

hairy vetch　［植］毛叶苕子，长柔毛野豌豆

organic cropping　有机种植

brassica　/'bræsɪkə/ *n.* 芸苔属植物，十字花科植物

glucosinolate　芥子苷，硫代葡萄糖苷

isothiocyanate　/'aisəuˌθaiəu'saiəneit/ *n.*［无化］异硫氰酸盐，［无化］异硫氰酸酯

glyphosate　/'glaɪfəseɪt/ *n.*［农药］草甘膦

Eleusine indica L.　牛筋草

Amaranthus palmeri S.　长芒苋

Ipomoea lacunose L.　野甘薯

Datura stramonium L.　曼陀罗

Digitaria sanguinalis L.　马唐

Amaranthus retroflexus L.　反枝苋

salient　/'seɪliənt/ *adj.* 显著的，突出的，跳跃的；*n.* 凸角，突出部分

triticale　/'traɪtɪkəl/ *n.* 黑小麦，黑小麦粒

E. crusgalli 稗草

Setaria verticillata（L.）P. Beauv.　狗尾草

Polygonum aviculare L.　扁蓄

Chenopodium album L.　藜

Notes

[1] organic cropping：有机种植，一种在生产中不使用化学合成的肥料、农药、植物生长调节剂，也不采用基因工程和离子辐射技术，而是遵循自然规律，采取农作、物理和生物的方法来培肥土壤、防治病虫害，以获得安全的生物及其产物的农业生产体系。其核心是建立和恢复农业生态系统的生物多样性和良性循环，以维持农业的可持续发展。

Exercises

Ⅰ Answer the following questions according to the text.

1. What is the difference between allelopathic cultivars and crop cultivars?
2. What measures can be used in allelopathic weed control? please give examples.
3. What are the objectives of allelopathic intercrops, cover crops, crop rotation?

Ⅱ Translate the following English phrases into Chinese.

organic farming　crop rotation　root exudation　integrated weed management
positive effects　weed control　crop cultivars　allelopathic intercrops

Ⅲ Translate the following Chinese phrases into English.

种子库　杂草侵染　土壤湿度　土壤肥力　化感栽培种
有机质　作物产量　覆土作物　轮作　农业生态系统

Ⅳ Choose the best answer for each of the following questions according to the text.

1. For instance, ________ maize and cowpea on alternate ridges helped reduce weed.
   A. allelopathic intercropping　B. allelopathic residues
   C. cover crops　D. crop rotation
2. What are the objectives using cover crops?
   A. improving soil fertility　B. soil quality
   C. crop productivity　D. all of the above
3. What are the objectives using crop rotation?
   A. maintain soil productivity and sustainability
   B. reduce pest infestations
   C. lowers weed infestation in crop fields
   D. all of the above
4. What is the most important measures for such allelopathic weed control?
   A. allelopathic intercropping　B. allelopathic residues
   C. cover crops　D. crop rotation
5. What are the objectives applying allelopathic plant residues?
   A. improve grain yield　B. enhanced moisture conservation
   C. nutrient availability　D. all of the above

Ⅴ Translate the following short passages into Chinese.

1. Allelopathic weed control can be implemented by growing allelopathic plants in close proximity to weeds which promote production of these chemicals; or by placing the allelopathic materials obtained from dead plants in close proximity to weeds.

2. Cover crops are grown with the aim of maintaining the sustainability of an agro-ecosystem. Various objectives of growing cover crops include improving soil fertility and soil quality, and suppressing weeds and plant pathogens.

3. Crop rotation lowers weed infestation in crop fields while it enhances the effectiveness of weed control when combined with other methods. Crop rotation becomes more effective when no weed seeds from the neighboring land invade the field under rotation.

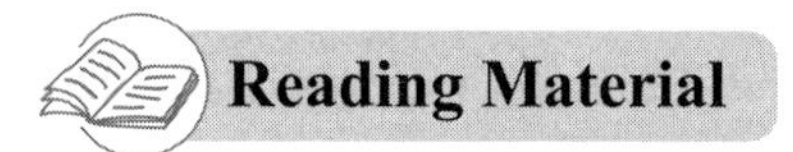

Allelopathy and Weed-Crop Ecology

Allelopathy was significant for weed-crop ecology in three ways:

1. As a factor affecting changes in weed species composition;
2. As an avenue of weed interference with crop growth and yield; and
3. As a possible weed management tool.

Allelopathy should not always be implicated when other explanations do not suffice, but it should not be overlooked because of the difficulty of establishing causality.

Effects on Weed Species

Why one species succeeds another is a question that has intrigued ecologists for many years. Weed scientists are interested in the same question but perhaps too often only for the life span of an annual crop. Weed scientists accept that plants change the environment and are changed by it. It is generally agreed that many early colonizers succeed by producing large numbers of seeds, whereas late arrivals succeed through greater competitive ability. This is true in old-field succession and in annual crops.

Ecologists have shown that successful plants may change the environment to their advantage through changes in soil nitrogen relationships caused by release of specific inhibitors of nitrogen fixation or nitrification.

Weed Interference

Weed seeds survive for long periods in soil and chemical inhibitors of microbial decay have been implicated in their longevity, but specific identification of inhibitors from weed seeds has not been accomplished. Allelochemicals have been implicated in the inability of some seeds to germinate in the presence of other seeds or in the presence of crop residues in soil. Although neither phenomenon has been exploited for weed management, there is little doubt that both occur. Eventual exploitation may depend on discovery of specific chemicals and their site of action. Because of the mass and volume of plant residue compared to the mass and volume of seed, the possibility of effects from plant residues is greater than the effects from seed.

The problems with replanting the same or different crops in a field have been cited to show the effect of allelochemicals on crop growth. the allelopathic potential of sorghum residues has been exploited for weed control in subsequent rotational crops. While there is little doubt that allelochemicals inhibit crop growth, a research challenge still exists to separate allelopathic effects from competition. Most greenhouse studies cannot be directly translated to the field because of different climatic, light, edaphic, and biological conditions, and possible effects of soil volume. Confirmation of allelopathy awaits development of appropriate experimental methods that verify its presence in field and greenhouse studies.

A fundamental assumption of biological control of weed is that damaged plants are less fit and compete poorly and therefore they will fail in the struggle for survival. That assumption, like so many in science, often is not borne out by research. When spotted knapweed control was attempted using larvae of two different root-boring insects and a parasitic fungus, its allelopathic potential increased significantly, and it had "more intense effects on native" vegetation. The authors conclude that while biological control can be very effective, it can often be less effective or fail. Without an understanding of the basic ecology of the area and the plants, it is not possible to know why success or failure occurred. "An invasive species that inhibits natives via unusually deep shade might be a more appropriate target for biological control than allelopathic invaders".

Weed Management

A living cover crop of spring-planted rye reduced early season biomass of common lambs quarters 98%, common ragweed 90%, and large crabgrass 42% compared to control plots with no rye. Wheat straw reduced populations of pitted morning glory and prickly sida in no-tillage culture. It was suggested that the wheat produced an allelochemicalthat inhibited emergence of several broadleaved species.

It is reasonable to assume that many plants have allelopathic potential or some susceptibility to allelochemicals when they are present in the right amount, form, and concentration at the appropriate time. It is equally reasonable to assume that allelopathy may have no role in the interference interactions of many species. However, enough work has been done to conclude that allelopathy could be utilized for development of new weed management strategies. Trials in South Dakota showed that fields planted with sorghum had 2～4 times fewer weeds the following year than similar fields planted with soybean or corn. It was proposed, although not proven, that reduced weed seed germination was due to phenolic acids and cyanogenic glucosides given off by sorghum.

Plant pathogens and allelochemicals from plant pathogens and other soil microorganisms can be used as bioherbicides. This possibility has been studied for more than three decades. Numerous pathogens and microbial allelochemicals have been isolated and studied for their bioherbicidal potential. A good example of a microbial product is the herbicide bialaphos (active ingredient phosphinothricin), which leads to accumulation of ammonium and disruption of primary metabolism. It is manufactured by fermentation as a metabolite of the soil microbe *Streptomyces viridochromeogenes*.

Pollen can also be allelopathic. Pollen can release toxins that inhibit seed germination, seedling emergence, sporophytic growth, or sexual reproduction. Two crops (timothy and corn) and four weeds (orange hawkweed, ragweed *parthenium*, yellow hawkweed, and yellow-devil hawkweed) are known to exhibit pollen allelopathy. Pollen allelopathy might be useful in biological weed management because the allelochemical is active in very low doses (as little as 10 grains of pollen per mm^2 on stigmas) and pollen is a small, naturally targeted distribution system.

Work to date has shown this to be an inconsistent effect, and, if developed, it might be useful when combined with other methods of weed management. Allelopathy is not, and will never be, a panacea for all weed problems. It is another weed management tool to be placed in the toolbox and used in combination with other techniques.

The second strategy where allelopathy may be used is weed-suppressing crops. This can be realized by discovering, incorporating, or enhancing allelopathic activity in crop plants. This technique would be most useful in crops maintained in high-density monocultures, such as turf grasses, forage grasses, or legumes.

The third area for allelopathic research and development includes the use of plant residues in cropping systems, allelopathic rotational crops, or companion plants with allelopathic potential. Many crops leave residues that are regarded as a necessary but not as a beneficial part of crop production, except as they contribute to soil fertility or tilth. Research indicates that plant residues have allelopathic activity, but the nature of this activity has not been explored sufficiently to permit effective use.

Rotation, a neglected practice in many agricultural systems, is being studied because of its potential for weed management through competition and allelopathy. Companion cropping is a relatively new technique for agricultural systems in developing countries. Multiple cropping is common in many developing countries where allelopathy may be operational without being obvious or defined. These systems may hold valuable lessons for further agricultural development of allelopathy as a useful weed management tool.

Weed scientists need to look beyond the immediate assumption that interference is always competition and see what they may not be looking for—an allelopathic effect—an unexpected, good thing. Perhaps there are expressions of allelopathy before our eyes that we don't see because we're not looking for them. If there are compounds in nature with such great specificity, they should be examined. The patterns of herbicide development point to greater specificity, and nature may have solutions in natural products if we recognize them, learn how they work, and exploit their capabilities.

Selected from: Zimdahl R L. Allelopathy. Fundamentals of Weed Science, 2018: 253-270.

Words and Expressions

causality /kɔːˈzæləti/ *n*. 因果关系

nitrogen fixation 固氮作用

nitrification /ˌnaɪtrəfəˈkeʃən/ *n*. [化学] 硝化作用

edaphic /ɪ'dæfɪk/ *adj*.［土壤］土壤的，与土壤有关的

bialaphos　双丙氨膦

phosphinothricin　草丁膦

Streptomyces viridochromeogenes　绿色产色链霉菌

sporophytic　孢子体的

orange hawkweed　山柳兰，橘色山柳兰

PART Ⅶ

GENETICALLY ENGINEERED PESTICIDE

Unit 13　Genetically Engineered Crops

New molecular tools provide opportunities for the development of genetically engineered (GE) pest-resistant crops that control key pests and require less input of foliar and soil insecticides. The first GE crops developed in the late 1980s expressed insecticidal proteins from the bacterium *Bacillus thuringiensis* Berliner (Bt) because of their known specificity and the excellent safety record of microbial Bt formulations. Pest-resistant Bt plants are now widely used on a global scale. There is evidence that Bt crops can reduce target pest populations over broad scales resulting in reduced damage on both GE and non-GE crops in the region. In addition, they have been shown to promote biological pest control in the system, if foliar insecticides are reduced. Host-plant resistance, whether developed through traditional breeding practices or genetic engineering, is an important tactic to protect crops against arthropod pests. The tools of genetic engineering have provided a novel and powerful means of transferring insect-resistance genes to crops, and there is evidence that those resistance traits have similar effects on natural enemies than resistance achieved by conventional breeding. GE insect resistant crops have been grown on a large scale for more than 20 years, and there is considerable experience and knowledge on how they can affect natural enemies and how their risks can be assessed prior to commercialization.

GE Plant Cultivation

Since the first GE plant was commercialized in 1996, the area grown with GE varieties has steadily increased. The two major traits that are deployed are herbicide-tolerance (HT) and resistance to insects. Here, we will focus primarily on insect-resistant GE crops. In 2017, GE varieties expressing one or several insecticidal genes from Bt were grown on a total of 101 million hectares worldwide, reaching adoption levels above 80% in some regions. Thus, Bt plants have turned what was once a minor foliar insecticide into a major control strategy.

The majority of today's insect-resistant GE plants produce crystalized (Cry) proteins from Bt. However, this bacterium possesses another class of insecticidal proteins, the vegetative insecticidal proteins (Vips), which are synthesized during the vegetative growth phase and have a different mode of action than Cry proteins. Vips are already deployed in some commercial maize hybrids and cotton. While the early generation of Bt crops expressed single cry genes, current varieties typically express two or more insecticidal genes. These so-called pyramid events are more effective in controlling the target pests and help to slow down the evolution of resistance. Currently, SmartStax® maize produces the most combined GE traits of any currently commercially cultivated GE crop, i. e. , six different cry genes to control lepidopteran and coleopteran pests and two genes for herbicide tolerance.

The application of the Bt technology, however, is currently largely limited to the three field crops maize, cotton, and soybean. Most of the Bt varieties target lepidopteran pests. This includes stemborers, such as *Ostrinia nubilalis* in maize, the pink bollworm *Pectinophora gossypiella* in cotton and the budworm/bollworm complex in cotton and soybean, including *Helicoverpa/Heliothis* spp. and other caterpillar pests. In the case of maize, traits are available that target the larvae of corn rootworms *Diabrotica* spp.. Recently, the technology has been applied to eggplant for protection against the eggplant fruit and shoot borer.

The adoption of the Bt technology differs among continents. While Bt-transgenic varieties are widely used in the Americas and in Asia, only few countries in Europe and Africa grow these crops. Bt maize is very popular in the Americas, often reaching>80% adoption. Bt cotton is also widely grown in the USA and Mexico, while Bt soybean remains at relatively low adoption levels (17%～34%) in South America with the exception of Brazil (58% adoption). In Chile, stacked Bt/HT maize and in Costa Rica, stacked Bt/HT cotton have been grown for seed export only. In several Asian countries and Australia, the technology is used to control lepidopteran pests in cotton with adoption levels>90%. Bt maize is grown at a significant level in The Philippines to control the Asian corn borer, *Ostrinia furnacalis* while Vietnam only introduced Bt-transgenic varieties in 2015 and their use is still limited. In Europe, the only product currently approved for cultivation is the Bt maize that produces the Cry I Ab protein and protects the plants from corn borers. The largest cultivation area is in Spain with an overall adoption level of 36% in 2017. In Africa, Bt crops are currently cultivated in only two countries. South Africa grows Bt maize to control stem borers and Bt cotton to control *Helicoverpa armigera*. Sudan deploys Bt cotton targeting the same pest. Use of Bt cotton has been temporarily halted in Burkina Faso after eight years. With the recent invasion of the fall armyworm in Africa, there is increased interest in using Bt maize as part of a management program.

Non-target Risk Assessment

Worldwide, GE plants are subject to an environmental risk assessment (ERA) before being released for cultivation. Growing insecticidal GE plants could harm natural enemies and biological control in three ways. First, the plant transformation process could have introduced potential harmful unintended changes. In the ERA, this risk is typically addressed by a weight-of-evidence approach considering information from the molecular characterization of the particular GE events and from a comparison of the composition and agronomic and phenotypic characteristics of the GE plant with its conventional counterparts. There is increasing evidence that the process of genetic engineering generally has fewer effects on crop composition compared with traditional breeding methods. The current approach is conservative, in particular because offtypes are typically eliminated over the many years of breeding and selection that happen in the process of developing a new GE variety. Second, the plant-produced insecticidal protein could directly affect natural enemies. Such potential toxicity is tested on a number of non-target species and these data are an important part of the regulatory dossier. Third, indirect effects could occur as a consequence of changes in crop management or arthropod food-webs. Such affects are addressed in the pre-market ERA but, because of the complexity of agro-ecosystems, potential impacts might only be visible once plants are grown in farmer fields.

Toxicity of the insecticidal protein to natural enemies is typically evaluated in a tiered risk assessment approach that is conceptually similar to that used for pesticides. Testing starts with laboratory studies representing highly controlled, worst-case exposure conditions and progresses to bioassays with more realistic exposure to the toxin and semi-field or open field studies carried out under less controlled conditions. From a practical standpoint, because not all natural enemies potentially at risk can be tested, a representative subset of species (surrogates) is selected for assessment. First, the species must be amenable and available for testing. This means that suitable lifestages of the test species must be obtainable in sufficient quantity and quality, and validated test protocols should be available that allow consistent detection of adverse effects on ecologically relevant parameters. Second, what is known about the spectrum of activity of the insecticidal protein and its mode of action should be taken into account to identify the species or taxa that are most likely to be sensitive. In the case of Bt proteins (and even more so in the case of insecticidal GE plants based on RNA interference) the phylogenetic relatedness of the natural enemy with the target pest species are of importance. Third, the species tested should be representative of taxa or functional groups that contribute to biological control and that are most likely to be exposed to the insecticidal compound in the field.

Safety

Studies to investigate the toxicity of the insecticidal compounds produced by Bt plants to natural enemies include direct feeding studies in which the natural enemies are fed artificial diet containing purified Bt protein, bitrophic studies where natural enemies are fed Bt plant tissue (e. g. , pollen), or tritrophic studies using a herbivore to expose the natural enemy to the plant-produced toxin. Numerous such studies have been conducted on a large number of Bt proteins, Bt crops and transformation events.

In summary, the available body of literature provides evidence that insecticidal proteins used in commercialized Bt crops cause no direct, adverse effects on non-target species outside the order (i. e. , Lepidoptera for Cry1 and Cry2 proteins) or the family (i. e. Coleoptera, Chrysomelidae for Cry3 proteins) of the target pests. This also holds true for Bt plants that produce two or more different insecticidal proteins. The available data indicate that these pyramided insecticidal proteins typically act additively in sensitive species and cause no unexpected effects in species that are not sensitive to the individual toxins. Recent studies have demonstrated that this is also true for a combination of Cry proteins and dsRNA.

Application

1. Bt Cotton

Bt crops has been well documented with Bt cotton in Arizona. In 1996, Cry Ⅰ Ac-cotton was introduced into Arizona to control the pink bollworm. Bt cotton led to dramatic reductions in the use of foliar insecticides for the target pest, all of them broad-spectrum in nature. The quickly increased adoption of Bt cotton led to broad, areawide control of the pest and opened the door for an opportunity to eradicate this invasive pest. Bt cotton became a cornerstone element in the pink bollworm eradication program initiated in 2006 in Arizona, and insecticide use for this pest ceased entirely by 2008.

Overall, the Arizona cotton IPM strategy has cumulatively saved growers over $500 million since 1996 in yield protection and control costs [$274/(ha · year)], while preventing over 25 million pounds of active ingredient from being used in the environment. While many components contributed to this transformative change that allowed conservation biological control to function at a high capacity in Arizona cotton production, Bt cotton was a keystone technology that eliminated the early season use of broad-spectrum insecticides for pink bollworm. Without this capstone event, it is unlikely this success would have been possible.

In China, a large-scale study demonstrated that the decline in insecticide sprays in Bt cotton resulted in an increased abundance of important natural enemies and an associated decline in aphid populations. More importantly, these effects were not only observed in the Bt crop itself but also in other (non-GE) crops within the region.

2. Bt Maize

As for cotton, studies have shown that using Bt maize (field corn) has resulted in large global reductions in the use of foliar insecticides for control of Lepidoptera. Studies on the widespread adoption of Bt maize in the Midwestern USA corn belt have demonstrated a dramatic decline in populations of *O. nubilalis*, and thus the need for insecticide treatments for this key lepidopteran pest. Furthermore, this decline occurred not only for those who adopted Bt maize, but also for surrounding maize farmers that did not. A similar "halo" effect of lepidopteran suppression by the widespread adoption of Bt maize in the eastern USA has also been documented, as well as the benefits of pest declines in surrounding vegetable fields. While these studies document lower pest pressure because of wide spread adoption of Bt maize and less need for insecticidal sprays, by implication they also suggest that widespread conservation of natural enemies may be occurring.

In the northeastern US where a considerable amount of sweet corn is grown, studies have shown that Bt sweet corn is far less toxic to the major predators in the system than the commonly used pyrethroid lambda cyhalothrin, spinosad, and indoxacarb. Furthermore, this study demonstrated that Bt sweet corn provided better control of lepidopteran pests, and did not negatively affect the predation rates of sentinel egg masses of the European corn borer, as did lambda cyhalothrin and indoxacarb.

Selected from: Romeis J, Naranjo S E, Meissle M, *et al*. Genetically engineered crops help support conservation biological control. Biological Control, 2019, 130: 136-154.

Words and Expressions

genetically engineered crops 基因工程作物
Bacillus thuringiensis 苏云金芽孢杆菌
insecticidal proteins 杀虫蛋白
pest-resistant 抗虫害的
host-plant [植] 寄主植物
genetic engineering 基因工程，遗传工程
plant cultivation 植物栽培学
herbicide-tolerance 除草剂耐性
insect-resistant 抗虫的
mode of action 作用方式，作用机理
stemborers 螟虫
Ostrinia nubilalis 玉米螟，欧洲玉米螟
Pectinophora gossypiella 棉红铃虫，红铃虫
Helicoverpa spp. 铃夜蛾属
Heliothis spp. 实夜蛾属
Diabrotica spp. 条叶甲属
Ostrinia furnacalis 亚洲玉米螟
Helicoverpa armigera 棉铃虫
environmental risk assessment 环境风险评价，环境风险评估
weight-of-evidence approach 证据权衡法
phenotypic /ˌfinoˈtɪpɪk/ *adj*. 表型的
offtype 变异型
dossier /ˈdɒsieɪ/ *n*. 档案，卷宗，病历表册
regulatory dossier 监管档案
bitrophic 二级营养，二重营养
tritrophic 三级营养，三重营养
pyramid /ˈpɪrəmɪd/ *n*. 金字塔；*vi*. 上涨，成金字塔状；*vt*. 使……上涨
halo effect 晕轮效应，光环效应
lambda cyhalothrin 高效氯氟氰菊酯
spinosad 多杀菌素
indoxacarb 茚虫威

Notes

[1] Genetically cngineered crops：基因工程作物。在重组 DNA 技术发明的背景下，人类培育转 Bt 基因作物和转基因耐除草剂等作物，通过改变作物的基因来提高作物抵抗生物胁迫或非生物胁迫的能力。其中以耐草甘膦转基因作物为典型代表，使用草甘膦杀除杂草的同时，可以保护作物免除药害，极大地缓解农民的除草压力，解决了大豆、玉米、棉花等农作物大面积种植的除草问题。

[2] Burkina Faso：布基纳法索，位于非洲西部沃尔特河上游的内陆国。1960 年 8 月 5 日独立，1984 年 8 月 4 日改为现名。布基纳法索是世界最不发达国家之一，也是周边非洲国家主要的外来劳工输出国。在经济上，以农牧立国，占全国近八成的劳动力。

[3] weight-of-evidence approach：证据权衡法，在某些情况下，在对所有现有资料进行评估后，证据权衡提示无遗传毒性危害。

Exercises

Ⅰ Answer the following questions according to the text.

1. What are the effects of Bt crops on chemical pesticides?
2. How about the safety of GE crops?
3. What kind of Bt crops have? and how about their application?
4. Talk about your views on GE crops.

Ⅱ Translate the following English phrases into Chinese.

agronomic characteristics　herbicide-tolerance insecticidal protein　insecticidal gene
phylogenetic relatedness　transgenic varieties　lepidopteran pests　sensitive species

Ⅲ Translate the following Chinese phrases into English.

表型特征　人工饲料　捕食率　抗虫性　光环效应　宿主
常规育种　间接影响　食物网　抗虫的　转基因作物　甜玉米

Ⅳ Choose the best answer for each of the following questions according to the text.

1. Currently，which crops can use the application of the Bt technology?

A. maize　B. cotton
C. soybean　D. all of the above

2. Which pests do most of the Bt varieties target?

A. Lepidoptera　B. Coleoptera
C. Diptera　D. Hymenoptera

3. Which ways could growing insecticidal GE plants harm natural enemies and biological control?

A. the plant transformation process could have introduced potential harmful unintended changes.
B. the plant-produced insecticidal protein could directly affect natural enemies.
C. indirect effects could occur as a consequence of changes in crop management or arthropod food-webs.
D. all of the above.

4. Which of the follow are belong to pesticides?

A. agrochemicals　　B. GE crops

C. natural enemies　　D. all of the above

5. How many different cry genes do control lepidopteran and coleopteran pests?

A. four　　B. five　　C. six　　D. seven

Ⅴ **Translate the following short passages into Chinese.**

1. The first GE crops developed in the late 1980s expressed insecticidal proteins from the bacterium *Bacillus thuringiensis* Berliner (Bt) because of their known specificity and the excellent safety record of microbial Bt formulations.

2. the first GE plant was commercialized in 1996, the area grown with GE varieties has steadily increased. The two major traits that are deployed are herbicide-tolerance (HT) and resistance to insects.

3. In summary, the available body of literature provides evidence that insecticidal proteins used in commercialized Bt crops cause no direct, adverse effects on non-target species outside the order or the family of the target pests.

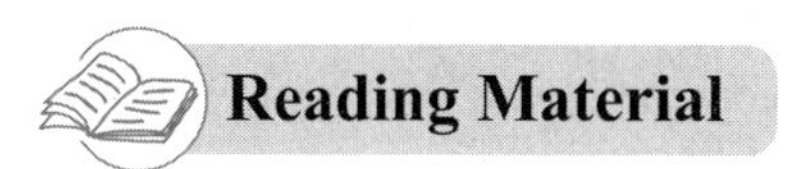

Transgenic Technology

Transgenic plants—plants carrying foreign genes—can be developed by inserting transgenes into any of the three genomes of plants, for example, nuclear, plastidal or mitochondrial. Traditionally, transgenic plants are made by inserting genes into the nucleus; however, to circumvent problems associated with the nuclear transformation, scientists have developed ways to insert genes into organelle genomes. Mitochondrial insertions have gained limited success, whereas chloroplast transformation is emerging as an alternative to nuclear transformation approach. Both approaches have their own strengths and limitations.

Agrobacterium-mediated Transformation

Agrobacterium tumefaciens is a gram-negative soil bacterium that causes crown gall in plants by inserting a part of its Ti (tumour-inducing) plasmid into the plant genome to produce a variety of amino acids, and sugar phosphate compounds collectively known as opines. The bacterium uses these compounds as an energy source to obtain carbons and nitrogen. Once the bacterium's sensory system detects acetosyringone one of the phenolic compounds secreted from plant wounds, the VirA protein is phosphorylated, which in turn phosphorylates VirG, a DNA binding protein. VirG switches on the expression of vir genes on Ti plasmid that are responsible for readying and delivering a 15 to 30 kb fragment of Ti plasmid, known as transfer DNA (T-DNA), into the plant cell to get integrated into plant chromosomal DNA through a mechanism similar to conjugation. Once inside the nuclear genome, it is then recognized by the host plant machinery to carry out its integration into the chromosomal DNA. After becoming a part of the host genome, the genes on the T-DNA start to express to produce auxins and cytokines, which result in an uncontrolled growth of the plant cell ultimately forming a tumour.

The genes on the T-DNA have regulatory elements that are eukaryotic in nature, and hence cannot be expressed by the bacterium's resident machinery. So the bacterium has to insert those genes into a plant cell to synthesize opines. The remaining Ti plasmid contains genes to take up and metabolize these opine molecules to produce energy. Therefore, any other bacterium lacking machinery for uptake and break down of opines cannot feed on the plant.

Scientists discovered that deleting genes involved in the production of opine molecules still resulted in transfer and integration of T-DNA. The realization that repeat regions of only 25 bases of T-DNA around a transgene are sufficient for integrating genes into a plant genome opened the doors of a new era in the history of plant biotechnology. Transformation of plants to introduce various desired genetic mutations using agrobacterium has become a routine now. Over the years, methods have been perfected by experimentation and now different strategies such as binary vector, co-integration system or floral dip methods, for example, are available to deliver transgenes into a plant genome using minimal T-DNA. Similarly, the Ri (root-inducing) plasmid of *Agrobacterium rhizogenes*, which causes a hairy root disease instead of crown gall, has also been attempted for transgene delivery. This could be useful for producing recombinant proteins for excreting in the medium through the root system. The molecular details of these strategies have been discussed elsewhere.

Using agrobacterium-mediated transformation procedure to develop transgenic plants is considered a "method of choice" when transgene insertions are intended for the nuclear genome because of simplicity and the ease of use. However, the narrow host range, which agrobacterium can infect, limits the usefulness of this method. For example, the bacterium has a limited host range and cannot be used to transform organelle genomes-chloroplast and mitochondria.

Biolistic Transformation

The impediments associated with the use of agrobacterium as a means of transmission of foreign genes was attempted to overcome by another method called biolistic transformation. In this system, DNA coated micro-projectiles like gold or tungsten are shot into plant cells using a biolistic particle delivery system such as a gene gun. Since such methods are based on the physical penetration of the transgenes, therefore, they are independent of cell type. Consequently, the biolistic transformation approach has broader transformation range, and can be used to transfer DNA into a variety of plant materials such as callus, embryos, leaves, and even the whole plant and organelles as well. Although the particle delivery system exhibits a broader transformation capacity, the transformation efficiency is significantly low as compared to agrobacterium-mediated methods. Other issues associated with this approach is the cost. The biolistic transformation is a costly technology, and hence is not preferred much. However, it is the only method for transforming organelle genomes like chloroplasts.

The process of recovering transgenic plants after delivery of the transgenes is essentially similar to that of agrobacterium-mediated transformation approach. Regeneration in plant in vitro cultures takes place via organogenesis or somatic embryogenesis. In organogenesis, unipolar bud-like structures are produced, which subsequently develop into distinct organs such as shoots or roots. Regenerated shoots are then cultured to produce root mass via root organogenesis. The shoots and roots can originate directly from the explant or indirectly from the undifferentiated callus. In organogenesis, the origin of bud primordia is of multicellular origin and the unipolar structure develops a vascular connection with the parent tissue or the cultured explant. On the other hand, somatic embryogenesis is the development of discrete embryo-like bipolar structures, referred to as somatic embryos, following a series of characteristic morphological stages. Somatic embryos arise from single cells and show no vascular connections with the parent tissue or explants. Somatic embryos can also originate directly from the explant or indirectly from the undifferentiated callus.

Briefly, a single transformed cell multiplies to give rise to a collection of undifferentiated transformed cells, referred to as callus, which is further cultured to ultimately give rise to plantlets through either organogenesis or embryogenesis. Since both the approaches produce a full plant from an individual transformed cell (direct) or from callus (indirect), the resulting plants therefore will be homozygous for that particular transgene. Organogenesis is a preferred method over embryogenesis for recovering transgenic plants because many plant species do not respond to embryogenesis, such as cereals.

For recovering transgenic plants, the putative transformants are first screened for resistance against a particular selection and then are identified using different molecular techniques; the most common ones are PCR (polymerase chain reaction) and Southern blotting. The expression of a particular gene is tested using reporter genes, whose expression is easy to monitor such as GFP or GUS.

Miscellaneous Methods

Other relatively less popular approaches to carry out the transformation of transgenes into plants include transduction and electroporation. Transduction uses plant viruses to introduce transgenes into plants. The transgenes are first packed into the genetic material of the virus. The modified virus is then allowed to infect plants to transport genes from the virus into plant cells. However, the plant viruses do not integrate their genes into the nucleus and therefore, are not suitable to make stable transformation events. Moreover, only two classes of plant viruses, caulimoviruses and geminiviruses, are able to infect higher plants; none of them is ideal for genetic modifications. Nevertheless, this method can be used to validate constructs quickly before embarking on stable transformation through either the agrobacterium or the particle delivery system. In addition, this method is quite useful for rapid production of important biomolecules like antibodies in plant leaves without going into stable transformation, which may take a long time to produce a desired biomaterial. Other than this, the viral vectors, particularly of geminiviruses are now being used for specialized applications such as VIGS (virus-induced gene silencing) and RNAi silencing to study the gene function and molecular mechanisms of viral replication in plants.

Electroporation uses electric current to make cell membrane permeable to allow DNA to enter into the cell. This method was used to overcome the problems associated with the agrobacterium method, physical penetration or viral-based strategies, which often result in cell toxicity, poor transformation efficiency or limited by a narrow range of transformable cell types. However, this method is also suitable for transient expression only. The DNA transferred by this method will not become a part of the genomic DNA of the host cell and therefore will not be transmitted to the progeny. Further, this procedure also requires isolation of protoplasts (plant cells without cell wall), which makes it labour-intensive and time-consuming practice with poor transformation efficiency. Although the transformation of intact plant cells using electroporation has been reported, such literature is scanty and the method was not found suitable for stable integration events. Therefore, such methods are now not in frequent use unless the plant material cannot be transformed otherwise. PEG-mediated transformation (also uses protoplasts) and microinjection are also not commonly used due to their unsuitability for high throughput applications, which often require higher transformation efficiencies, speed and a broad range of transformable material with few drawbacks.

Selected from: Ahmad N, Mukhtar Z. Genetic manipulations in crops: Challenges and opportunities. Genomics, 2017, 109 (5-6): 494-505.

Words and Expressions

transgenic plants 转基因植物
mitochondrial /ˌmaɪtəʊˈkɒndriəl/ *adj*. 线粒体的
circumvent /ˌsɜːkəmˈvent/ *v*. 包围；智取，绕行，规避
organelle /ˌɔːgəˈnel/ *n*. [细胞] 细胞器，细胞器官
Agrobacterium tumefaciens 根癌农杆菌，农杆菌
crown gall [植保] 冠瘿病
opine /əʊˈpaɪn/ *v*. 以为，表示意见，说明；*n*. 冠瘿碱
acetosyringone 乙酰丁香酮，丁香酮
auxin /ˈɔːksɪn/ *n*. [生化] 植物生长素；植物激素
cytokine /ˈsaɪtəkaɪn/ *n*. [细胞] 细胞因子，细胞激素
eukaryotic /juːˌkærɪˈɒtɪk/ *adj*. 真核的，真核生物的
Agrobacterium rhizogenes 毛根农杆菌，发根土壤杆菌
biolistic transformation 基因枪转化
tungsten /ˈtʌŋstən/ *n*. [化学] 钨
organogenesis /ˌɔːgənəˈdʒenɪsɪs/ *n*. [胚] 器官发生，[胚] 器官形成
somatic embryogenesis 体细胞胚胎发生

miscellaneous /ˌmɪsəˈleɪniəs/ *adj*. 混杂的，多方面的，多才多艺的
electroporation /ɪˌlektrəpəˈreɪʃ(ə)n/ *n*. [遗] 电穿孔
caulimovirus 花椰菜花叶病毒
geminivirus 双粒病毒，双生病毒

Unit 14 RNAi

RNA interference (RNAi), usually referring to the small interfering siRNA pathway, is promising in the development of a new generation of insect pest control products. RNAi is a conserved mechanism for posttranscriptional sequence specific gene silencing and is initiated by the introduction of doublestranded RNA (dsRNA). RNAi can be sorted into three pathways depending on the type of small RNA biogenesis and the Argonaute protein involved: microRNA (miRNA), small interfering RNA (siRNA) and piwi-interacting RNA pathways. Among them, the siRNA pathway has shown the most potential in insect pest control. Briefly, when exogenous dsRNA designed to target an endogenous gene is delivered to the target pest, it triggers the slicing activity of Dicer, resulting in the long dsRNA being processed into short 21-22nt siRNAs. These siRNAs are then incorporated into a complex together with Argonaute and other proteins, forming the RNA-induced silencing complex. In this complex, one of the small RNA strands is selected as the guide strand and serves as a sequence-specific guide to cleave the target endogenous mRNA through sequence-specific complementary binding. By targeting the expression of a vital endogenous gene necessary for survival, RNAi can be exploited to induce target gene silencing and pest mortality.

Prior to the exploitation of RNAi in pest management, several important aspects about the nature of RNAi in insects need to be understood: (i) siRNA is one of the key antiviral pathways in insects. Although insects are known to lack an adaptive immunity system, there are crosstalks between RNAi and other non-RNAi immune mechanisms, reflecting the sophisticated immunity network of insects. (ii) The fitness costs of RNAi in antiviral immunity. The activity of the RNAi pathway like other immune pathways could be significantly induced upon viral infections, which could cause a disturbance in the existing homeostasis, resulting in a trade-off with other individual parameters. (iii) RNAi-based pest control relies on the silencing of crucial genes for survival in the target insect pest, ultimately causing mortality or inhibition of the pest populations. However, the potential of RNAi (siRNA) in inducing gene silencing may be limited within certain thresholds, resulting in poor gene silencing efficiency. (iv) Direct degradation of dsRNA in the insect gut as well as in the hemolymph by nucleases can decrease gene silencing efficiency. (v) Virus-encoded suppressors of RNAi factors can also inhibit the activity of the RNAi pathway.

dsRNA Delivery

The efficient delivery of dsRNA is crucial in moving RNAi-based insect pest control from the lab to the field. From an ecological viewpoint, dsRNAs not only move within an organism but they can also transfer from the environment to the organism (environmental uptake) and between interacting organisms (cross-kingdom dsRNA trafficking), thereby subsequently inducing gene silencing in the targeted organism.

1. Environmental Uptake

The environmental RNAi approach can be exploited in the application of sprayable dsRNA formulations for insect pest control, similar to the application of classical chemical insecticides. The delivery and uptake of insecticides in insect pests mainly happen through the insect exocuticle or ingestion during feeding on host plants. In the potential RNA imediated control of chewing pests, the spray-dsRNA formations can be successful and such applications, for example against *L. decemlineata*, are on their way to be commercialized. However, sucking-piercing pests, such as aphids, whiteflies as well as spider mites, which are either feeding on the plant phloem or cells, require the dsRNA uptake via the insect integument directly or via ingestion of the dsRNA from plant phloem or cells. For sprayable dsRNA formulations, this presents an extra challenge.

Using a topical delivery approach, significant gene silencing of five *CYP4* genes was achieved and resulted in suppressed insecticide resistance in *Diaphorina citri*. A soaking-based dsRNA delivery was also used successfully in targeting genes in piercing-sucking pests. If dsRNA could be taken up by plant leaves then these sucking pests could be targeted via ingestion of dsRNA during feeding on the plant phloem or cells.

Therefore, studies elucidating the mechanism of dsRNA uptake by insect cuticle or plant leaf are required in the development of dsRNA-sprayable pest control strategies. Two aspects of fundamental studies are currently needed: (i) the mechanism of dsRNA uptake by insects or plant leaves, which can be achieved by using labelled dsRNA/siRNA to visualize the mobility of dsRNA/siRNA and (ii) exploring the ingredients ormethods that could enhance the stability and the absorbing efficiency of dsRNA by insects and plants. For example, adjuvants used for chemical pesticides can be evaluated for their ability to improve the uptake of dsRNA by the insect/plant leaf. Also, liposomes, polymers and peptides have been successfully used to increase persistence and cellular uptake of dsRNA in insects. Nanoparticles, such as nanosheets, have potential for applications of dsRNA by topical delivery, thus it may also be useful in delivery insect pest-specific dsRNAs.

2. Cross-Kingdom dsRNA Trafficking

Based on the concept of cross-kingdom RNAi, plants and microbes can be exploited to mediate dsRNA production and delivery. Transgenic plants expressing dsRNA have been shown to be useful in insect pests control. In the pioneering work for the control of *D. virgifera*, Snf7-dsRNA was transcribed and delivered via transgenic corn. The engineered transgenic plants expressing sequence-specific dsRNA to target pests are currently investigated in various organisms. Several studies have found that dsRNA produced by nuclear gene expression in these transgenic plants, is being processed by the plant's own RNAi machinery, resulting in mainly small RNAs being available for the insects. This could potentially affect the efficacy of these host plants in achieving a high degree of gene silencing in pest insects. Therefore, the expression of long dsRNAs in plastids where the RNAi machinery is absent, offers a particularly promising strategy that could lead to a more efficient silencing and therefore a more efficient management of the pest species.

Microbes, another important ecological interacting factor of insect pests, also provide an alternative way for dsRNA production and delivery. The bacterial model *Escherichia coli* is not only used for large-scale production of dsRNA but also the ingestion of engineered *E. coli* in expressing dsRNA targeting chitinase directly, increased the mortality and reduced the weight of the feeding larvae of *Mythimna separata*. In addition, the engineered fungi *Saccharomyces cerevisiae*, expressing specific dsRNA targeting *Drosophila suzukii*, led to a significant decrease of fitness of the targeted pest. Besides these model microbes, an ecological design focused on exploiting the insect symbiotic bacteria for trauma-free sustainable delivery of insect pest-specific dsRNA. Recent studies also identified a great number of insect viruses, especially RNA viruses. Based on plasmid-based reverse genetics systems, engineered viruses could also produce host-specific dsRNA, thus providing potential in pest control. Viruses are obligated to use the host nucleic acid and protein synthesis machinery and dsRNA is produced as an intermediate molecule during RNA virus replication, while RNA hairpins are present after transcription with DNA viruses. A first proof of concept was recently provided by Taning et al. who engineered a Flock House Virus to produce *D. melanogaster* specific dsRNA and managed to cause a high degree of RNAi silencing upon viral infection, leading to a strong mortality.

Studying the nature of cross-kingdom dsRNA movement is also useful in obtaining an optimal pest control strategy. For instance, Varroa mites are one of the major drivers in bee decline, and the control of this pest is crucial to protect bees from mites transmitted bee viruses. Intriguingly, dsRNA targeting mite genes can be transferred to mites through feeding of dsRNA-supplemented syrup to bees. In addition, the dsRNA can also be horizontally transferred between individuals and also across generations in bees.

Synergistic Effects of RNAi

RNAi technology is a manipulator of gene expressions, implying that the key genes in insect pests, required for maintaining homeostasis with the interacting host plants and pathogens, could be targeted, presenting an alternative approach in RNAi-based pest control strategy. For instance, silencing of genes involved in feeding in aphids, such as effectors to plant defences, resulted in significant mortality. Similarly, targeting the detoxification gene, glutathione *S*-transferase gene, by phloem-mediated RNAi in the whitefly, *Bemisia tabaci*, showed long-term inhibition of the population. Moreover, by depleting insect immunity genes, such as toll receptors, the virulence of pathogens can be increased.

Thus, the concept of combining RNAi with plant resistance or pathogens termed as RNAi plus, could be helpful in solving current challenges of RNAi in insect pest control: (i) RNAi could be used to assist pest control strategies such as plant-produced toxins, insecticides, pathogens, etc. (ii) Neither RNAi nor plant-produced toxins, insecticides and pathogens are fully efficient in controlling pests, however, the combination could lead to efficient pest control. (iii) The combination of RNAi and other pest control strategies could largely delay the development of resistance. For instance, variable combinations could be used, taking into account synergistic effects of various pathogens or various insecticides or plant toxins, with the RNAi-based control strategy. In such scenario, the RNAi-based strategy will be designed to target various genes involved in maintaining homeostasis between the plant host, insect pests and insect pathogen, thereby following an ecological principle rather than targeting a single factor which can easily pressure the insects to develop resistance. In addition, with environmental RNAi or cross-kingdom RNAi, the delivery of RNAi could also be enhanced either through the host plants or the insect pathogens.

Challenges in Applying RNAi

Until now, the approach of using RNAi in insect pest control has only been reported for commercialization in the western corn rootworm. While most RNAi-based insect pest control studies are still in the lab stages, others are only meaningful for functional gene analyses, far away from pest control in practice.

1. dsRNA Delivery Methods

The overall success of using RNAi for pest control is dependent on the mode of delivery of dsRNA. Since gene silencing is only limited to cells that take up the target gene dsRNAs, dsRNA delivery remains a major consideration in the development of an RNAi-based insecticide. Different delivery systems have been studied in various insect species and the main dsRNA delivery methods tested so far include soaking, feeding, injection and transgenic plants expressing dsRNA.

dsRNA uptake in insects could be divided into two levels: (i) dsRNA uptake from the environment into the insect body. This could either be from the ingestion of dsRNA from diet or direct uptake via the insect exocuticle. Current practical methods of dsRNA delivery include either transgenic plants modified to express insect pest-specific dsRNAs or sprayable dsRNA formations. The transgenic approach of dsRNA delivery offers the best potential for practical field application and has been widely studied in various key insect pests and spider mites. Alternatively, with the development of microbial production systems for exogenous dsRNAs, pest-specific dsRNAs can now be produced in large scales and at affordable prices for direct application in the field, by spraying. (ii) Cellular uptake and the stability of dsRNA in insects after uptake of the dsRNA from the environment into the insect body. Limiting factors, such as direct degradation and entrapment of internalized dsRNA in endosomes, have recently been shown to be crucial in achieving a successful RNAi during this step of dsRNA delivery.

2. Efficiency of RNAi

The success of RNAi in pest control depends on its efficiency in causing mortality to the targeted pests, ultimately leading to a decrease in the insect pest population under the economic threshold. The choice of a target gene that could lead to such a level of pest control is based on: genotype—the relative gene silencing level of the targeted gene via RNAi; phenotype—the function of the targeted gene, which implies that the role of this gene must be crucial for the survival of the insect. However, gene silencing efficiency is not only influenced by the target gene but also by the method of dsRNA delivery as mentioned earlier. Other associated factors, based on the nature of the target gene that could affect gene silencing efficiency include: the absolute expression level of the target gene, the relative expression level of the target gene at various time points, tissue-specific expression pattern of the target gene as well as the expression of paralogous genes with similar functions.

3. Resistance Development to RNAi

As observed with most insecticides, resistance to RNAi-based insecticides is bound to happen. Therefore, the key questions are what potential mechanisms could be involved in resistance development against RNAi based insecticides and which possible sustainable control strategies could be explored to delay these resistance challenges. In a recent study, the first case of field-resistant strains of *D. virgifera* against dsRNA (dsDvSnf7) was reported. A 130-fold resistance was observed in the strain feeding on transgenic plants expressing dsDvSnf7 in comparison with the susceptible strain reared on nontransgenic plants.

Selected from: Niu J, Taning C N T, Christiaens O, *et al*. Rethink RNAi in Insect Pest Control: Challenges and Perspectives. Advances in Insect Physiology, 2018, 55: 1-17.

Words and Expressions

RNA interference　RNA 干扰
posttranscriptional　/ˌpəusttræns'kripʃənəl/ *adj*. 转录后的
gene silencing　基因沉默
doublestranded RNA (dsRNA)　双链的 RNA
microRNA (miRNA)　微 RNA，微小 RNA
small interfering RNA (siRNA)　小干扰 RNA
exogenous　/ek'sɒdʒənəs/ *adj*. 外生的，外因的，外成的
endogenous gene　内源基因
homeostasis　/ˌhəumiə'steɪsɪs/ *n*. [生理] 体内平衡，[自] 内稳态
hemolymph　/'hiːməlɪmf/ *n*. 血淋巴
environmental uptake　环境吸收
trafficking　/'træfɪkɪŋ/ *n*. 非法交易（尤指毒品买卖）；*vi*. 交易
L. decemlineata　马铃薯甲虫，马铃薯叶甲
Diaphorina citri　柑橘木虱
D. virgifera　玉米根虫，玉米根萤叶甲
Escherichia coli　大肠杆菌
chitinase　/'kaitiˌneis/ n. [生化] 几丁质酶，[生化] 壳多糖酶，甲壳质酶
Mythimna separata　黏虫；东方黏虫，水稻黏虫
Saccharomyces cerevisiae　酿酒酵母
Drosophila suzukii　樱桃实蝇，樱桃果蝇
trauma　/'trɔːmə/ *n*. [外科] 创伤（由心理创伤造成精神上的异常），外伤
hairpin　/'heəpɪn/ *n*. 簪，束发夹，夹发针
D. melanogaster　黑檀体果蝇
viral infection　病毒性感染，病毒感染

Varroa mite 蜂螨，蜂虱

synergistic effects 协同作用，协同效应

exocuticle /'eksəuˌkjuːtikəl/ *n*. 外角皮，外角质层，外小皮

cellular uptake 细胞摄取，细胞摄入

entrapment /ɪn'træpmənt/ *n*. 诱捕，圈套，截留

economic threshold 经济阈值，经济限界

expression level 表达水平

paralogous genes 同源基因

detoxification gene 解毒基因

Bemisia tabaci 烟粉虱

Notes

[1] RNA interference (RNAi)：RNA 干扰，指在进化过程中高度保守的、由双链 RNA (doublestranded RNA，dsRNA) 诱发的、同源 mRNA 高效特异性降解的现象。

[2] Argonaute protein：Argonaute 蛋白，RNA 诱导沉默复合物 (RISC) 的催化成分，这种蛋白复合物负责 RNA 干扰 (RNAi) 的基因沉默现象。Argonaute 蛋白结合小干扰 RNA (siRNA) 片段，并具有针对与其结合的 siRNA 片段互补的信使 RNA (mRNA) 链的内切酶活性。这些蛋白也部分负责 siRNA 底物的导向链选择和乘客链的破坏。

[3] piwi-interacting RNA：piRNA，从哺乳动物生殖细胞中分离得到的一类长度约为 30nt 的小 RNA，并且这种小 RNA 与 PIWI 蛋白家族成员相结合才能发挥它的调控作用。目前，越来越多的文献表明 piRNA 在生殖细胞的生长发育中的调控是由 PIWI-piRNA 复合物引起的基因沉默导致的。

Exercises

Ⅰ Answer the following questions according to the text.

1. What is RNA interference? And how many pathways are RNAi sorted into?
2. Please describe the RNAi process.
3. What aspects of the RNAi properties of insects need to be understood before RNAi can be used to control pests
4. Please explain the synergistic effects of RNAi.
5. What are the challenges of applying RNAi?

Ⅱ Translate the following English phrases into Chinese.

doublestranded RNA	endogenous gene	gene silencing	topical delivery
sucking-piercing pest	silencing complex	viral infection	model microbes

Ⅲ Translate the following Chinese phrases into English.

咀嚼式害虫	靶标害虫	敏感菌株	体内平衡	基因表达
逆向遗传学	协同效应	关键基因	环境摄取	细胞吸收
转基因植物	RNA 干扰	免疫系统	表达水平	靶标基因

Ⅳ Choose the best answer for each of the following questions according to the text.

1. RNAi can be divided into three pathways according to the types of small RNA biogenesis and the Argonaute proteins involved ________.

   A. miRNA　　B. siRNA

   C. piwi-interacting RNA　　D. all of the above

2. Which pathway shows the greatest potential for pest control?

   A. miRNA　　B. siRNA

   C. piwi-interacting RNA　　D. all of the above

3. What are the methods for dsRNA delivery?

   A. soaking　　B. feeding

   C. injection　　D. all of the above

4. Which of the following processes is the delivery of dsRNA by insect exocuticle or ingestion during feeding on host plants?

   A. environmental uptake　　B. cross-kingdom dsRNA trafficking

   C. A and B　　D. all of the above

5. Microbes, another important ecological interacting factor of insect pests, also provide an alternative way for ________.

   A. environmental uptake　　B. cross-kingdom dsRNA trafficking

   C. A and B　　D. all of the above

Ⅴ Translate the following short passages into Chinese.

1. RNAi can be sorted into three pathways depending on the type of small RNA biogenesis and the Argonaute protein involved: microRNA (miRNA), small interfering RNA (siRNA) and piwi-interacting RNA pathways.

2. RNAi-based pest control relies on the silencing of crucial genes for survival in the target insect pest, ultimately causing mortality or inhibition of the pest populations.

3. The environmental RNAi approach can be exploited in the application of sprayable dsRNA formulations for insect pest control, similar to the application of classical chemical insecticides. The delivery and uptake of insecticides in insect pests mainly happen through the insect exocuticle or ingestion during feeding on host plants.

4. RNAi could be used to assist pest control strategies such as plant-produced toxins, insecticides, pathogens, etc. dsRNA uptake from the environment into the insect body. This could either be from the ingestion of dsRNA from diet or direct uptake via the insect exocuticle.

Methodology of dsRNA Uptake

Methods of dsRNA uptake in insects can greatly vary and strongly influence the efficiency of gene silencing, thus their potential as insect pest control agent. It is important to note that since gene silencing is only limited to cells that are infected, the main challenge is the selection of the delivery system. In both types, methods of delivery must be first defined, being effectively easier and better understood for cell-autonomous RNAi machinery. The main delivery methods include microinjection, soaking, feeding, transgenic technique, and viral infection.

Microinjection

Microinjection, i. e. the direct injection of dsRNA into the body of insects, has been one of the most effective delivery methods for systemic RNAi types. Short dsRNA have had the most success with this mechanism. In addition, the 5' end of the dsRNA can affect the effectiveness of RNAi; a phosphorylated 5' end exhibits better gene silencing rate than does a hydroxylated 5' end. The major advantage of injecting dsRNA into the insect body is the high efficiency of inhibiting gene expression. There are however some limitations with micro-injection. First, the cost for in vitro synthesis and storage of dsRNA is relatively high, and the steps are complicated. In addition, injection pressure and the wound generated inevitably affect the insects. It has been shown that skin damage stimulates the immune response. In practice, this delivery method would have very limited application as pest control agent.

Soaking

Soaking *D. melanogaster* embryos in a dsRNA solution can inhibit gene expression, and its effectiveness is comparable to the injection method in that it requires a higher concentration of dsRNA. Soaking *D. melanogaster* S2 cells in CycE and ago dsRNA solutions has been shown to effectively inhibit the expression of these two genes for cell cycle, thereby elevating levels of protein synthesis. The soaking method is suitable only for certain insect cells and tissues as well as for specific insects of developmental stages that readily absorb dsRNA from the solution, and therefore, it is rarely used.

Feeding

Compared to other methods, dsRNA feeding is the most attractive primarily because it is convenient and easy to manipulate. Since it is a more natural method of introducing dsRNA into insect body, it causes less damage to the insect than microinjection. It is especially popular in very small insects that are more difficult to manipulate using microinjection. Early insect RNAi feeding studies were frustrating; for example, the injection of dsRNA effectively silenced the aminopeptidase gene *slapn*, which is expressed in the midgut of *Spodoptera littoralis*, but feeding with dsRNA did not achieve RNAi.

Fortunately, there are other studies showing that dsRNA feeding can be successful for RNAi studies in insects. Feeding dsRNA to *E. postvittana* larvae has been shown to inhibit the expression of the carboxylesterase gene *EposCXE1* in the larval midgut and also to inhibit the expression of the pheromone-binding protein EposPBP1 in adult antennae. dsRNA feeding also inhibite the expression of the nitrophorin 2 (*NP2*) gene in the salivary gland of *Rhodnius prolixus*, leading to a shortened coagulation time of plasma. dsRNA feeding has also been successful in many other insects, including insects of the orders Hemiptera,

Coleoptera, and Lepidoptera.

The main challenge remains that there needs to be a greater amount of material for delivery as silencing has been shown to be incomplete. This phenomenon has been observed after ingestion of *CELL-1* dsRNA by the termite *Reticulitermes flavipes*, TPS dsRNA in *N. lugens* nymphae, Nitrophorin 2 dsRNA by *Rhodnius prolixus*. In addition, different species of insects have different sensitivities to RNAi molecules when delivered orally. For example, *Glossina morsitans* fed with dsRNA may effectively inhibit the expression of *TsetseEP* in the midgut, but cannot inhibit the expression of the transferrin gene *2A192* in fat bodies due to lack of transfer capacity between tissues. The mechanisms associated with the transfer of gene expression through feeding delivery method still need further study.

In addition, one method that may be better than direct feeding with dsRNA is the use of transgenic plants to produce dsRNA. The advantage of this method is the generation of continuous and stable dsRNA material. Genetically engineered dsRNA-producing yeast strains have also been developed to feed *D. melanogaster*, but gene silencing was not successful. However, dsRNA produced in bacteria is effective in *C. elegans*. Therefore, the use of bacteria, especially insecticidal microorganisms, to produce dsRNA for insect RNAi merits further study.

Transgenic Insects

The advantage of using transgenic insects that carry the dsRNA is that as it is inheritable, the expression can be stable and continuous. The technique has been proposed to help either reduce population through introduction of sterile insects or for population replacement. In this case, dsRNA must be first injected in the host insect. Tests are being conducted on several species with promising results but as stated by Scolari et al., there is a need to understand environmental and genetic influences when assessing the potential use of such transgenics. The transgenic method has been first used in *D. melanogaster* with the *GAL4/UAS* transgenic system that leads to the expression of hairpin RNA. Subsequently, transgenic technology has generated transgenic *Aedes aegypti* that produces dsRNA. Through the use of a U6 promoter in *D. Melanogaster*, S2 cells can generate short hairpin RNA (shRNA) to inhibit gene expression. RNAi molecules targeting the circadian clock gene per have also introduced into *Bombyx mori* embryos by a piggyback plasmid to obtain genesilenced transgenic individuals. The transfection technique has been used to silence the *D. melanogaster* mitochondrial frataxin gene *dfh*, generating large-sized, long-lived larvae and short-lived adults. The GAL4/UAS transgenic system has also been used in *B. mori* to allow for induction of the transgenic construct. Therefore, gene function can be studied within a certain time period, and the study of gene functions in development, physiology, and the nervous system is possible.

Virus-mediated Uptake

Virus-mediated RNAi methods involve the infection of the host with viruses carrying dsRNA formed during viral replication and targeting the gene of interest in the host. For example, recombinant Sindbis virus introduced into *B. mori* cells through electroporation can produce dsRNA to inhibit *BR-C* gene expression, causing the larvae not to pupate or leading to adult defects. Virus-mediated RNAi studies are still rare. However, this method takes advantage of the infection and ability of the virus to spread rapidly in a host population. Virus-mediated RNAi does not require screening for transgenic insects or tissues, and thus, it has unique advantages.

Selected from: Yang G, You M, Vasseur L, *et al*. Stoytcheva M. Pesticides in the Modern World—Pests Control and Pesticides Exposure and Toxicity Assessment. Development of RNAi in Insects and RNAi-Based Pest Control, 2011: 27-35.

Words and Expressions

microinjection /ˌmaɪkroɪn'dʒɛkʃən/ *n*. 显微注射法，微注射，微量注射

aminopeptidase /əˌmino'pɛptɪˌdes/ [生化] 氨肽酶，[生化] 氨基肽酶

Spodoptera littoralis 海灰翅夜蛾，棉贪夜蛾，棉叶虫，灰翅夜蛾

E. postvittana 苹果褐卷蛾

antennae /æn'teniː/ *n*. [电讯] 天线（等于 aerial），[昆] 触须，[植] 蕊喙，直觉

nitrophorins 硝基磷蛋白

salivary /sə'laɪvəri/ *adj*. 唾液的，分泌唾液的

Rhodnius prolixus 长红锥蝽，长红猎蝽

coagulation /kəuˌægju'leɪʃn/ *n*. 凝固，凝结，凝结物

Reticulitermes flavipes 黄肢散白蚁，北美散白蚁

N. lugens 褐飞虱，稻褐飞虱

nymphae /'nimfiː/ *n*. 蛹，若虫

C. elegans 秀丽隐杆线虫

Aedes aegypti 埃及伊蚊

Bombyx mori 家蚕

frataxin 共济蛋白

Sindbis virus 辛德毕斯病毒，辛德比斯病毒

PART Ⅷ

BIOASSAYS

Unit 15 Experimental Design

The experimental design must be considered in terms of the experimental unit —the entity actually receiving the treatment. Examples of experimental units are single insects, or cages containing plants or fruit. Treatments frequently involve use of groups, rather than individuals, exposed to each treatment level. For some species, a standard procedure is to select individuals from a laboratory colony, place them in groups of 10 into a petri dish or other container, and treat each insect with a measured drop. In this instance, the individual insect is the experimental unit. Suppose that the entire group is sprayed simultaneously or presented with treated diet. In this situation, the group is the experimental unit.

In a quantal response bioassay, nothing except the treatment level or intensity can distinguish one group from another: the experimental unit must be a constant. Occasionally, an unforeseen event (such as cannibalism or a simple mistake in counting) can occur without affecting the integrity of the bioassay. In experiments to select arthropods for resistance to a treatment, a preplanned minor modification of the initial experimental unit (for example, from use of two insects in each container) to another unit (for example, use of one insect in solitary confinement) may also be necessary as the insects grow and develop.

Randomization

The process of assigning experimental units to treatments and controls at random is known as randomization. Use of this process is one way to avoid bias so that results represent responses of the units. A formal randomization procedure is best to use even if more record keeping is required and the experiment is more expensive to undertake. If randomization is not formalized, for example, large insects might be selected preferentially, as might larvae that move most slowly. Results of the data analysis would then pertain to subsets of the population and not to the population as a whole. Logistical convenience can result in bioassays with invalid designs.

In the laboratory, for example, insects are often held in stacks of petri dishes or rearing containers placed on shelves where gradients in light intensity and temperature might significantly affect response. A sample consisting only of insects selected from the top dishes at the front of shelves is not a random sample of the population. Instead, containers from which individuals are removed should be randomly selected to account for location on shelves and position within stacks. When wild populations are tested, individualized randomization procedures suited to each population are usually necessary. For example, natural units such as leaves, branches, or cones might be numbered, and insects on each unit can then be selected at random for testing. If the arthropod remains in or on the natural unit during testing, then these numbered units can be randomly assigned to treatment levels. In the laboratory or greenhouse, light intensity and temperature gradients must be considered in the randomization process for these units.

Treatments

Dr. Tarleton may be vehement about the importance of randomization in experimental design, but he is an absolute fanatic about treatment methods remaining constant during a statistically valid bioassay. In a quantal response bioassay with a single explanatory variable, only the intensity of treatment can be subject to change. Thus, group 1 could change the distance to the target, but not the fact that they were shooting paintballs; groups 2 and 3 were stuck with using their water pistols. When multiple explanatory variables are investigated, the identity of variables also cannot be changed in the midst of the bioassay. Group 2 could change the distance to the target and continue their measurements of wind speed, but not change the fact that they were shooting water at the target. Finally, the treatments in a quantal response bioassay with multiple explanatory variables and multiple response variables cannot be changed midway through the experiment: group 3 could not change the variables it measured or the effects it recorded. When an investigator changes the exposure method (that is, the treatment) midway through an experiment with one explanatory variable, the two (or more) exposure methods become additional explanatory variables. Paula Maven remembers Dr. Tarleton's warning: "In the end, you have destroyed your experiment. Remember that!" Suppose an investigator also changes the times at which an effect (such as mortality) is assessed midway through the experiment. Without knowing the relationship between time and mortality, versus exposure method, Dr. Tarleton describes the subsequent bioassay as a fruitcake: "A mixture of apples, oranges, walnuts, those strange candied fruits, and flour; who knows what you have?"

Controls

A control group should be exposed to everything except the treatment, and must be included in each replication of a bioassay. Dr. Tarleton explains that he cannot foresee a circumstance in which the inclusion of controls is similar to a clinical trial of a cancer drug with human subjects. The rationale for control groups is simple: without them, it will be difficult to attribute an observed effect or response to the treatment with any certainty. Suppose that test subjects are exposed to a chemical in water. In this case, the control group should be treated only with water. Excess test subjects should not be used in control groups that are assembled after selection of treatment units. Instead, the controls must represent a randomly selected treatment unit.

Use of a common control group for several treatments in a quantal response bioassay is not desirable even if the same solvent is used. Each treatment tested in a quantal response bioassay must have its own control. To illustrate what might happen if a common control is used, suppose a control group is treated first, followed by groups treated with several levels of treatment A, several levels of treatment B, and several levels of treatment C. None of the controls respond, the test subjects treated with treatment A respond in relation to the dose applied, and all test subjects treated with treatments B and C respond regardless of dose. What does this bioassay reveal about the effects of treatments B and C? The answer is, unfortunately, not very much. Possibly, these two treatments are very effective, and lower rates should be tested. Equally possible is contamination of the application device with the highest level of treatment A and resultant contamination of treatment B, followed by contamination of treatment C with treatment B. Given the choice of testing more treatments at the expense of testing a control with each one, the wisest option is to test fewer treatments.

Is an "untreated" control a real control? The answer is no, unless the experiment is done with the treatment in its pure form (without a carrier or solvent) or unless the fumigation activity of a chemical is being tested. In tests with fumigants, air is the carrier and an untreated control is appropriate. However, when a treatment is applied as an emulsion in water, the appropriate control is water containing all of the emulsifying agents in the undiluted formulation, then diluted to the same concentrations present in the aliquots actually tested.

Replication

Garland Tarleton next discusses a common tendency among scientists never to replicate a successful experiment. Reports of unreplicated bioassays occasionally appear in the published literature. Without exception, all experiments must be replicated. A replication is repetition of a bioassay at a different time, but under the same conditions (as much as possible) as the first test. One purpose of replication is to randomize effects related to uncontrollable procedures and conditions. These include effects caused by workers or by time. Another reason for replication is to detect errors in formulation; for this reason, the best way to ensure true replication is to prepare a fresh series of doses diluted from a new stock solution.

Some investigators refer to subsets within a replication as replications. For example, 20 insects might be treated in groups of 10 with concentrations of 0 (the control), 1, 2, 5, 7, and 10 mg/mL of chemical X. Unless each group of 10 treated with any concentration was treated with a fresh formulation and at a different time, it is not a separate replicate. For this example, each group of 20 treated with the same concentration is a replicate and each of the groups of 10 treated with that concentration is a subset of that replicate. When the subset is called a replicate, the investigator has done pseudoreplication. Pseudoreplication is not unique to quantal response bioassays, but it occurs with alarming frequency.

Ideally, replications should be done on different days within a relatively short period of time. This recommendation assumes that day-to-day variability is not in itself a source of error in the experiment. However, time constraints related to the condition of test subjects, use of the application equipment, or varying personnel may require modifications of the ideal method. Decisions about the minimum time that should pass between replications must be made by investigators based on their common sense. Just as no specific guidelines are available to define the lower time limit for replication, none are available to identify the maximum time that can pass after which a treatment becomes an altogether different experiment. Certainly, the same bioassay done with the same species, but in different years, cannot be considered a replication in the usual sense. Because of these problems, administration of a series of fresh solutions or treatment levels is more specific than time as a criterion for true replication.

Suppose, for example, that chemical A is to be tested on a group of arthropods collected from the field, and a shipment of another group from a different site will arrive the next day. Time available to do the bioassay is obviously limited. One set of treatment solutions is prepared, but three different groups will be tested one hour apart. Is this true replication? The answer is no, not because only an hour separates the experiments, but because a possible error in preparation of the test solutions cannot be detected. If the experiments were done with fresh solutions, the biosassay would be truly replicated.

Now suppose that fresh solutions are used. In the first replication, the concentrations that are applied kill between 5% and 95% of the insects. In the second, all of the controls live, but all of the insects treated with the chemical die regardless of concentration applied. Although these results may be valid, a procedural error may have occurred; but in which replicate? Results of the third and fourth replications might suggest that a formulation error occurred in one of the first two replicates. Results from a replicate can be disregarded if an error is known to have occurred in that replicate. Data must never be discarded without extremely strong justification, such as a known procedural error or an outlier that is detected as part of the data analysis. If an error does not seem likely, all data should be used for statistical analyses. Inconsistency from replication to replication may be the result of extrabinomial variability or outliers and various other reasons.

Order of Treatments within a Replication

On both practical and theoretical grounds, Dr. Tarleton suggests that the order in which treatment rates are applied in a quantal response bioassay should be from lowest to the highest, never vice versa. If the treatment is highest to lowest in a bioassay with a chemical, for example, results can be spurious because the test subjects may actually be exposed to a higher rate than intended. The impracticality of cleaning an application device and letting it dry completely between applications of different concentrations is obvious. But such cleanings are necessary even if concentrations are applied in random order. Theoretically, some residual solvent will remain in the application device after cleaning without subsequent drying. If the solvent is pure, the concentration of the chemical tested next will be diluted more than if solvent plus a lower concentration of chemical remains in the needle, syringe, spray reservoir, or nozzle. In the treatment sequence of lowest to highest rate, the application device need not be cleaned.

When more than one treatment is tested as part of each replication, the order in which the treatments are applied should be randomized among replications. Suppose that, unknown to the investigator, small amounts of chemical A interact with chemical B to cause a greater effect than would be expected if the joint effects of the two chemicals were independent and additive. Each time chemical B was tested, observed effects will not just be due to B if it is always tested after A. Randomization of the treatment sequence of chemicals helps minimize the possible experimental bias and error that may occur.

Selected from: Robertson J L, Russell R M, Preisler H K, *et al*. Bioassays with arthropods, 2nd. Florida: CRC Press LLC, 2007: 13-18.

Words and Expressions

petri dish *n*. 有盖培养皿，皮氏培养皿
quantal /'kwɒntəl/ *adj*. 局量子的
intensity /ɪn'tensəti/ *n*. 强烈，剧烈，强度，亮度
cannibalism /'kænɪbəlɪzəm/ *n*. 嗜食同类，自相残杀
resistance /rɪ'zɪstəns/ *n*. 反抗，抵抗，阻力，电阻
solitary /'sɒlətri/ *adj*. 孤独的
sluggish /'slʌgɪʃ/ *adj*. 行动迟缓的
randomization /'rændəmai'zeiʃən/ *n*. 随机化，随机选择
bias /'baɪəs/ *n*. 偏见，偏爱，斜线；*vt*. 使存偏见
pertain /pə'teɪn/ *v*. 适合，属于
logistical /lə'dʒɪstɪkl/ *adj*. 推理的，逻辑的，计算的
stack /stæk/ *n*. 堆，一堆，堆栈；*v*. 堆叠
gradient /'greɪdiənt/ *adj*. 倾斜的；*n*. 梯度，倾斜度，坡度
cone /kəun/ *n*. [数、物] 锥形物，(松树的) 球果；*vt*. 使成锥形
vehement /'viːəmənt/ *adj*. 激烈的，猛烈的，(情感) 热烈的
fanatic /fə'nætɪk/ *n*. 狂热者，入迷者；*adj*. 狂热的，盲信的
explanatory /ɪk'splænətri/ *adj*. 说明的，解释的
paintball /'peintbɔːl/ 彩弹球
walnut /'wɔːlnʌt/ *n*. 胡桃，胡桃木
replication /ˌreplɪ'keɪʃn/ *n*. 复制
rationale /ˌræʃə'nɑːl/ *n*. 基本原理
repetition /ˌrepə'tɪʃn/ *n*. 重复，循环，副本
resultant /rɪ'zʌltənt/ *adj*. 作为结果而发生的，合成的
fumigation /ˌfjuːmɪ'geɪʃn/ *n*. 烟熏法，熏烟消毒法
aliquot /'ælɪkwɒt/ *n*. [数] 能整除；*adj*. 能整除的，部分的
pseudoreplication *n*. 假重复法

discard /dɪ'skɑːd/ *vt*. 丢弃，抛弃；*v*. 放弃

criterion /kraɪ'tɪəriən/ *n*.（批评判断的）标准，准据，规范

inconsistency /ˌɪnkən'sɪstənsi/ *n*. 不一致，易变

extrabinomial *adj*. 二项的，二项式的；*n*. 二项式

spurious /'spjʊəriəs/ *adj*. 伪造的，假造的，欺骗的

impracticality /ɪmˌpræktɪ'kæləti/ *n*. 不切实际，办不到

syringe /sɪ'rɪndʒ/ *n*. 注射器，洗涤器；*v*. 注射，冲洗，灌洗

reservoir /'rezəvwɑːr/ *n*. 水库，蓄水池

Notes

[1] Pseudoreplication：假重复法。使用推理统计来测试处理效果的数据来自实验，其中要么是处理没有重复（尽管样本可能是这样），要么是重复在统计学上不是独立的。

Exercises

Ⅰ Answer the following questions according to the text.

1. What is the meaning of bioassay?

2. What should be considered first when designing a bioassay?

3. What is the best way to ensure that the assigning experimental units are random? And what is the importance of randomization in experimental design?

4. Why is use of a common control group for several treatments in a quantal response bioassay undesirable even if the same solvent is used?

5. What is the purpose of replication?

Ⅱ Translate the following English phrases into Chinese.

control group	experimental unit	pseudoreplication	treatments and control
randomization	laboratory colony	temperature gradient	dose-response curve
quantal response	experiment design	explanatory variable	fumigation activity

Ⅲ Translate the following Chinese phrases into English.

重复	处理方法	方差分析	生长速率	随机数字表
致死中量	实验偏差	统计分析	逻辑安排	致死中浓度

Ⅳ Choose T if you think the statement is true according to the text and F if it is false.

1. The entities actually receiving the treatment of bioassay always involve use of individuals, rather than groups.

2. In a quantal response bioassay, nothing except the treatment level or intensity can distinguish one group from another: the experimental unit must be a variable.

3. Randomization is one way to avoid bias so that results represent responses of the units.

4. When wild populations are tested, individualized randomization procedures suited to each population are usually unnecessary.

5. In a quantal response bioassay with a single explanatory variable, only the intensity

of treatment can be subject to change. When multiple explanatory variables are investigated, the identity of variables also cannot be changed in the midst of the bioassay.

6. The rationale for control groups is simple: without them, it will be difficult to attribute an observed effect or response to the treatment with any certainty.

7. In tests with fumigants, water is the carrier and an untreated control is appropriate.

8. When a treatment is applied as an emulsion in water, the appropriate control is water containing all of the emulsifying agents in the diluted formulation, then diluted to the same concentrations present in the aliquots actually tested.

9. Ideally, replications should be done on different days within a relatively short period of time.

10. On both practical and theoretical grounds, Dr. Tarleton suggests that the order in which treatment rates are applied in a quantal response bioassay should be from highest to the lowest.

Ⅴ Translate the following short paragraphs into Chinese.

1. Without exception, all experiments must be replicated. A replication is repetition of a bioassay at a different time, but under the same conditions (as much as possible) as the first test. One purpose of replication is to randomize effects related to uncontrollable procedures and conditions. These include effects caused by workers or by time. Another reason for replication is to detect errors in formulation; for this reason, the best way to ensure true replication is to prepare a fresh series of doses diluted from a new stock solution.

2. The process of assigning experimental units to treatments and controls at random is known as randomization. Use of this process is one way to avoid bias so that results represent responses of the units. A formal randomization procedure (such as the use of a random number table) is best to use even if more record keeping is required and the experiment is more expensive to undertake.

Factors Affecting the Bioassay

Larval Instar

Larval sensitivity to bacterial toxins is reduced as the larvae develop; 2nd instar *Ae. vexans* were about 10 times more sensitive than 4th instars at a water temperature of 25℃. It is therefore essential that larvae of more or less the same age are always used for the bioassay.

Insect Species

Large differences in sensitivity among mosquito species can be found due to differences in their feeding habit, ability to activate the protoxin, and toxin binding to midgut cell receptors. For example, larvae of *Culex pipiens* were 2 ~ 4 times less susceptible to *B. thuringiensis* subsp. *israelensis* than Aedes species of the same instar.

Temperature

The feeding rate of *Aedes vexans* decreases with temperature, and this effect leads to a reduction in consumption of bacterial toxins. With a rise in water temperature from 5 to 25℃, sensitivity of 2nd instars of this mosquito species to *B. thuringiensis* subsp. *israelensis* increases by more than 10 times. The influence of water temperature on the efficacy of the bacterium against the insects differs also with larval instar. Second-instar *Ae. Vexans* are twice as sensitive to the microbe as 4th instar larvae at a water temperature of 15℃. This difference in sensitivity was 10 times higher at 25℃.

Larval Density

At a larval density of ten *Ae. vexans* 4th instars per cup with 150 mL water the LC_{50} was (0.0162 ± 0.004) mg/L; with 75 larvae per cup the LC_{50} increased by about 7 times [(0.1107 ± 0.02) mg/L]. This indicates that when the number of larvae increases, larger amounts of *B. thuringiensis* subsp. *israelensis* have to be used to reach mortality levels obtained at lower mosquito populations.

State of Nutrition

Feeding the larvae before starting the bioassay caused an increase in the LC_{50}. In contrast, the LC_{50} of larvae that were fed sparsely throughout their growth prior to the bioassay, and not at all on the bioassay day itself, was very low. These results were obtained in a comparative study conducted among several laboratories. When water polluted with bacteria was used in the test cups, 2~3 times more *B. thuringiensis* had to be applied in order to achieve the same effect as was obtained when clean water was used.

Types of Larval Feeding

Aedes aegypti larvae prefer to feed from the bottom of the container, whereas *Cx. pipiens* and even more so Anopheles species are surface feeders, although all three mosquito species will also feed on microbe particles in suspension. Wettable powders commonly consist of large particles and will precipitate within minutes to the bottom of the test cups. In contrast, fluid formulations that are based on minute particles will disperse homogeneously in water. Thus, the availability to and the efficacy of the spore-crystal mixtures of the product against the larvae will depend on the feeding behavior of the mosquito species. Also, the type of container may affect the bioassay results, as protoxins will stick to vial surfaces that are electrically charged.

Sunlight

Ultraviolet sunlight destroys the bacterial toxins and kills the vegetative cells and spores. Therefore, bioassay cups with water should not be exposed to the sun or to irradiation by UV lamps. In small-scale experiments, a severe sunlight inactivation of the microbial product in clean water has been reported.

Heat

Exposing the vegetative cells, live spores and toxins for 10 min to 80℃ will inactivate them. Therefore, microbial products should not be stored close to heat sources. Furthermore, to prolong viability, fluid products should be stored at 5℃ and the standard powders at −18℃. Liquid products should not be stored below the freezing point, as that will cause the proteins to aggregate and thus will increase particle size. To overcome this problem, a thorough homogenization of the product is made before use in the bioassay.

Selected from: Navon A, Ascher K R S. Bioassays of Entomopathogenic Microbes and Nematodes. UK: CABI Publishing, 2000: 44-46.

Words and Expressions

2nd instar　二龄
Ae. vexans　刺扰伊蚊
protoxin　/prəʊ'tɒksɪn/ [药] 原毒素，强亲和毒素
midgut　/'mɪdgʌt/ [解] 中肠
Culex pipiens　尖音库蚊
aedes　/ei'i:di:z/ *n.* [昆虫] 伊蚊（一种传染黄热病的蚊子）
sparsely　/'spɑ:sli/ *adv.* 稀疏地，稀少地
Aedes aegypti　埃及伊蚊
Cx. pipiens　淡色库蚊
anopheles　/ə'nɒfə,lɪz/ 按蚊，疟蚊
vegetative　/'vedʒɪtətɪv/ *adj.* 有关植物生长的，植物的

Unit 16 Bioassays of *B. thuringiensis*

Artificial Diets

The aim of using bioassays based on an artificial diet was to provide the worker with a rapid, standardized and simple procedure for estimating the activity of a microbial strain. The nutrients in the diet are a substitute for the natural food and the agar gel provides a texture similar to that of plant tissues but devoid of their undesirable side effects due to plant allelochemicals and microorganisms. Initially, a diet was proposed for the single bioassay insect, *Trichoplusiani*, or *Anagasta kuehniella*. However, with the growing international interest in *B. thuringiensis* as a useful substitute for chemical insecticides, efforts were made to select the most effective *B. thuringiensis* strains against specific insect pests. This change required the use of more than one bioassay diet. Navon et al. proposed a standardized diet that, with additions of feeding stimulants available from processed food fractions, would be suitable for almost any lepidopterous species. In addition, a wide choice of diets that could be adapted for a bioassay diet is available from rearing manuals. Even a diet based on a calcium alginate gel that is prepared without a heating step has been developed. In this diet, heat-labile components of *B. thuringiensis* and enzymes can be used.

A standardized bioassay diet can be used for any target insect provided that the specific phagostimulants are included for specialist feeders. Also, to preserve spore activity, inclusion of antibiotics in the diet should be avoided. However, diet preservatives with bacteriostatic effects, such as methyl-*p*-hydroxybenzoate, can be used. One standardized diet has been made suitable for rearing neonate larvae. The other diet formula is for third instars. Other food sources with similar nutritional ratios can be used for any lepidopterous species, provided that the nutrients contain the necessary phagostimulants to induce insect feeding.

To prepare the diet dissolve the agar and the methyl-*p*-hydroxybenzoate in half of the water by heating in an autoclave. Use the other half of the water to homogenize the remaining ingredients, except the ascorbic acid, in a blender. Mix together the hot agar/methyl-*p*-hydroxybenzoate solution and the nutrient homogenate. Add the L-ascorbic acid at 50℃. Keep the diet at this temperature by holding it in a container in a hot water bath at approximately 70℃ or in an electric heating basket. Weigh diet portions from this container for each of the microbe dilution series, mix in the aqueous microbe mixture, stir with a laboratory stirrer for 30 s and pour the diet into the rearing cavities or cells. The microbial mixture should not exceed 5% of the diet's volume; larger microbial portions may dilute the nutrients' concentration and cool the diet so that it will set prematurely. Mix the dilute microbe mixtures first and then the concentrated ones.

Use the standard reference microbes obtained from the international *B. thuringiensis* collections. Store all the standard microbes at −10℃. If you use the same insect species in most of your bioassays and you have small amounts of the microbial standard, select an internal standard. Then calibrate the insecticidal power of the internal standard against the international standard to express the microbial activity in potency units. The tested *B. thuringiensis* products are available mostly as wettable powders and liquid concentrates. The dilution solution used for the microbe preparations is saline buffer solution (8.5 g NaCl, 6.0 g K_2HPO_4 and 3.0 g KH_2PO_4 per litre, pH 7.0) together with 0.05% polysorbitan monooleate (Tween 80) as a surfactant. To test potencies of unknown *B. thuringiensis* preparations, conduct a preliminary assay with a tenfold dilution series of the experimental microbial powders and the standard. The results of these assays are used to select a narrower dilution in which the LC_{50} will fall approximately midway in the series.

Mixing the *B. thuringiensis* insecticidal components (spores, crystals, protoxin and more) with the diet renders the microbe available not only to defoliators but also to larvae that penetrate into the diet and feed on inner layers of the medium. In addition, by using this mixing procedure undesired effects of the adjuvants of the *B. thuringiensis* product, mostly fermentation residues, on larval feeding will be minimized. In contrast to this, if commercial products are applied to the diet's surface, the fermentation adjuvants will accumulate there and may introduce dose-dependent errors in the bioassay at high *B. thuringiensis* concentrations.

In bioassays with 3rd-instar larvae, diet portions in the cells should be about 1 mL or 1 g. In a 1st-instar bioassay less than 1 g can be used, provided the diet does not dehydrate within the bioassay period. Weigh diet portions for a single *B. thuringiensis* concentration, mix with the aqueous microbe mixture in a blender and pour into the cavities. In the grid cells, pour the diet into the petri dish first and then fit the grid. For neonate bioassays, keep the diet at room temperature in a hood for 1 h to evaporate any condensed water. Put a single larva in each of the cavities. To transfer neonates use a camel-hair brush. Avoid touching the larvae with the brush; instead, let the larva "parachute" on its spinning thread. Hold the thread with the brush and let the larva touch the diet. You may hold several larvae together with the brush and save inoculation time. Special attention should be paid to closing the rearing units. In the grid cells, a filter paper of 9 cm diameter is placed on the grid and a 5~10 mm thick plastic sponge is placed on top of the filter paper. The petri dish lid is put on top of the plastic sponge and rubber bands are used to close the cells tightly. In this way, the sponge is pressed against the grid so that neonate larvae cannot escape from the cells. Neonate larvae are used instead of 3rd instars for several reasons: (i) this instar is available from the insect colony in much larger numbers than any other instar and with less input of labor and materials; (ii) the bioassay period is shorter; and (iii) precision is higher because larval mortality is more uniform and confidence intervals of the LC_{50} are smaller.

In "official bioassays" take measurements of larval weight, size and head capsule to describe the instar. In 3rd-instar bioassays, postecdysed larvae (after moult) are preferable as "standardized larvae". The length of this bioassay is based on 3rd-instar larvae surviving for 7 days. Touch the larvae with a needle to confirm mortality. For the neonate bioassays use larvae that are 0～12 h old and deprived of food; mortality is counted after 48 or 96 h.

Natural Food

Leaf and greenhouse-plant bioassays are an intermediate step between dietary (artificial diet) bioassays and field assays. Whereas dietary bioassays were designed to determine the activity of the spores and crystals accurately in an artificial medium, plant bioassays have two purposes. First, they consist of plant tissues with most of the chemical and physical barriers presented by the agricultural crop. For example, the alkaline leaf surface produced by the epidermal glands is present in both greenhouse and field plants. However, other barriers may not be expressed in the sheltered plants; the trichome density in greenhouse seedlings is significantly lower than in field plants. Second, they allow the evaluation of effects of formulation adjuvants on the phylloplane. Feeding stimulants, fermentation residues, surfactants, rain-fasting materials and stickers may affect the larval feeding behavior and thereby the ingestion of the spore crystal mixture. Cotton, tomato, maize and cruciferous species are among the most common crops raised in the greenhouse for bioassays. Cotton is a useful bioassay plant because it is a host of several major lepidopterous insect pests. Also, within 2～3 weeks, the seedlings already have eight to ten leaves that can be used either for detached-leaf or for potted plant bioassays. However, leaf bioassays with bollworm and borer larvae cannot substitute for assays with flower buds and fruit that are the natural target organs of these insects in the field.

Leaf Bioassays

One of the common leaf bioassays uses leaf discs. The disc test is based on a standardized size of leaf tissue and therefore application of the *B. thuringiensis* in aqueous mixture per unit area is simple and accurate. However, a leaf disc cannot be infested in bioassays with more than one neonate larva of lepidopterous species with cannibalistic tendencies. An alternative method is to use the entire leaf with a water supply. The whole leaf provides hiding places for the larvae, so that the physical contact among larvae that occurs in the disc bioassay is reduced. In addition, the water balance of the plant tissue is maintained better when the leaf has a water supply. The method is as follows: pour a 2 cm layer of 1% agar solution in a 15～20 cm^3 glass vial. Dry the condensation water by exposing the vial to reduced air pressure in a hood for 1～2 h. Pipette 50 μL of the test solution on each 10～15 cm^2 sized leaf side. Let the mixture dry on the leaves. Cut the petiole at 2 cm distance from the leaf. Hold the petiole with forceps and insert it into the agar layer. Put 5 neonate larvae in each vial. Close the vial with a cotton cloth held tightly with a rubber band. Use five vials per treatment (25 neonates). In this bioassay, leaf freshness is preserved for 3 days whereas the bioassay period is 1～2 days only. Flower bud bioassays are conducted in a similar manner with one or two larvae; the bud petiole is inserted into the agar layer as for the leaves.

A conventional mixing of two *B. thuringiensis* subspecies in the formulation is one of the means to widen the insect host range of the microbial product. When combining subspecies *kurstaki* and *aizawai* in an aqueous mixture or in the tank mix, different insect species should be used in parallel, for example *Helicoverpa armigera* and others which are susceptible to subsp. *kurstaki* strains and *Spodoptera* species for subsp. *aizawai* strains. Such a combination of *B. thuringiensis* strains has been developed in a granular feeding bait formulation.

The leaf and flower bud bioassays are also suitable for assaying granular formulations. An accurate quantitative application of granules on leaves and flower buds was developed using a dispersion tower. With this tower, effects of granular sizes on the larvae were determined. In the agar vials, granules larger than 250 μm were not suitable for the bioassay because they dropped off the leaf and were not available to the larvae, whereas granules of less than 150 μm adequately adhered to the leaf surface.

Potted-plant Bioassays

Potted-plant bioassays have several uses that cannot be provided with the leaf bioassays. These include: (i) testing the activity of *B. thuringiensis* with intact plant organs; (ii) applying the microbial product by spraying or dusting the plant, where an accuracy of dosing exceeding that of a field application can be achieved; (iii) extending the bioassay time for neonates exposed to the intact leaves to more than 3 days (until the leaf cage area is totally consumed); and (iv) assaying the residual effect of the microbial preparation. Potted-plants bioassays, conducted by caging 1st-, 2nd- or 3rd instar larvae on leaves of potted plants, are useful for assaying mortality, leaf consumption and inhibition of larval weight gain under greenhouse conditions. The cages consist of two plastic cylinders, one attached to each side of the leaf. Each half cage is closed on its outer side by a 150～224 μm mesh metal screen. The screen is attached by heating it and pressing it against the plastic cylinder until sealed. The two half-cage units are held together by an uncoiled wire paper clip.

The leaf area consumed within the cage can be measured with a leaf area meter as a parameter for assessing the efficacy of the *B. thuringiensis* preparation against 3rd instar larvae. This cage is also used in bioassays with bollworms on cotton flower buds and fruits in potted plants. For this purpose, a nick measuring 3～4 mm^2 is cut out from the half-cage surface as a space for the insertion of the petiole of the flower bud or fruit and elastic filler is used to prevent larval migration through this hole in the cage. For neonate bioassays, a ring of polyethylene sponge is glued to the perimeter of each of the half-cage units. In this way, when the cage is closed, the sponge is pressed against the leaf surface preventing the 1st-instar larvae from escaping.

In order to determine the EC_{50}, the larvae are weighed every 1～2 days. In a 7～8 day bioassay, larvae have to be transferred to new leaves because 3rd-instar larvae will consume the whole leaf area in the cage within 3～4 days.

Field Plants Bioassays

This type of bioassay is conducted in natural plant and environmental conditions. Spraying volume and formulation affect the effectiveness of pest management. The leaf and flower bud cages are also used for the field bioassay. 2nd and 3rd instar larvae are used in the bioassay. Smaller screen meshes (100～200 μm) are used for field bioassays with neonate larvae. It should be noted that neonate mortality in the control insects brought about by natural entomopathogenic microorganisms on the phylloplane is often unacceptably high.

Portable meteorological stations are useful in the experimental plots to record the circadian temperature and humidity during the bioassay periods. Bioassays for recording the residual effect of *B. thuringiensis* are conducted by sequential caging of treated plants at 0, 2, 4 and 6 days from the time of the initial microbe application.

Selected from: Navon A, Ascher K R S. Bioassays of Entomopathogenic Microbes and Nematodes. UK: CABI Publishing, 2000: 3-12.

Words and Expressions

nutrient /'njuːtriənt/ *adj*. 有营养的
agar gel 琼脂凝胶
devoid /dɪ'vɔɪd/ *adj*. 全无的，缺乏的
Trichoplusiani 粉纹夜蛾
Anagasta kuehniella 地中海粉螟
strain /streɪn/ *n*. 血统，种，（品）系，菌株，变种
lepidopterous /ˌlepɪ'dɒptərəs/ *adj*.［昆］鳞翅类的
calcium /'kælsiəm/ *n*.［化］钙（元素符号 Ca）
alginate /'ældʒineit/ *n*. 藻酸盐
labile /'leɪbɪl/ *adj*. 不安定的，易发生变化的
phagostimulant /'feɪgəstɪmjulənt/ *n*. 诱食剂
spore /spɔːr/ *n*. 孢子；*vi*. 长孢子
antibiotics /ˌæntɪbaɪ'ɒtɪks/ *n*. 抗生素，抗生学
preservative /prɪ'zɜːvətɪv/ *n*. 防腐剂
bacteriostatic /bækˌtiəriə'stætik/ *adj*. 细菌抑制的，阻止细菌繁殖法的
methyl-*p*-hydroxybenzoate 对羟基苯甲酸甲酯
neonate /'niːəuneɪt/ *n*.（尤指出生不满一个月的）婴儿
autoclave /'ɔːtəukleɪv/ *n*.（烹调用）高压锅，高压灭菌器
homogenize /hə'mɒdʒənaɪz/ *vi*. 均质化；*vt*. 使均匀
ascorbic acid /əsˌkɔːbɪk'æsɪd/ *n*.［生化］抗坏血酸维生素 C
calibrate /'kælɪbreɪt/ *v*. 校准
saline /'seɪlaɪn/ *adj*. 盐的，苦涩的；*n*. 盐湖，盐田
buffer solution 缓冲溶液
polysorbitan monooleate 聚氧乙烯己六醇油酸酯
protoxin /prəu'tɒksɪn/ *n*. 原毒素，强亲和毒素
defoliator /ˌdiː'fəulieitə/ *n*. 除叶剂，食叶昆虫
adjuvant /'ædʒuvənt/ *adj*. 辅佐的；*n*. 助理员，助剂
dehydrate /ˌdiːhaɪ'dreɪt/ *vt*.（使）脱水

parachute /'pærəʃu:t/ *n.* 降落伞
inoculation /ɪˌnɒkju'leɪʃn/ *n.* 接木，接种，接插芽
postecdysis /pəust'ekdɪsɪs/ 蜕皮后期
moult /məult/ *n.* 换毛，脱毛；*v.* 脱毛，换毛
alkaline /'ælkəlaɪn/ *adj.*［化］碱的，碱性的
epidermal /ˌepɪ'dɜ:məl/ *adj.*［解］［生］表皮的，外皮的
gland /glænd/ *n.*［解剖］腺，［机械］密封管
trichome /'traɪkəum/ *n.*［植］毛状体
crucifer /'kru:sɪfə/ *n.* 十字花科植物，执十字架的人
bollworm /'bəulwɜ:m/ *n.* 一种蛾的幼虫，螟蛉
infest /ɪn'fest/ *v.* 大批滋生
cannibalistic /ˌkænibə'listik/ *adj.* 食人肉的，同类相食的
vial /'vaɪəl/ *n.* 小瓶；*vt.* 装入小瓶
pipette /pɪ'pet/ *n.* 吸液管
petiole /'petɪəul/ *n.* 叶柄，柄部
forceps /'fɔ:seps/ *n.* 镊子，钳子
Helicoverpa armigera 棉铃虫
intact /ɪn'tækt/ *adj.* 完整无缺的
cylinder /'sɪlɪndər/ *n.* 圆筒，圆柱体，汽缸，柱面
uncoil /ˌʌn'kɔɪl/ *vt.* 解，解开；*vi.* 展开，解开
insertion /ɪn'sɜ:ʃn/ *n.* 插入
elastic /ɪ'læstɪk/ *adj.* 弹性的
polyethylene /ˌpɒli'eθəli:n/ *n.*［化］聚乙烯
perimeter /pə'rɪmɪtər/ *n.* 周长，周界
circadian /sɜ:'keɪdiən/ *adj.* 生理节奏的，以 24h 为周期的

Notes

［1］agar：琼脂，又名洋菜、琼胶、石花胶、燕菜精、洋粉、寒天、大菜丝，是植物胶的一种，常用海产的麒麟菜、石花菜、江蓠等制成，为无色、无固定形状的固体，溶于热水，为最常用的微生物培养基的固化剂。

［2］potency：效价，抗生素的计量单位。

Exercises

Ⅰ Answer the following questions according to the text.

1. What is the aim of using bioassays based on an artificial diet?
2. Why neonate larve used instead of 3rd instars in dietary bioassays?
3. What are the purposes of plant bioassays?
4. Why is cotton a useful bioassay plant?
5. In contrast with the leaf bioassays, what are the advantages of potted-plant bioassays?

Ⅱ Translate the following English phrases into Chinese.

agar gel	spore activity	rearing manual	cruciferous species
field assays	plastic sponge	experimental plot	fermentation residues
artificial diet	buffer solution	feeding stimulants	meteorological stations

Ⅲ Translate the following Chinese phrases into English.

叶碟	生测周期	效价单位	鳞翅目害虫	田间植株生物测定
初孵幼虫	靶标昆虫	微生物菌种	叶片生物测定	盆栽植株生物测定

Ⅳ Choose T if you think the statement is true according to the text and F if it is false.

1. Selecting the most effective *B. thuringiensis* strains against specific insect pests required the use of more than one bioassay diet.

2. In order to preserve spore activity, inclusion of antibiotics in standardized artificial bioassay diet should be avoided.

3. In dietary bioassay, the tested *B. thuringiensis* products are available scarcely as wettable powders and liquid concentrates.

4. In dietary bioassay, for the neonate bioassays, larvae that are 0～12 h old and deprived of food are used as experimental units; mortality is counted after 48 or 96 h.

5. Dietary bioassays are designed to determine the activity of the spores and crystals accurately in an artificial medium.

6. Leaf bioassays with bollworm and borer larvae can substitute for assays with flower buds and fruit that are the natural target organs of these insects in the field.

7. In leaf bioassays, the leaf disc test is based on a standardized size of leaf tissue and therefore application of the *B. thuringiensis* in aqueous mixture per unit area is simple and accurate.

8. The leaf and flower bud bioassays are suitable for assaying granular formulations. Because accurate quantitative application of granules on leaves and flower buds was developed using a dispersion tower in which effects of granular sizes on the larvae are determined.

9. Field plant bioassay is conducted in natural plant and environmental conditions. Only formulation of pesticides affects the effectiveness of pest management.

10. Field plant bioassays for recording the residual effect of *B. thuringiensis* are conducted by disordered caging of treated plants from the time of the initial microbe application.

Ⅴ Translate the following short paragraphs into Chinese.

1. The aim of using bioassays based on an artificial diet was to provide the worker with a rapid, standardized and simple procedure for estimating the activity of a microbial strain.

2. A standardized bioassay diet can be used for any target insect provided that the specific phagostimulants are included for specialist feeders. Also, to preserve spore activity, inclusion of antibiotics in the diet should be avoided.

3. Let the mixture dry on the leaves. Cut the petiole at 2 cm distance from the leaf. Hold the petiole with forceps and insert it into the agar layer. Put 5 neonate larvae in each vial. Close the vial with a cotton cloth held tightly with a rubber band.

4. Potted-plant bioassays have several uses that cannot be provided with the leaf

bioassays. These include: (i) testing the activity of *B. thuringiensis* with intact plant organs; (ii) applying the microbial product by spraying or dusting the plant, where an accuracy of dosing exceeding that of a field application can be achieved; (iii) extending the bioassay time for neonates exposed to the intact leaves to more than 3 days; and (iv) assaying the residual effect of the microbial preparation.

Types of Bioassays

Leaf Assays

Most initial screening assays of maize, potatoes and cotton are no-choice tests that use leaves, or leaf sections, from young plants which typically have the highest concentration of *B. thuringiensis* proteins in leaf tissues. Leaves are detached, placed on top of moist filter paper in petri dishes and infested with one or more neonate larvae of the appropriate test species. An alternative approach is to cut out circular leaf discs with a cork borer and place the discs on top of agar, containing antimicrobials, within individual wells of microtitre assay plates. Neonate larvae are added and the wells are covered with plastic film, heat sealed using a tacking iron and ventilated using pin holes. In leaf assays, larval growth (weight, instar) and survivorship is measured after 2～4 days; damage to the leaf tissue is most easily scored using an index comparing transgenic with control leaf tissue. If transgenic events cannot be ranked using these assays, other approaches become necessary. Later instar larvae of the test species are relatively less susceptible to transgenic proteins and can be used instead of neonates. Also, species with different susceptibility can be used in the assay.

Pollen Assays

Evaluating expression of *B. thuringiensis* protein in pollen is important for ecological risk assessment because many non-target beneficial insects use pollen as a food source. Also, pollen is critical for larval growth of some pest species. The presence of *B. thuringiensis* protein in maize pollen, for example, is an important aspect of plant resistance to second-generation *O. nubilalis* since neonate larvae often develop exclusively on pollen prior to attaining a size large enough to penetrate sheath collar tissue. The procedure described below for *B. thuringiensis* protein expression in maize pollen illustrates one bioassay approach. Pollen was collected from maize plants during the first 7 days of pollen shed. To prevent protein degradation, pollen was stored in tightly capped vials at −80℃ until tested. Microtitre plates (24-well) with agar and mould inhibitor were used. Each well received 50～100 μg of pollen and two neonate *O. nubilalis* larvae. Completed wells were sealed with Mylar film and ventilated with pin holes. Test duration was 4 days at 26℃. Larval survivorship and stunting were scored. At 4 days, there was ≥85% survival on the control pollen and ≥85% of surviving larvae were second instar.

Root Assays

Maize is grown in 7.6 cm diameter pots for 7～8 days until foliage height is approximately 7.6 cm. Then the soil is gently washed off the roots using running water and individual plants are placed on top of the filter paper/agar base. Ten neonate *D. undecimpunctata howardi* larvae (～24 h old) are added and the top is replaced. Test duration is 7 days at L : D (light : dark) 14 : 10, 26℃, and 70% RH (relative humidity). Under these conditions, control mortality is low (≤10%) and larvae feed only on the roots. A soil assay useful for root resistance evaluation uses 7.6 cm diameter plastic pots with drainage holes covered with polyurethane foam and commercial potting soil such as MetroMix®. Plants are grown as for the root assay described above. At 7.6～10.1 cm of leaf growth, the soil at the base of the maize is lightly "cultivated" using a metal spatula and the soil is inoculated with 10 neonate larvae (24～48 h old) of *D. undecimpunctata howardi* pre-fed on the roots of non-transgenic germinated maize seeds. The test is scored after 7 days by washing the soil off the roots into two stacked US Standard sieves, No. 40 (425 μm opening) on the bottom and No. 20 (850 μm opening) on top. After washing most soil through the sieves, living rootworm larvae are floated out of the remaining soil and roots by immersing the sieves into a tray of saturated magnesium sulphate solution.

Callus Assays

The precedent for callus bioassay comes from work on non-transgenic plants. Williams et al. demonstrated that callus derived from insect-resistant maize genotypes was also resistant. Callus can be evaluated by either direct infestation with insects or diet incorporation. Callus (0.2～0.6 g) can be tested within wells of 24-well microtitre plates. Because callus is largely (> 80%) water, preblotting callus on sterile filter paper and placement of the callus on top of a sterile filter paper disc, such as a 1/20 Difco Concentration Blank, reduces excess free moisture. Transfer of callus to wells using sterile technique and covering the bottom of wells with antimicrobial agar (propionic+phosphoric acid at 3.0 mL/L) minimizes microbial contamination. One or more neonate larvae are added to each well, which is then sealed with Mylar and ventilated with pin holes. Test duration is 4～7 days at 23～28℃ after which larval survivorship and size/weight relative to controls are recorded. Because of its extreme sensitivity to Cry Ⅰ Ab protein, *Manduca sexta* (L.) is useful for assay of transgenic maize callus to verify transformation and protein expression. However, *M. sexta* sensitivity is a drawback when an attempt is made to prioritize or rank the relative potency of different transgenic constructs or prioritize the relative potency of different transgenic events sharing the same construct. Although *O. nubilalis* is less sensitive to Cry Ⅰ Ab protein than *M. sexta*, many *B. thuringiensis* gene constructs in maize callus produce 100% mortality of neonate *O. nubilalis* larvae. Such constructs are difficult to rank regarding relative insecticidal activity. As with leaf assays, one solution is to use other species of Lepidoptera that will consume maize callus but are less sensitive to the insecticidal proteins used in the constructs. *H. zea*, *Diatraea grandiosella* (Dyar) and *S. frugiperda* are extremely useful in this regard since all survive and develop normally on maize callus and moult to second instars within 5 days at 26℃ on non-transgenic callus. Potent *B. thuringiensis* maize constructs can be evaluated with later instar larvae of *O. nubilalis* since they are less sensitive to insecticidal proteins than neonates. A growth inhibition assay uses third instar *O. nubilalis* larvae (4～8 mg) fed for 7 days on rootworm diet. A sample of at least 12 larvae is weighed, added to callus from one line (one larva per well), allowed to feed for 4 days, and then reweighed. The percentage growth inhibition (= growth reduction) of treatment groups compared with non-transgenic controls can be calculated using the following equation:

$$\text{growth inhibition}(\%)=\left(1.0-\frac{F_t \cdot I_c}{F_c \cdot I_t}\right)\times 100\%$$

where F = final mean weight, I = initial mean weight, t = treatment, c = control.

Selected from: Navon A, Ascher K R S. Bioassays of Entomopathogenic Microbes and Nematodes. UK: CABI Publishing, 2000: 27-29.

Words and Expressions

survivorship /sə'vaɪvəʃɪp/ *n.* 生存
transgenic /ˌtrænz'dʒenɪk/ *adj.* 转基因的，基因改造的
pollen /'pɒlən/ *n.* 花粉；*vt.* 传授花粉给
O. nubilalis 欧洲玉米螟
microtitre plate 微孔板
D. undecimpunctata howardi 黄瓜甲虫
humidity /hjuː'mɪdəti/ *n.* 湿气，潮湿，湿度
polyurethane /ˌpɒli'juərəθeɪn/ *n.* [化] 聚氨酯
spatula /'spætʃələ/ *n.* (涂油漆，涂药等) 抹刀，压舌板
sieve /sɪv/ *n.* 筛，滤网，不会保密的人；*v.* 筛，过滤
callus /'kæləs/ *n.* 胼胝，(植物伤口的) 愈合组织，老茧
regeneration /rɪˌdʒenə'reɪʃn/ *n.* 再生，重建
genotype /'dʒenətaɪp/ *n.* 基因型
propionic /prɒpa'ɪɒnɪk/ *adj.* 丙酸的
Manduca sexta 烟草天蛾
Diatraea grandiosella 玉米秆草螟
S. frugiperda 草地贪夜蛾

PART Ⅸ
RESISTANCE

Unit 17 Resistance

The development of resistance to pesticides is generally considered to be one of the most serious obstacles to effective pest control today. The first case was recognized in 1908 by Melander, who noted an unusual degree of survival of San Jose scale (*Quadraspidiotus perniciosus*) after treatment with lime sulfur in Clarkston Valley of Washington.

A panel of World Health Organization (WHO) experts defined resistance as "the development of an ability in a population of a pest to tolerate doses of toxicants that would prove lethal to the majority of individuals within the same species". The term *behavioristic resistance* describes the development of the ability to avoid a dose that would prove lethal. Resistance is distinct from the *natural tolerance* shown by some species of pests. Here a biochemical or physiological property renders the pesticide ineffective against the majority of normal individuals.

Cross-resistance is a phenomenon whereby a pest population becomes resistant to two or more pesticides as a result of selection by one pesticide only. It must not be confused with *multiple resistance*, which is readily induced in some species with simultaneous or successive exposure to two or more pesticides. Cross-resistance is caused by a common mechanism.

Resistance is an Inevitable Result of Evolution

Populations are polymorphous and show genetic variability between individuals in the same population. Even if they have been inbred for some time, the genetic difference of the individuals may be considerable. Every gene can occur as different versions, and these are known as alleles. New techniques in molecular genetics make it possible to study these differences with great precision. One insect specimen, the fruit fly, has 13,601 genes, and each of them can have hundreds of alleles. New alleles can be formed by mutations, and genes may also be duplicated to increase the total gene pool of the species. Most alleles are very rare, but if conditions change so that an allele becomes advantageous for survival and reproduction, it will in a few generations become the main allele in the population. One enzyme family, the CYP enzymes, often referred to as cytochrome P450 or mixed-function oxidases, is often involved in resistance because they are able to catalyze oxidation and detoxication of a wide variety of substances. The fruit fly has 90 different genes that code for these enzymes. Just one may be a rare allele of one of these genes, with a code for just one different amino acid, and may make an enzyme that is more active in degrading a specific pesticide (e. g., for a pyrethroid or a carbamate). This rare variant makes it easier to survive and reproduce in a pyrethroid or carbamate sprayed field. Other enzymes, such as the glutathione transferases, are important for detoxication of xenobiotics. They are also coded for by numerous genes that have many alleles.

In plants, nematodes, and microorganisms, the situation is similar, although resistance development to herbicides, fungicides, and nematicides appeared later. The small plant *Arabinopsis thaliana* has, for instance, 25,498 genes, and the free-living nematode *Caenorhabditis elegans* has 19,099 genes. It is not very surprising that an allele of one or other of all these genes may make the organism less sensitive to an herbicide or nematicide.

Lethal toxicants in the environment will, of course, have a dramatic effect on the population. Only those individuals that for some reason survive are able to reproduce. An individual with alleles or gene duplications that make it less sensitive to the toxic environment will have much better opportunities to reproduce. The next generation of the pest will therefore have a higher frequency of these alleles. If the pest organism cannot be completely wiped out by the pesticide or by other means, resistance will appear sooner or later. Pesticides can therefore be regarded as consumable with a restricted time of usefulness. After having been used some years, the development of resistance may render them useless.

Physiological and Biochemical Factors

During the 1950s, an era when biochemical knowledge developed very fast, there was a very strong belief that by finding the biochemical mechanism for resistance, it should be easy to find some substance that counteracted it, for instance, inhibitors of enzymes that detoxicate the pesticide or a new pesticide that shows higher activity toward the resistant insects.

A priori, i. e., without doing any empirical research, the following mechanisms have been postulated for insect resistance to insecticides. Most of the points also apply to weeds and fungi:

(1) Behavior: Insects may have modified their behavior so that they avoid the areas sprayed with the insecticide. Such behavior may be genetically determined.

(2) Reduced penetration of the pesticide through the cuticle or the intestine.

(3) Lower transport into the target sites.

(4) Lowered bioactivation: Some pesticides such as the sulfur-containing organophosphates may often be bioactivated.

(5) Increased storage in fat depots or other inert organs.

(6) Increased excretion of active ingredients.

(7) Increased detoxication or decreased bioactivation.

(8) Less sensitive receptors or enzymes that are inactivated or hyperactivated by the

pesticides.

(9) The development of alternative physiological pathways so that those disturbed by the pesticides are not so important.

(10) More robust or bigger organisms so that they can tolerate bigger doses.

Extensive research has shown that points (7) and (8) are almost always involved, but with the other factors playing a modifying or additional role. Enhanced detoxication of the insecticide is often found in the resistant insect, or a modification of the bimolecular that is its target.

How to Delay Development of Resistance

Many scientists and administrators have addressed this problem, but there is no clear-cut simple method for preventing resistance. The only method to delay development of resistance is to slow down the evolutionary process, which means providing better reproduction possibilities to susceptible individuals. P. Richter argued that the aim of pesticide treatment is to reduce the target population substantially. The population recovers from three sources of founder individuals: immigrants, survivors in refuges, and resistant individuals. There is a short period after treatment during which the survivors are solitary. The population density may be too small for successful breeding. A rapid influx of immigrants will greatly enhance the chance of the resistant trait being conserved. Once a certain quantity of resistant insects escapes extinction, the *R*-allele can quickly increase in frequency during following treatments. If this argument holds, resistance development from a very low frequency of *R*-alleles is dependent on survivors of susceptible insects in refuges or fast immigration. Otherwise, the few surviving *R*-individuals are not able to breed.

1. Refuge Strategy

The strategy implies that some areas are not treated with pesticides, but are kept as a refuge for susceptible individuals. These insects can then mate without any selection pressure from pesticides. The surviving, more or less resistant, individuals in the nearby treated fields will mate with the individuals from the refuge area and the resultant offspring will be heterozygotes and susceptible to the pesticide. The refuge strategy has two critical assumptions: that inheritance of resistance is recessive, and that random mating occurs between resistant and susceptible individuals. If resistance is recessive, hybrid first-generation (F_1) offspring produced by mating between *S*- and *R*-insects are killed by the pesticide. If mating is random, initially rare homozygous *R*-individuals are likely to mate with the more abundant homozygous susceptible insects that have not been exposed to the insecticide. In most cases, where the genetics of resistance have been studied, the resistance alleles are recessive or semidominant.

The primary strategy for delaying insect resistance to transgenic *B. thuringiensis* (Bt) cotton plants is to provide such refuges of cotton plants that do not produce Bt toxins. Pink bollworm resistance to Bt cotton is recessive. Survival of hybrid F_1 progeny is not higher than survival of the susceptible strain, but is markedly lower than the resistant strain. However, resistant larvae on Bt cotton require an average of 5.7 days longer to develop than larvae on non-Bt cotton. Bt-resistant moths from Bt plants will therefore probably mate more often with other *R*-moths, and the refuge method may not work. To achieve random mating, resistant adults from Bt plants and susceptible ones from refuge plants must emerge synchronously.

2. Mixing Pesticides with Different Modes of Action

Two pesticides where the mechanism of resistance is likely to be caused by different genes may be mixed. It is highly unlikely that individuals with resistance to one will also be resistant to the other. Using such high doses of both pesticides such that they will kill heterozygote resistance will probably be prevented when the frequency of *R*-alleles for both is low, and none of the insects are homozygotes for mechanisms of resistance for both pesticides. Sometimes insecticides with widely different modes of action and structure may give resistance through a common mechanism. One example is parathion-methyl resistance in houseflies, which may be caused by a certain glutathione transferase. This same enzyme, however, is also able to detoxicate lindane.

Sequential or rotational use of pesticides will not prevent development of resistance, at least not if the survival of heterozygotes is better than that of normal susceptible ones. It may just make a delay because the selection pressure for each is lower.

3. Switching Life-stage Target

In many pests, for instance, in cabbage flies, the larvae, which are the real pest, will probably be much more robust and have a better repertoire of detoxifying enzymes in order to detoxify the isothiocyanates and other natural pesticides present in the host plant, while the adults may not eat anything during their short life span and hence do not have detoxication enzymes. The adults therefore will be more sensitive to pesticides and less likely to develop resistance. If pesticides are used against the adult stage and not against the real pest, the larvae, resistance development may be slowed down.

4. Increased Sensitivity in Resistant Pests

If it were possible to develop two pesticides such that increased resistance to one of them led to increased susceptibility to the other, resistance development would at least be delayed. There is one example of this principle. The systemic fungicide *diethofencarb* is particularly effective against *Botrytis* spp. , which are resistant against benzimidazole. Benzimidazoles like *carbendazim* and *thiophanate* bind to a site on the tubulin protein and inhibit mitosis. Resistant *Botrytis* has a tubulin that does not bind benzimidazoles, but may bind diethofencarb better.

5. Inhibition of Detoxication Enzymes

After DDT dehydrochlorinase was found as one of the causes of DDT resistance in flies, it was subsequently found that *N*, *N*-dibutyl-*p*-chlorobenzene sulfonamide inhibits the enzyme. This substance, called WARF-Antiresistant, was tried, and although it increased the sensitivity of resistant insects, it was of no durable practical usefulness because either the flies developed other mechanisms of resistance or the amount of DDT dehydrochlorinase increased to overcome the action.

Selected from: Stenersen J. Chemical Pesticides: Mode of Action and Toxicology. Florida: CRC Press, 2004: 188-206.

Words and Expressions

obstacle /'ɒbstəkl/ *n*. 障碍，妨害物

Quadraspidiotus perniciosus 梨园盾蚧

lime sulfur 石硫合剂

behavioristic /bi,heivjə'ristik/ *adj*.［心理］行动主义的

lethal /'liːθl/ *adj*. 致命的；*n*. 致死因子

render /'rendə(r)/ *vt*. 呈递，归还，实施；*vi*. 给予补偿

polymorphous /,pɒli'mɔːfəs/ *adj*. 多形的，多形态的

inbred /,ɪn'bred/ *adj*. 天生的，先天的，内在的

allele /ə'liːl/ *n*.［遗传学］等位基因

specimen /'spesɪmən/ *n*. 范例，标本，样品

mutation /mjuː'teɪʃn/ *n*. 变化，转变，(生物物种的) 突变

duplicate /'djuːplɪkeɪt/ *adj*. 复制的，两倍的；*n*. 复制品；*vt*. 复写，复制

oxidase /'ɒksɪdeɪz/ *n*.［生化］氧化酶

oxidation /,ɒksɪ'deɪʃn/ *n*.［化］氧化

glutathione transferase *n*.［生化］谷胱甘肽转移酶

xenobiotics /,zenə,bai'ɔtiks/ *n*. 外源性物质，［生物］异种生物学；有害异物

consumable /kən'sjuːməbl/ *adj*. 可消费的；*n*. 消费品

Arabinopsis thaliana 拟南芥

Caenorhabditis elegans 秀丽隐杆线虫

counteract /,kaʊntər'ækt/ *vt*. 抵消，中和，阻碍

susceptible /sə'septəbl/ *adj*. 易受影响的，容许……的；*n*. 易得病的人

cuticle /'kjuːtɪkl/ *n*. 角质层，表皮，护膜

intestine /ɪn'testɪn/ *adj*. 内部的，国内的；*n*.［解，动］肠

excretion /ɪk'skriːʃn/ *n*.(动植物的) 排泄，排泄物

hyperactivate 超活化

influx /'ɪnflʌks/ *n*. 流入，汇入

extinction /ɪk'stɪŋkʃn/ *n*. 消失，消灭，废止，［物］消光

heterozygote /,hetərə'zaɪgəut/ *n*. 异质接合体，异形接合体

inheritance /ɪn'herɪtəns/ *n*. 遗传，遗产

recessive /rɪ'sesɪv/ *adj*. 退行的，逆行的，［遗］隐性的；*n*. 隐性性状

hybrid /'haɪbrɪd/ *n*. 杂种，混合物；*adj*. 混合的，杂种的

homozygous /,həuməu'zaigəs/ *adj*.［生］同型结合的，纯合子的

semidominant /'semɪ'dɒmɪnənt/ *adj*.［生］半显性的

progeny /'prɒdʒəni/ *n*. 后裔

synchronously /'siŋkrənəsli/ *adv*. 同时地，同步地

detoxicate /diː'tɒksɪkeɪt/ *vt*. 使解毒

repertoire /'repətwɑː(r)/ *n*.（准备好演出的）节目，指令系统

isothiocyanate /'aisəuˌθaiəu'saiəneit/ *n*.［化］异硫氰酸盐（或酯）

dicthofencarb 乙霉威

Botrytis spp. 葡萄孢真菌

benzimidazole /ˌbenzimi'dæzəul/ *n*.［化］苯并咪唑

carbendazim /kɑː'bendeɪzɪm/ 多菌灵

thiophanate /ˌθaiəu'fæneit/ 硫菌灵

tubulin /'tjuːbjulin/ *n*.［生化］微管蛋白

mitosis /maɪ'təusɪs/ *n*.有丝分裂，间接核分裂

dehydrochlorinase /diːhaɪdrə'klɔːrɪneɪs/ *n*.［生化］脱氯化氢酶

Notes

［1］Bt（*Bacillus thuringiensis*）：苏云金杆菌，是一种细菌杀虫剂，属芽孢杆菌。1911年，德国人贝利纳（E. Berliner）在苏云金这个地方的一家面粉厂里，发现有一种寄生在昆虫体内的细菌，有很强的杀虫力。于是，人们称这种细菌为苏云金杆菌。

［2］refuge strategy：庇护所策略。以美国、澳大利亚等国为例，每个农场主每种植80亩的Bt棉，必须同时种植常规棉花品种20亩，作为“庇护所”养活不产生Bt抗性的害虫，不产生抗性的害虫相对数量很大，通过交配便可以稀释抗性基因来延缓抗性发展。

［3］allele：等位基因，是指在一对同源染色体上，占有相同座位的一对基因，它控制一对相对性状。

Exercises

Ⅰ Answer the following questions according to the text.

1. What are the definitions of resistance，cross-resistance and multiple-resistance?

2. What's the distinction between cross-resistance and multiple-resistance?

3. Please list some priori that are postulated for insect resistance to insecticides and give a brief explanation to each of them.

4. What does refuge strategy imply? What are the two critical assumptions about refuge strategy?

5. List at least four methods used to delay development of resistance mentioned in this text.

Ⅱ Translate the following English phrases into Chinese.

cross-resistance	lethal toxicant	multiple-resistance	biochemical mechanism
first-generation	hybrid progeny	molecular genetics	susceptible individual
target population	tolerate dose	genetic variability	behavioristic resistance

Ⅲ Translate the following Chinese phrases into English.

族群密度	抗性个体	果蝇	杂合子抗性	抗性发展	生化/生理特征
微管蛋白	成虫阶段	寄主	农药穿透性	解毒酶	谷胱甘肽转移酶

Ⅳ Choose the best answer for each of the following questions according to the text.

1. ________is easily caused in some species by simultaneously or successively exposing to two or more pesticides.

A. Cross-resistance　　B. Multiple-resistance

C. Natural-resistance　　D. Behavioristic resistance

2. Which of the following statement is not true?

A. Allele is any of a group of genes that a gene occur as.

B. New alleles can be formed by mutations.

C. If an allele becomes advantageous for survival and reproduction, it will in a few generations become the main allele in the population.

D. The CYP enzymes is easily involved in resistance because they can reproduce very quickly.

3. Which of following statement is true?

A. Two pesticides where the mechanism of resistance is unlikely to be caused by different genes may be mixed.

B. It is highly likely that individuals with resistance to one will also be resistant to the other.

C. Sequential or rotational use of pesticides will surely prevent development of resistance.

D. Sequential or rotational use of pesticides may just make a delay of resistance development.

4. The method is called ________ that pesticides are used against the adults that may not eat anything during their short life span and not against the real pest, the larvae, in order to slow down resistance development

A. refuge strategy　　B. increased sensitivity in resistant pests

C. switching life-stage target　　D. inhibition of detoxication enzymes

5. How to delay development of resistance?

A. refuge strategy and mixing pesticides with different modes of action

B. switched life-stage target and increased sensitivity in resistant pests

C. inhibition of detoxication enzymes

D. all of the above

Ⅴ Translate the following short passages into Chinese.

1. The term *behavioristic resistance* describes the development of the ability to avoid a dose that would prove lethal. Resistance is distinct from the *natural tolerance* shown by some species of pests. Here a biochemical or physiological property renders the pesticide ineffective against the majority of normal individuals.

2. *Cross-resistance* is a phenomenon whereby a pest population becomes resistant to two or more pesticides as a result of selection by one pesticide only. M*ultiple resistance* is readily induced in some species with simultaneous or successive exposure to two or more pesticides.

3. Lethal toxicants in the environment will have a dramatic effect on the population. Only those individuals that for some reason survive are able to reproduce. An individual with

alleles or gene duplications that make it less sensitive to the toxic environment will have much better opportunities to reproduce. The next generation of the pest will therefore have a higher frequency of these alleles. If the pest organism cannot be completely wiped out by the pesticide or by other means, resistance will appear sooner or later.

4. The strategy implies that some areas are not treated with pesticides, but are kept as a refuge for susceptible individuals. These insects can then mate without any selection pressure from pesticides. The surviving, more or less resistant, individuals in the nearby treated fields will mate with the individuals from the refuge area and the resultant offspring will be heterozygotes and susceptible to the pesticide.

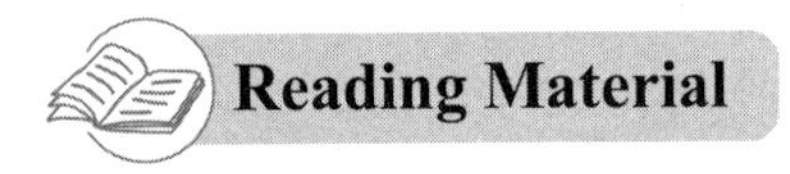

Classification of Resistance

Pesticide resistance evolved in insects, fungi, and bacteria long before it was observed in weeds. Resistance evolves following persistent selection for mutant genotypes that may be pre-existing or arise *de novo* in weed populations. Herbicide resistant weeds were predicted shortly after the introduction of herbicides. Given the examples of pesticide resistance in insects and fungi it seemed inevitable that the continuous or frequent use of the same herbicide against the same populations of weeds would eventually result in resistant weeds. The first herbicides to be used persistently over large areas were 2,4-D and MCPA. Fortunately, these auxinic herbicides are not prone to rapid selection for resistance and, with the exception of 2,4-D-resistant *Daucus carota*, resistance was not reported until the appearance of triazine herbicide-resistant weeds in the early 1970s.

Resistance, unless otherwise stated, denotes the evolved capacity of a previously herbicide-susceptible weed population to withstand a herbicide and complete its life cycle when the herbicide is used at its normal rate in an agricultural situation.

Target-site resistance is the result of a modification of the herbicide-binding site (usually an enzyme), which precludes herbicides from effectively binding. This is the most common resistance mechanism.

Cross-resistance occurs where a single resistance mechanism confers resistance to several herbicides. *Target-site cross-resistance* can occur to herbicides binding to the same target site (enzyme). Good examples are the two classes of herbicide chemistry (aryloxyphenoxypropionates and cyclohexanediones). While chemically dissimilar, both inhibit the enzyme acetyl coenzyme A carboxylase and resistant biotypes frequently exhibit varying levels of target-site cross-resistance in both. Some target sites may have more than one domain, e. g., the targeted protein in photosystem Ⅱ has separate domains that bind triazine-type herbicides and phenolictype herbicides.

Nontarget-site resistance is resistance due to a mechanism other than a target site modification. Nontarget-site resistance can be endowed by mechanisms such as enhanced metabolism, reduced rates of herbicide translocation, sequestration, etc. Such mechanisms reduce the amount of herbicide reaching the target site.

Nontarget-site cross-resistance occurs when a single mechanism endows resistance across herbicides with different modes of action. Such mechanisms are usually unrelated to the herbicide target site. Examples are cytochrome P450-based nontargetsite cross-resistance, and glutathione transferase-based resistances, which degrade a spectrum of herbicides that have different sites of action.

Multiple-resistance occurs when two or more resistance mechanisms are present within individual plants or a population. Depending on the number and type of mechanisms, a population and/or individual plants within a population may simultaneously exhibit multiple-resistance to many different herbicides.

Selected from: Powles S B, Shaner D L. Herbicide resistance and world grains. Florida: CRC Press, LLC, 2001: 2-3.

Words and Expressions

de novo /ˌdiːˈnəuvəu/ *n*. 重新，更始

MCPA (2-methyl-4-chloro-phenoxyacetic acid) 2-甲基-4-氯苯氧基乙酸

Daucus carota 胡萝卜

preclude /prɪˈkluːd/ *vt*. 排除，妨碍，阻止

aryloxyphenoxypropionate 芳氧苯氧丙酸酯

cyclohexanedione /saɪkləuˈheksəniːdiən/ 环己二酮

translocation /ˌtrænsləuˈkeɪʃən/ *n*. 迁移，移动，置换

Unit 18 Mechanisms of Resistance

Similar mechanisms for resistance to pesticides have been observed in insects, fungi, bacteria, plants, and vertebrates. These include changes at target sites, increased rates of detoxification, decreased rates of uptake, and more effective storage mechanisms. Most resistance to pesticides is inherited in a typical Mendelian fashion, but in some cases, resistance can be attributed to, or influenced by, relatively unique genetic and biochemical characteristics, e. g. , extranuclear genetic elements in bacteria and higher plants. A thorough understanding of the genetic, biochemical and physiological mechanisms of pesticide resistance is essential to the development of solutions to the pesticide-resistance problem.

Genetic Mechanisms

Insects, vertebrates, most higher plants, and fungi of the class Oomycetes are diploid, and some fungi are dikaryotic. Therefore, the genes responsible for resistance may exist in duplicate. Multiple allelic forms are known for many resistance genes. These alleles often produce an effect that is greater than additive. In some cases a resistance gene may exist in multiple copies, a condition called gene amplification. This is known to occur, for example, in the insects *Myzus* and *Culex*. Several genes at different loci also can be involved in resistance.

Most fungi are haploid in their vegetative state, as are bacteria generally, although multiple genomes are found in actively growing cultures. In a haploid state, the expression of each gene involved in resistance is not modified by another allele as in the case of the diploid organism. Many fungal cells, however, are multinucleate and heterokaryotic with respect to resistance genes, and these genes can produce a modification of resistance expression analogous to that found in diploid organisms. Furthermore, the resistance level of the organism is frequently the result of the interaction of alleles of several genes at different loci. This interaction is known as polygenic resistance. An additional complication in bacteria is the existence of extrachromosomal genes, which can act alone, or in concert with chromosomal genes, to confer resistance. In plants, herbicide resistance can be inherited in the plastid genome.

Genes that can mutate to confer resistance to a pesticide may be either structural or regulatory. Some structural genes are translated into products (enzymes, receptors, and other cell components, such as ribosomes and tubulin) that are targets for pesticides. The mutation of structural genes can result in a critical modification of the gene products, such as decreases in target site sensitivity or increased ability to metabolize pesticides. Regulatory gene products may control rates of structural gene transcription. They may also recognize and bind pesticides and thereby control induction of appropriate detoxifying enzymes.

A clear and detailed understanding of the molecular genetic apparatus of the resistant organism can provide essential information for devising tools and strategies for avoidance and management of practical pesticide resistance problems. Some examples include: (i) the construction of genetically defined organisms for investigation of the biochemical mechanism of pesticide action and for studies on population dynamics of biotypes that are heterozygous or polygenic for pesticide resistance; (ii) the rational design of synthetic antagonists to combine with regulatory proteins and block the induction of detoxifying enzymes; (iii) genetic engineering of herbicide-resistant plants, insecticide resistant beneficial insects, and microbial antagonists; and (iv) preparation of monoclonal antibodies for rapid and specific detection of resistance in a pest population. Ideally, this research should lead to the isolation, cloning, and sequencing of alleles conferring resistance and elucidation of their structure relative to their susceptible alleles.

Physiological and Biochemical Mechanisms

In insects and plants the principal biochemical mechanisms of resistance are: (i) reduction in the sensitivity of target sites; (ii) metabolic detoxication of the pesticide by enzymes such as esterases, monooxygenases, and glutathione-sulfotransferases; and (iii) decreased penetration or translocation of the pesticide to the target site in the insect. Alleles involving alteration of target sites include altered acetylcholinesterase resistance to organophosphates and carbamates, alterations in the gene for the receptor protein target of DDT and pyrethroids, and changes in the receptor protein target for cyclodiene insecticides. Metabolic resistance in the house fly seems to be under the control of a single gene whose product is a receptor protein. This protein binds insecticides, and the protein: insecticide complex induces synthesis of multiple detoxifying enzymes. Whether or not similar metabolic receptor proteins exist in other insects is not known. Decreased penetration has a minor or modifying effect on the level of resistance. A minor change in penetration, however, may have a profound effect upon the pharmacokinetics of a toxicant.

In plant pathogenic fungi, resistance has been attributed mainly to single gene mutations that (i) reduce the affinity of fungicides for target sites (e. g., ribosomes, tubulin, enzymes); (ii) change the absorption or excretion of the fungicides; (iii) increase detoxication. Most cases of practical fungicide resistance can be attributed to the first mechanism, which often results in a striking increase in resistance level brought about by mutation of a single gene. For this reason, fungicides that act at a single target site are at great risk with respect to the possibility of resistance development.

Resistance to other fungicides, such as ergosterol biosynthesis inhibitors and polyene antibiotics, occurs through a polygenic process. Each gene mutation produces a relatively small, but additive, increase in resistance. When many mutations are required to achieve a significant level of resistance, there is an increased likelihood for a substantial loss of fitness in the pathogen. There have been no major outbreaks of resistance to these fungicides in the field, but this situation is changing rapidly and problems are beginning to occur with the ergosterol biosynthesis inhibitors.

Three bactericides are used to control plant diseases in the United States: copper complexes, streptomycin, and oxytetracycline. Resistance to streptomycin in *Erwinia amylovora*, the pathogen of fireblight disease of pear and apple trees, has been a widespread problem. Resistance appears to be controlled by alteration of a structural chromosomal gene that reduces the affinity of the bacterial ribosome for streptomycin, an inhibitor of protein synthesis. In contrast, the most common mechanism of streptomycin resistance in human bacterial pathogens is mediated by an extrachromosomal gene that regulates the production of an enzyme (phosphorylase) that detoxifies streptomycin. The application of oxytetracycline to control streptomycin-resistant strains of *Erwinia amylovora* on pear trees is a relatively new practice, and reports of tetracycline resistance have not yet appeared. Oxytetracycline has been injected into palm trees and stone fruit trees for several years to control mycoplasmalike organisms, apparently without the development of resistance.

Plants utilize the same general resistance mechanisms as insects. The efficacious use of herbicides on crops is made possible because many crop plants are capable of rapid metabolic inactivation of the chemicals, thereby avoiding their toxic action. Target weeds are notably deficient in this capacity. It is apparent, though, that the capability to metabolize herbicides to innocuous compounds constitutes a potentially important basis of evolved resistance to herbicides in weeds. Documented cases of resistance have been due to other mechanisms, however, such as alteration of the herbicide's target site. For example, newly appearing *s*-triazine-resistant weeds have plastid-mediated resistance that involves a reduced affinity of the thylakoids for triazine herbicides.

The herbicide paraquat disrupts photosynthesis in target weeds by intercepting electrons from photosystem I, part of the metabolic cycle that fixes energy from sunlight into plant constituents via a complicated flow of electrons. Transfer of electrons from paraquat to oxygen gives rise to highly reactive oxygen radicals that damage plant membranes. Paraquat-resistant plants have higher levels of the enzyme superoxide dismutase, which quenches the reactive oxygen radicals.

The mechanisms of weed resistance to the dinitroaniline herbicides and to diclofop-methyl have not yet been identified. A number of herbicides act on the photosynthetic mechanism in the chloroplasts. Although the frequency of resistant plants arising from plastid mutations would normally be very low, a plastome mutator gene has been recognized that increases the rate of plastome mutation in weeds. This factor could be largely responsible for the plastid-level resistance to herbicides that have emerged in some weeds.

Resistance to anticoagulants is the most widespread and thoroughly investigated heritable resistance in vertebrates. Warfarin resistance in rats has been observed in several European countries, and in 1980 more than 10 percent of rats were resistant to warfarin in 45 out of 98 cities surveyed in the United States.

Research and Implementation

Despite the continual threat of resistance, we may still be able to exploit our expanding knowledge of the genetic and biochemical makeup of pests by designing pesticides that can circumvent existing resistance mechanisms, at least long enough to provide chemical manufacturers a reasonable rate of financial return on the investment needed to develop a new pesticide. Realistically, though, it is difficult to be optimistic on this point in practical situations where a synthetic pesticide is applied repeatedly to the same crop or environment to control a well-adapted pest. History promises no encouragement, at least for most pests, for the discovery of a "silver bullet". On the other hand, it is indeed encouraging that there are examples of pesticides, both selective and nonselective, that have been used for years in certain situations without setting off rapid, extensive resistance. The phenoxy herbicides (e. g. , 2,4-D) and the broadspectrum fungicides have been used successfully for decades without serious resistance problems. Still, the wisest course for future research appears to be the integration of a diversity of approaches to pest control chemical, biological, and cultural, because an integrated application of multiple methods will produce minimum selection pressure for development of resistance to pesticides. Evolution of resistance to minimally selective or multi target synthetic chemicals might be delayed indefinitely if the selection pressure were kept within "reasonable" limits. The pressure might be reduced with crop rotations and careful management, but may be virtually impossible in agricultural areas typified by repeated monocultures.

The development of resistance is encouraged by pesticides that act upon single biochemical targets. Unfortunately, the modes of action of many systemic plant fungicides, and most modern synthetic insecticides and herbicides, are biochemical site-specific. Many of these fungicides and insecticides have produced a rapid, major buildup of resistance genes in pest populations after just a few seasons of use. Undoubtedly, the potential for resistance development to such compounds will continue to be a limiting factor in the widespread use of these compounds, although compounds differ in the degree of risk for rapid development of resistance. In addition, some compounds lend themselves to relatively effective resistance management strategy.

Selected from: Glass E H. Pesticide Resistance: Strategies and Tactics for Management. Washington: National Academy Press, 1986: 45-49.

Words and Expressions

vertebrate /'vɜ:tɪbrət/ *n*. 脊椎动物；*adj*. 有椎骨的，有脊椎的

inherited /ɪn'herɪtɪd/ *adj*. 通过继承得到的，遗传的，继承权的

Mendelian /men'di:liən/ *adj*.（奥地利遗传学家）孟德尔的；*n*. 孟德尔学派的人

extranuclear /ˌekstrə'nju:klɪə/ *adj*.［核］核外的，胞质的

oomycetes /u:mɪ'sets/ *n*. 卵菌

diploid /'dɪplɔɪd/ *adj*. 双重的，倍数的，双倍的；*n*. 倍数染色体

dikaryotic /dɪkeəri':ɒtɪk/ *adj*. 双核的，双核细胞的

amplification /ˌæmplɪfɪ'keɪʃn/ *n*. 扩大

Myzus 瘤蚜属，苑瘤属，桃蚜

Culex /'kju:leks/ *n*.［动］库蚊属（*Culex*）蚊子
loci /'ləʊkaɪ/ *n*.部位，部，地
genome /'dʒi:nəʊm/ *n*.［生］基因组，染色体组
haploid /'hæplɔɪd/ *n*.［生物］单倍体；*adj*.单一的
heterokaryotic /'hetərəʊkærɪ'ɒtɪk/ *adj*.［生］异核体的
polygenic /ˌpɒlɪ'dʒenɪk/ *adj*.［医］多基因的
extrachromosomal /'ekstrəˌkrəʊmə'səʊməl/ *adj*.［生］染色体外的
chromosomal /ˌkrəʊmə'səʊməl/ *adj*.染色体的
plastid /'plæstɪd/ *n*.［医］质体，成形原体
ribosome /'raɪbəsəʊm/ *n*.［生化］核糖体
transcription /træn'skrɪpʃn/ *n*.抄写，抄本，信使核糖核酸的形成
antagonist /æn'tægənɪst/ *n*.敌手，对手
esterase /'estəreɪz/ *n*.［生化］酯酶
monooxygenase /ˌmɔnəu'ɔksidʒineis/ *n*.［生化］单（加）氧酶
pharmacokinetics /ˌfɑ:məkəʊkɪ'netɪks/ *n*.药物（代谢）动力学
affinity /ə'fɪnəti/ *n*.密切关系，吸引力，姻亲关系，亲和力
ergosterol /ɜ:'gɒstərɒl/ *n*.［生化］麦角固醇
copper complex 铜络合物
streptomycin /ˌstreptə'maɪsɪn/ *n*.链霉素
oxytetracycline /ˌɒksɪtetrə'saɪkli:n/ *n*.［微］土霉素，氧四环素
Erwinia amylovora ［医］解淀粉欧文（氏）菌，梨水疫病欧文（氏）菌
fireblight 火疫病
phosphorylase /fɒs'fɒrɪleɪz/ *n*.［生化］磷酸化酶
tetracycline /ˌtetrə'saɪkli:n/ *n*.［微］四环素
mycoplasma /ˌmaɪkə'plæzmə/ *n*.［微］支原体，支原菌
innocuous /ɪ'nɒkjuəs/ *adj*.无害的，无毒的，无伤大雅的，不得罪人的
thylakoid /'θaɪləkɒɪd/ *n*.［植］类囊体
paraquat /'pærəkwɒt/ *n*.百草枯（除草剂）
photosynthesis /ˌfəʊtəʊ'sɪnθəsɪs/ *n*.光合作用
superoxide dismutase ［化］超氧化物歧化酶
quench /kwentʃ/ *vt*.结束，熄灭，淬火；*vi*.熄灭，平息
dinitroaniline /daɪnaɪt'rəʊnɪli:n/ *n*.［化］硝基苯胺
diclofop-methyl /dɪk'lɒfɒpm'eθɪl/ 禾草灵
plastome mutator 质体突变体
anticoagulant /ˌæntikəʊ'ægjələnt/ *n*.［药］抗凝血剂
heritable /'herɪtəbl/ *adj*.可遗传的，可继承的
warfarin /'wɔ:fərɪn/ 灭鼠灵
chromosome /'krəʊməsəʊm/ *n*.［生物］染色体
circumvent /ˌsɜ:kəm'vent/ *vt*.围绕，包围，智取
phenoxy /fɪ'nɒksɪ/ *adj*.［化］含苯氧基的
broadspectrum *adj*.广谱的

recombinant /rɪˈkɒmbɪnənt/ *n*. [遗] 重组 [复合] 器官，重组细胞，重组体 [子]

enzymology /ˌenzaɪˈmɒlədʒɪ/ *n*. [生化] 酶学

Notes

[1] Mendelian：奥地利生物学家孟德尔创立的一种遗传学说。孟德尔是“现代遗传学之父”，是遗传学的奠基人。

[2] gene amplification：基因扩增。细胞内选择性复制 DNA，产生大量的拷贝。

[3] structural gene & regulatory gene：结构基因与调节基因，它是对基因的功能所作的区分，是以直线形式排列在染色体上。结构基因是决定合成某一种蛋白质分子结构相应的一段 DNA。调节基因是调节蛋白质合成的基因。

[4] warfarin：华法林，又名灭鼠灵，为香豆素类口服抗凝血药。

[5] silver bullet：银色子弹，或者称“银弹”，指由纯银质或镀银制成的子弹。在欧洲民间传说及 19 世纪以来哥特小说风潮的影响下，银色子弹往往被描绘成具有驱魔功效的武器，是针对狼人等超自然怪物的特效武器。后来被比喻为具有极端有效性的解决方法，作为“杀手锏、最强杀招、王牌”等的代称。

Exercises

Ⅰ Answer the following questions according to the text.

1. Please list mechanisms for resistance to pesticides.

2. How do the fungi produce resistance?

3. Why is a clear and detailed understanding of the molecular genetic apparatus of the resistant organism very important? Give some examples of the essential information it can provide for devising tools and strategies for avoidance and management of practical pesticide resistance problems.

4. Which are the principal biochemical mechanisms of resistance in insects and plants?

Ⅱ Translate the following English phrases into Chinese.

structural gene	copper complex	population dynamics
molecular genetic apparatus	resistant gene	resistance expression
metabolic inactivation	metabolic detoxication	regulatory gene
polygenic resistance	monoclonal antibody	superoxide dismutase

Ⅲ Translate the following Chinese phrases into English.

基因工程　基因突变　质体突变体　最小选择压力　基因转录

解毒系统　基因扩增　遗传适应性　染色体外基因　生物合成抑制剂

Ⅳ Choose the best answer for each of the following questions according to the text.

1. With regard to structural genes and regulatory genes, which of the following statement is not true?

A. The mutation of structural genes can result in decreases in target site sensitivity.

B. Structural gene products may control the rates of regulatory gene transcription.

C. The mutation of structural genes can result in increased ability to metabolize pesticides.

D. Regulatory gene may control induction of appropriate detoxifying enzymes.

2. Which of the following is not the main biochemical mechanism of resistance in insects and plants?

A. reduction in the sensitivity of target sites

B. metabolic detoxication of the pesticide by enzymes

C. decreased penetration and/or translocation of the pesticide to the target site in the insect

D. decrease storage in depots or other inert organs

3. As to the three bactericides use to control plant diseases in the United States, which of the following statement is true?

A. The most common mechanism of streptomycin resistance in human bacterial pathogens is mediated by a chromosomal gene that regulates the production of phosphorylase.

B. Resistance to streptomycin appears to be controlled by alteration of a structural chromosomal gene that reduces the affinity of the bacterial ribosome for streptomycin.

C. There are no reports of tetracycline resistance since oxytetracycline is used to control streptomycin resistance on pear trees.

D. Phosphorylase, whose production is regulated by an extrachromosomal gene, can detoxicate streptomycin.

4. In plant pathogenic fungi, the followings are the traits of single gene mutations that contribute to resistance except ________.

A. reduced affinity of fungicides

B. changing the absorption or excretion of the fungicide

C. increased penetration

D. increased detoxication

5. According to the text, which of the following examples of pesticides have been used for years in certain situations without setting off very fast and widespread resistance?

A. ergosterol biosynthesis inhibitors and polyene antibiotics

B. phenoxy herbicides and the broadspectrum fungicides

C. antioagulants and warfarin

D. triazine herbicides and stretomycin

6. The wisest course for future research appears to be the integration of a diversity of approaches to pest control chemical, biological, and cultural, because ________.

A. an integrated application of multiple methods will not produce any selection pressure for development of resistance to pesticides

B. an integrated application of multiple methods will kill much more pests than any other methods

C. an integrated application of multiple methods will produce minimum selection pressure for development of resistance to pesticides

D. an integrated application of multiple methods will not produce resistance to pesticides

Ⅴ **Translate the following short passages into Chinese.**

1. A thorough understanding of the genetic, biochemical and physiological mechanisms of pesticide resistance is essential to the development of solutions to the pesticide-resistance problem.

2. Plants utilize the same general resistance mechanisms as insects. The efficacious use of herbicides on crops is made possible because many crop plants are capable of rapid metabolic inactivation of the chemicals, thereby avoiding their toxic action. Target weeds are notably deficient in this capacity. It is apparent, though, that the capability to metabolize herbicides to innocuous compounds constitutes a potentially important basis of evolved resistance to herbicides in weeds.

3. Synthetic chemicals will probably continue for some time as the major weapon against most pests because of their general reliability and rapid action, and their ability to maintain the high quality of agricultural products that is demanded by urban consumers today. Although new chemicals offer a short-term solution, this approach to pest control alone will rarely provide a viable, long-term strategy.

Resistance Management

The global economic impact of pesticide resistance has been estimated to exceed $4 billion annually. Other estimates have been lower, but most scientists, agrochemical technical personnel, and agricultural workers agree that resistance is a very important driver of change in modern agriculture.

Resistance management attempts to ameliorate the development of resistance through strategies, tactics, and tools that reduce selection pressure. Management steps are deployed to reduce resistance evolution by:

1. Diversifying mortality sources with strategies of managing resistance such as sequencing, rotating, or alternating pesticides with differing modes of action and the use of other strategies of integrated pest management including biological control, resistant varieties, cultural control, and pheromone disruption, among others

2. Monitoring to detect low frequency resistant alleles

3. Modeling to predict resistance development or

4. Facilitating the survival or immigration of susceptible individuals that will dilute the frequency of homozygous resistant individuals in pest populations

Resistance exhibits many of the characteristics described by Garret Hardin in his article "Tragedy of the commons". His concept relates to a public animal grazing area known as a "commons". Many families could benefit from this single resource by careful management and equal sharing. However, overgrazing by even a single user could upset the balance of regrowth and destroy it for all. Hardin's argument, oversimplified, is that individuals are compelled to do this. Much like the grass in those fields, the proportion of individual pests in a population that is susceptible to a pesticide is a precious commodity held in common. Such a statement may sound surprising, but the susceptible genes can be "overgrazed" by a single individual who continues to apply an insecticide that only serves to establish a resistant population. The now abundant resistant individuals will disperse and establish in other fields. In short order this pesticide would no longer be effective in that region. Very little incentive exists for an individual producer to manage resistance on his or her farm if a neighbor ignores resistance management principles and thus selects a resistant strain, especially if in practice this results in increased crop losses.

To complicate the resistance management issue, very little resistance reporting has not been anecdotal. Early on, many resistance episodes were attributed to poor spray coverage, ineffective timing, and rain wash-off. Therefore resistance evolution from the early 1950s to the 1980s was often described as a pesticide applicator problem. Various stakeholders, including industry, government and state agencies, and university representatives, sought other explanations for insecticide failure. Because resistance monitoring was difficult, expensive, and of questionable value, widespread and effective monitoring programs have not generally been supported by the private or public sectors. Ironically, monitoring had been suggested by scientists and government agencies and welcomed as a resistance management strategy. This contrast reflects the uncertain nature of deploying a monitoring strategy with adequate efficiency to allow the implementation of alternative resistance management tactics. As a result, resistant pest populations have become established before pest managers have even suspected a problem; thus their reporting has been anecdotal. Some might say that for implementation of resistance management in the field, it is better to assume that resistance must be present rather than to waste time and money in monitoring because it can be economically impractical. Rather than taking action only after monitoring procedures declare that the pest population is resistant, it is not unreasonable to recommend the prevention of resistance by implementing a resistance management strategy whenever pesticides are used.

Selected from: Wheeler W B. Pesticides in Agriculture and the Environment. New York: Marcel Dekker, Inc., 2002: 242-245.

Words and Expressions

diversify /daɪ'vɜːsɪfaɪ/ *v*. 使多样化，作多样性的投资
tragedy /'trædʒədi/ *n*. 悲剧，惨案，悲惨，灾难
overgraze /ˌəuvə'greɪz/ *v*. 过度放牧
anecdotal /ˌænɪk'dəutl/ *adj*. 轶话的，多轶事趣闻的
episode /'epɪsəud/ *n*. 一段情节，［音］插曲，插话
coverage /'kʌvərɪdʒ/ *n*. 覆盖
stakeholder /'steɪkhəuldər/ *n*. 赌金保管者，利益相关者
ironical /ai'rɔnik/ *adj*. 讽刺的，用反语的

PART X
PESTICIDE RESIDUES

Unit 19 Pesticide Residues in Food and Trade

In their efforts to supply a safe and abundant food supply, the world's farmers must cope with a variety of production challenges. To face threats posed by insect pests, weeds, and fungi during the growing season and post-harvest, a variety of tactics, including the use of pesticides, may be necessary as part of an integrated pest management (IPM) approach. If it is necessary to use pesticides, the potential presence of trace concentrations of pesticide residues in food commodities at harvest and after processing poses a dilemma. Consumers generally would prefer to eat food free of pesticide residues, yet pesticides are often integral components of IPM programs. To resolve this situation, a "food-chain compromise" has been reached in practice to meet the needs of both farmers and consumers.

The globalization of the food chain means that consumers and the farmers who supply them may reside in different regions separated by great distances and political boundaries. For any given meal, a fresh banana or apple or mango may have been grown half a world away. This brings up the question of how pesticide residues in food and the food-chain compromise are regulated at the international level.

Regulation of Pesticide Residues

The primary regulatory standard employed to control pesticide residues in food is the maximum residue limit (MRL). The MRL has been defined as "the maximum concentration of pesticide residue that is legally permitted or recognized as acceptable on a food, agricultural commodity, or animal feed". MRLs are not set on the basis of toxicology data, but once proposed based on good agricultural practice (GAP) they must be evaluated for safety. This is generally accomplished through a risk assessment process that compares dietary intakes estimated from expected residue concentrations in foods consumed with the relevant health-related regulatory endpoints, the acceptable daily intake (ADI) and the acute reference dose (ARfD).

1. The World Food Code and Codex MRLs

Food standards elaborated by Codex include harmonized MRLs for pesticide residues in food. These standards are developed through activities coordinated by the Codex Committee on Pesticide Residues (CCPR). The scientific evaluations upon which Codex MRLs are based result from the FAO/WHO Joint Meeting on Pesticide Residues (JMPR), active since 1963. As part of the JMPR, a WHO panel reviews pesticide toxicology data to estimate the ADI and the ARfD. A FAO panel reviews pesticide GAP and residue chemistry data to estimate MRLs. Following adoption of the JMPR recommendation by CCPR, the Codex Alimentarius Commission (CAC) formally promulgates the MRLs as Codex standards.

The importance of Codex standards is that they offer a globally harmonized, unbiased and authoritative source of MRLs that take into account the various national GAP for a particular pesticide-commodity as well as available residue trial data. The authoritative nature of Codex MRLs has, in fact, been recognized and agreed in principle by the majority of important trading countries. The World Trade Organization (WTO), through a 1995 agreement on the Application of Sanitary and Phytosanitary Measures (SPS), identified Codex MRLs as the official reference for food safety issues which affect international food trade and the basis for resolution of trade disputes. Thus, it would appear that with respect to management of residues and MRL issues associated with global trade, the mechanism for preempting potential national differences in GAP is neatly in place.

Two primary factors, however, have served to retard the universal implementation of Codex MRLs for worldwide regulation of pesticide residues on food moving in international trade. The first is that Codex MRLs have not been established for all important pesticides and crops. Although more than 700 pesticide active ingredients are authorized in one country or another on a worldwide basis, as of 2006, Codex MRLs had only been established for around 180 pesticides. The primary causes for this incomplete set of Codex MRLs include the historically slow nature of the Codex standard elaboration process, limited JMPR resources to complete evaluations and failure of members to submit sufficient field residue trials at GAP for some crops. Thus, farmers may use many pesticides on crops for which no Codex MRLs are available. The second factor hindering the effective regulation of pesticide residues in world food trade by Codex MRLs is the incomplete recognition of their applicability for trade by several major food-importing regions including the EU, Japan, and the U.S.. In these regions, legislation mandates the development of a specific set of national or regional pesticide MRLs based primarily on locally approved GAP. Although Codex MRLs may be considered in the development of such MRL systems, in practice the MRL promulgation process strongly favors local GAP as the basis for standard-setting. As might be expected, this approach leads to national or regional MRLs which may differ in some cases from Codex MRLs.

2. U.S. Tolerances

U.S. MRLs, referred to as "tolerances", are established by the U.S. Environmental Protection Agency (EPA) under auspices of the Federal Food, Drug, and Cosmetic Act (FFDCA). Tolerances are established on raw agricultural commodities (RAC) and also on processed commodities if the residue level in the process fraction will be greater than that for the RAC. Tolerances for more than 300 active ingredients have been established by EPA.

Enforcement of U.S. tolerances is the responsibility of the U.S. Food and Drug Administration (FDA). In the absence of a specific tolerance, residues must be below detectable levels. Based on modifications to FFDCA mandated by the Food Quality Protection Act (FQPA) of 1996, several new elements were introduced to the EPA tolerance process. These include the need to consider the special sensitivity of infants and children, the potential exposure via multiple routes of exposure, and the potential for exposure to other pesticides and chemicals with a common mechanism of toxicity. Under FQPA, the EPA was also required to complete a reevaluation of all existing tolerances during a 10-year period. Domestically established MRLs apply also to imported commodities, but there is an established process for evaluation of residue data from other countries in support of import tolerances.

3. Japan MRLs

Japan MRLs are established by the Ministry of Health, Labor, and Welfare (MHLW) in consultation with the Food Safety Committee (FSC) under auspices of the Food Sanitation Law. Until recently, specific with-holding limits (WHLs) were set under the Agricultural Chemical Control Law to govern residue limits associated with approved GAP, but these were applicable only for domestically grown agricultural commodities. For some pesticides, MRLs were also established by MHLW to govern the residue levels on both domestic and imported food commodities.

The importance of Japan MRLs stems from the highly influential role of Japan as a major food importer from neighboring countries within Asia as well as the broader Pacific Rim and beyond. The recent move to adopt a comprehensive set of "positive list" MRLs will greatly increase the importance of Japan MRLs for world trade, and exporting nations whose GAP was not specifically considered in development of the positive list may be most impacted. Another factor which increases the impact of Japan MRLs is the strict system of compliance monitoring and enforcement which is implemented by the MHLW and local government. In addition to random monitoring, targeted and mandatory monitoring of 50% or 100% of certain commodities may be required following one or two MRL violations, respectively. Continued violations may result in targeted import bans for the problem commodities.

4. EU MRLs

At present, complete harmonization of MRLs across the European Union (EU) member states has not yet been accomplished. Thus, most regulation of pesticide residues in food is based on MRLs established by national legislation in each member state. These un-harmonized MRLs may reflect different GAP and thus may differ between members. A program for creation of a harmonized set of EU MRLs applicable across all member states has been making slow but significant progress since the early 1990's. Harmonized EU MRLs are established by the European Commission. Harmonized EU MRLs have so far been established for around 150 pesticide active ingredients. New legislation was approved by the European Parliament during 2005 which established an accelerated program for achieving a single, harmonized set of EU-wide MRLs for all crop/pesticide combinations. This harmonized listing will be based on (1) existing EU MRLs, (2) MRLs currently in force within the 25 members' states, and (3) Codex MRLs. Promulgation of a complete set of

EU MRLs may take several years to occur based on the complexities of selecting the most appropriate value to reflect differences in GAP among member states and requirements for safety determination via dietary intake assessment. The new European Food Safety Agency (EFSA) is expected to play a major role in implementation of the accelerated EU MRL process. A process for establishment of an EU import MRL based on overseas GAP and data will also be available in the future.

Disharmonized MRLs, Monitoring, and Consumer Safety

The existing world situation of partially harmonized regulation of pesticide residues in food, with influential MRL systems including those of Codex, the EU, Japan, and the U.S. , has led to negative consequences for growers and consumers alike. First, mismatches between GAP and applicable MRLs of food-exporting and food-importing regions may lead to the creation of trade barriers and irritants. Farmers in one country may not be able to employ authorized GAP for certain crops and pesticides in their own country because of such discrepancies due to fears of or actual import violations. Such fears may be theoretical because in many cases the actual residues present in food moving in international trade are much lower than established MRLs. Such trade-related concerns are often high for minor crops, which may lack specific data or grouping with major crops for purposes of MRL establishment. A second consequence of MRL disharmony is the retarded adoption by farmers of many of the newer, reduced risk pesticides which may take several years to achieve worldwide approvals and all required MRLs. Although the evaluation policies of key organizations such as U.S. EPA and Codex have accelerated the introduction of new pesticides with more favorable human health and environmental safety profiles, farmers in food-exporting countries may be forced to continue to use older pesticides while they await establishment of all applicable MRLs. Third, there may be significant economic impacts which may result from disharmonized MRL standards among trading partners. A World Bank case study of divergent banana MRLs indicated that a 1% increase in regulatory stringency for one key pesticide could decrease world banana imports by 1.6%. If the lowest existing national MRL rather than the higher Codex MRL had to be observed by all banana growers, who might not know the destination of their harvest, an estimated 5.5 billion USD in lost exports would be annually predicted. Finally, discordant MRLs and the trade violations they may yield have spawned sensational and inaccurate publicity regarding pesticide residues in food and decreased consumer confidence.

Actual pesticide monitoring programs from key food-importing countries indicate that in many instances no pesticide residues are detected and in the vast majority of instances where detectable residues of pesticides occur, these levels are well below established MRLs. For example, monitoring of domestic and imported foods in the U.S. during 2003 by the U.S. Department of Agriculture Pesticide Data Program found that 43% of fresh fruits and vegetables had detectable residues. In 0.3% of the samples, U.S. tolerances were exceeded and, in 1.6% of the samples, residues were detected for which no U.S. tolerance existed. Similarly, monitoring of foods in the UK by the Pesticide Residues Committee during 2004 found that 31% of food commodities had detectable residues and the established MRL was exceeded in 1% of the samples. Compliance monitoring in Japan has revealed similar levels of detection and MRL violation rates < 1%, although implementation of the positive list system of comprehensive MRLs has been predicted to increase the violation rate by 5- to 6-fold.

What about the dietary intake and consumer safety relevance of detected residues and the low incidence of MRL violations? First, for residues at or below the MRL it should be mentioned that dietary intake assessments are conducted in setting the MRL to ensure that cumulative residues which may be present in all food sources are below toxicological endpoints. The endpoints employed for the dietary risk assessments, such as the ADI, are conservative in nature and generally established at levels 100-fold lower than those found to cause no adverse effects in test animals. Thus, human exposures many times the level of the MRL would be required to reach even those levels which may have minimal biological impacts. Although the MRL is not a health-based or toxicological standard, the favorable comparison of estimated food intake containing residues at the MRL level gives confidence that food with residues at or below the MRL poses no human health concern. Second, for residues present above the established MRL, it must be kept in mind that this is only an indication that either GAP has not been followed or, for food imports, that GAP in the country of origin may differ from that of the importing country. Thus, an MRL exceedence should be considered as a trade violation and not as a human safety risk. In fact, the vast majority of MRL violations constitute a negligible level of exposure and health risk despite news media headlines to the contrary.

Selected from: Ohkawa H, Miyagawa H, Lee P W. Pesticide Chemistry. Weinheim: WILEY-VCH Verlag GmbH, 2007: 29-37.

Words and Expressions

residue /'rezɪdjuː/ *n*. 残余，滤渣，残数，剩余物

integrated pest management 有害生物综合治理

maximum residue limit 最大残留限量

unbiased /ʌn'baɪəst/ *adj*. 没有偏见的

preempt /prɪ'empt/ *v*. 先占

retard /'rɪtɑːd/ *vt*. 延迟，使减速，阻止，妨碍

mandate /'mændeɪt/ *n*.（书面）命令，训令，要求；*vt*. 委任统治

auspices /'ɔːspɪsɪz/ *n*. 由…主办及赞助

risk assessment 风险评估

acceptable daily intake 每日允许摄取量

acute reference dose 急性参考剂量

codex /'kəʊdeks/ *n*.（圣书、古代典籍的）抄本，法律，规则，药典

promulgate /'prɒmlgeɪt/ *vt*. 发布，公布，传播

tolerance /'tɒlərəns/ *n*. 宽容，容忍，(食物中残存杀虫剂的)(法定) 容许量

discordant /dɪs'kɔːdənt/ *adj*. 不调和的，不和的，不和谐的

spawn /spɔːn/ *n*. (鱼等的) 卵，(植物) 菌丝，产物；*v*. 产卵

sanitation /ˌsænɪ'teɪʃn/ *n*. 卫生，卫生设施

mandatory /'mændətəri/ *adj*. 命令的，强制的，托管的

harmonization /ˌhɑːmənaɪ'zeɪʃn/ *n*. 调和化，一致，融洽

parliament /'pɑːləmənt/ *n*. 国会，议会

irritant /'ɪrɪtənt/ *n*. 刺激物；*adj*. 刺激的

discrepancy /dɪ'skrepənsi/ *n*. 相差，差异，矛盾

divergent /daɪ'vɜːdʒənt/ *adj*. 分歧的

stringency /'strɪndʒənsi/ *n*. 严格，紧迫，说服力

disharmony /dɪs'hɑːməni/ *n*. 不调和

inaccurate /ɪn'ækjərət/ *adj*. 错误的，不准确的

cumulative /'kjuːmjələtɪv/ *adj*. 累积的

Notes

[1] the Application of Sanitary and Phytosanitary Measures：《实施动植物卫生检疫措施的协议》，简称 SPS 协议。该协议的目标是："维护任何政府提供其认为适当健康保护水平的主权，但确保这些权利不为保护主义目的所滥用并不产生对国际贸易的不必要的障碍。"

[2] acute reference dose (ARfD)：急性参考剂量。它用来评价外来化学物短时间急性暴露造成的健康损害。

[3] acceptable daily intake (ADI)：每日容许摄入量。每日通过各种途径，摄入某种物质对人体健康不会产生已知的任何不良影响的剂量，用每千克体重摄入的毫克数表示。

[4] the Codex Alimentarius Commission：国际食品法典委员会，由联合国粮食及农业组织 (FAO) 和世界卫生组织 (WHO) 共同建立，以保障消费者的健康和确保食品贸易公平为宗旨的一个制定国际食品标准的政府间组织。

[5] Codex Committee on Pesticide Residues (CCPR)：国际食品法典农药残留委员会。

[6] Joint Meeting on Pesticide Residues (JMPR)：农药残留联席会议。

[7] Federal Food, Drug and Cosmetic Act (FFDCA)：《美国联邦食品、药品与化妆品法案》。

[8] U.S. Food and Drug Administration (FDA)：美国食品药品管理局。

[9] the Ministry of Health, Labor and Welfare (MHLW)：日本厚生省，主管医疗卫生、食品安全、劳动及就业的部门。

Exercises

Ⅰ Answer the following questions according to the text.

1. What is the definition of MRL?

2. What are the two main factors that withhold the universal implementation of Codex MRLs for worldwide regulation of pesticide residues on food moving in international trade?

3. Some new elements were added to the EPA process on the basis of modifications to FFDCA mandated by FQPA of 1996. Please give some examples of these new elements.

4. Why do the Japan MRLs have such impact in the world?

5. What are the negative consequences for growers and consumers?

Ⅱ Translate the following English phrases into Chinese.

GAP　residue trial　pesticide residue　cumulative residue
trade dispute　post-harvest　detectable levels　pesticide toxicology
agricultural commodity　sensitivity of infants and children

Ⅲ Translate the following Chinese phrases into English.

食品安全　痕量浓度　农药主成分　最大残留限量　每日允许摄取量
毒性机理　风险评估　食物链全球化　急性参考剂量　有害生物综合治理

Ⅳ Choose the best answer for each of the following questions according to the text.

1. Which of the following was not the main reason for the incomplete set of Codex MRLs?

A. historically slow nature of the codex standard elaboration process

B. limited JMPR resources to complete evaluations

C. failure of members to submit sufficient field residue trials at GAP for some crops

D. disagreements among members of Codex MRLs

2. In the course of the formal promulgation of Codex standards, ________ reviews pesticide toxicology data.

A. a FAO panel　B. CAC　C. a WHO panel　D. CCPR

3. WTO identified ________ as the official reference for food safety issues which influence global food trade and the basis for resolution of trade disputes.

A. U.S. Tolerances　B. Codex MRLs　C. Japan MRLs　D. EU MRLs

4. With regard to U.S. Tolerances, which of the following statements is not true?

A. Tolerances for more than 300 active ingredients have been established

B. Tolerances are established on raw agricultural commodities (RAC) but not on processed commodities

C. Enforcement of U.S. tolerances is the responsibility of FDA

D. In the absence of a specific tolerance, residues must be below detectable levels

5. The harmonized EU MRLs will be established on the following list except ________.

A. existing EU MRLs

B. MRLs currently in force within the 25 members' states

C. MRLs in some other countries

D. Codex MRLs.

6. According to the text, which of the following statements is not correct?

A. Residues present above the established MRL indicates that GAP has not been followed.

B. MRL exceedence should be considered as a trade violation.

C. MRL exceedence should be considered as a human safety and health risk.

D. The vast majority of MRL violations constitute a negligible level of exposure.

Ⅴ Translate the following two short passages into Chinese.

1. Consumers generally would prefer to eat food free of pesticide residues, yet pesticides

are often integral components of IPM programs. To resolve this situation, a "food-chain compromise" has been reached in practice to meet the needs of both farmers and consumers.

2. The importance of Codex standards is that they offer a globally harmonized, unbiased and authoritative source of MRLs that take into account the various national GAP for a particular pesticide-commodity as well as available residue trial data.

3. Actual pesticide monitoring programs from key food-importing countries indicate that in many instances no pesticide residues are detected and in the vast majority of instances where detectable residues of pesticides occur, these levels are well below established MRLs.

Sample Collection, Preparation, and Analysis

There are frequently large differences in residue levels between residues on raw agricultural commodities and those in a form actually consumed. The FDA's regulatory monitoring program focuses on the raw commodities while PDP samples are often washed, presumably just as typical consumers would do, Surveillance for pesticide residues 281 prior to analysis. In FDA's Total Diet Study, foods are prepared in institutional kitchens to be 'ready for consumption' before analysis. Residues encountered following this approach more closely resemble residues to which consumers may be exposed, as this approach takes into account such factors as washing, cooking, peeling, processing, evaporation, and the passage of time.

Samples collected for pesticide residue analysis are subject to several laboratory steps before pesticides can be detected and the residue levels are determined. Food samples are usually subject to an initial blending step followed by extraction using organic solvents or newer solvent-minimizing steps to isolate the pesticides from many other components in the foods. Additional unwanted components that may have been extracted along with the pesticides may be removed using clean-up procedures often involving column chromatography, volatilization, liquid-liquid partitioning, or chemical degradation. Such procedures often result in the fractionizing of the analyte extract into different subgroups; the properties of the individual pesticides will determine the particular subgroup into which each pesticide will primarily reside.

Chemical modification steps frequently follow the clean-up procedures and may result in the formation of pesticides modified to make them easier to separate, detect, or quantify. Analytical instruments in the laboratory such as high performance liquid chromatographs and gas chromatographs are frequently used to separate individual pesticides found in the cleaned-up extracts from other pesticides and other components of the extract. At the end of the instrument is a detector that takes advantage of the properties of the analyte to detect its presence once it has been separated from other components in the extract. Both qualitative and quantitative detection measurements are used to identify the analyte and determine how much of the analyte is present.

When violative residues are determined in tolerance enforcement programs, the commodities from which the violative residues reside are subject to seizure and possible destruction. To provide assurance that the samples are indeed violative, a second, independent analytical method is frequently used to confirm the findings of the original method.

While single residue methods (SRMs) for individual pesticides are frequently developed by the pesticide manufacturers and submitted to the EPA, US regulatory agencies normally use multi-residue methods (MRMs) capable of detection of a large number of different pesticides. Such methods frequently involve fractioning the original sample extract into many different subgroups and analyzing each subgroup separately. The MRM procedures used in FDA's regulatory monitoring program, for example, are capable of detecting more than 200 individual pesticides, representing approximately half of the pesticides with EPA tolerances and many others that have no tolerances. Occasionally, regulatory agencies must rely on other SRMs or MRMs capable of detecting pesticides that cannot be determined using the more common MRMs.

Analytical advances are leading to the development of more specific and more sensitive methods to separate, identify, and quantify pesticide residues from foods. The use of mass spectrometers, either alone or coupled to high performance liquid chromatographs and gas chromatographs, provides even greater sensitivity and analyte selectivity. In addition, a technique known as immunoassay, in which sensitive antibodies are developed from laboratory organisms and used to identify pesticide residues, is gaining widespread popularity due to its relatively low cost and low sample preparation requirements.

Selected from: Watson D H. Pesticide, Veterinary and other Residues in Food. Florida: CRC Press LLC, 2004: 281-283.

Words and Expressions

surveillance /sɜːˈveɪləns/ *n*. 监视，监督
peeling /ˈpiːlɪŋ/ *n*. 剥皮，剥下的皮
extraction /ɪkˈstrækʃn/ *n*. 抽出，[化] 提取（法）萃取法，抽出物
chromatography /ˌkrəʊməˈtɒgrəfi/ *n*. [化] 色谱法
partition /pɑːˈtɪʃn/ *n*. 分割，分开，隔离物；*vt*. 区分，隔开，分割
degradation /ˌdegrəˈdeɪʃn/ *n*. 降级，降格，退化
fractionize /ˈfrækʃəniz/ *v*. 化为分数，细分，分为小部分
analyte /ˈænəlait/ *n*.（被）分析物
high performance liquid chromatograph 高效液相色谱仪
gas chromatograph 气相色谱仪
multi-residue method 多残留方法
mass spectrometer 质谱仪
immunoassay /ˌɪmjunəʊˈæseɪ/ *n*. 免疫测定

Unit 20 Analysis of Pesticide Residues

Pesticides have helped us to grow food in abundance and eliminate pests. Unfortunately, many pesticides can also have negative effects both on the environment and on humans. The use of pesticides must consequently be carefully controlled and closely monitored to maximize their benefits and minimize harmful effects. To support good stewardship of pesticide uses, many analytical methods have been developed to measure levels of specific pesticide residues in food and in the environment.

Analytical methods for a pesticide may vary depending on the sample type and the purpose of the analysis. In the United States, over 700 pesticides are registered for use in food production, and many analytical methods for pesticides are described in the literature. Farmers can choose from many different pesticides to control the multitude of insect pests, fungi, and weeds that attack their crops. Rotations of different pesticides on a crop are recommended to reduce the buildup of resistance by pests, potentially further increasing the number of residues that may be found on a commodity. Finally, mixtures of pesticides are often used for more effective control of pests. A greater variety of pesticides are used in growing fruits and vegetables than for any other food items. Because it is not possible to know which pesticide residues you might find on a given crop, samples need to be screened for all possible residues.

The single-residue methods (SRMs) describe analysis of a single pesticide (or a group of related compounds derived from it) in a specific crop because they have been developed to register particular pesticides for particular applications or crops. The multiresidue methods (MRMs) can be used for assaying a wide range of pesticides in many different types of samples.

The presence of residues in fruits and vegetables makes pesticide residue testing a real challenge. Furthermore, rapid analysis is essential for assaying perishable samples such as lettuce, strawberries, and cucumbers. It is challenging for any chemist and for any method to analyze for unknown pesticides in a variety of matrices in a short time. For many regulatory laboratories, it is often the goal to complete the analysis the same day the samples are received. Even though MRMs may sometimes provide less method sensitivity or analytical precision than SRMs, they are the methods of choice for regulatory pesticide residue analysis because of their ability to detect a large number of pesticides, their applicability to a wide range of matrices, and the relative ease and speed of sample analysis. Here are five fundamental steps:

Sample Processing

Samples submitted to laboratories may consist of several individual fruits or vegetables. The exact numbers and sizes of samples vary depending on each nation's regulations. In general, the samples range from five to 20 individual fruits or plants or from 10 to 20 kg in total weight depending on the particular commodity. Some sample manipulation, such as the removal of outer layers of leafy vegetables, removal of cores of fruits, and washing, may be required by regulations. In the United States, unless otherwise indicated in the Code of Federal Regulations (CFR-40), regulatory samples cannot be manipulated through brushing, washing, peeling, removing outer leaves, or any other procedure that could affect the magnitude of pesticide residues. Samples may require further preparation for analysis such as cutting and chopping coarsely prior to extraction. Most laboratories chop and homogenize entire samples unless the applicable government regulation requires the preservation of an unaltered portion of the submitted sample. Samples are often homogenized by using common commercial food processors, providing both maceration and mixing at the same time. A subsample, usually in the range of 25～200 g, is taken for extraction.

Extraction

Water-miscible solvents such as methanol, acetonitrile, and acetone are the most common extracting solvents, along with ethyl acetate, which also extracts significant amounts of water. Much of the weight of fruits and vegetables (80%～95%) is due to water, and this water derived from the commodity mixed with the solvent becomes an efficient pesticide extraction medium. For example, a 50 g apple sample (80% moisture content) combined with 100 mL of acetonitrile yields an extracting solvent that is 70% acetonitrile in water. If necessary, additional water can be added to compensate for the low moisture status of some samples. For example, 10～20 mL of water is often added to low-moisture samples (e.g., wheat, rice, soybeans) to increase the aqueous proportion of the extraction solvent. Extraction of pesticides into organic solvent is often enhanced by further blending and shearing of the homogenate. Several types of blenders are used. Most common extraction devices have a rotating blade mounted at either the top or bottom of the vessel. Two to five minutes of blending at a moderate speed (2000～5000 rpm) normally suffices for the extraction of pesticides. To accomplish a more thorough extraction, MRMs can specify a device that disrupts samples through the generation of cellrupturing ultrasound in addition to mechanical mixing and shearing. Repeated extractions to ensure complete recovery of residues are often omitted from MRMs to save time and effort. Immediately after extraction, solvent is separated from nonextractable plant material. Simple filtering to remove plant material may be accomplished by using Sharkskin—filter paper, which is designed for quick filtration. Centrifugation is also used to separate insoluble materials from soluble extracts.

Purification

The filtrate resulting from sample extraction is a complex mixture that contains organic solvents, water, biochemicals (lipids, sugars, amino acids, and proteins), and secondary metabolites (organic acids, alkaloids, terpenoids, etc.) at high concentrations with very minute amounts of the pesticide residues of interest. It is a challenging task to isolate and detect pesticide residues of interest in the presence of high levels of background chemicals, often called matrix interferences. Most crude extracts require some purification prior to analysis.

Purification involves the removal of water, evaporation of excess organic solvent, and selective trapping to separate pesticides from interferences. The greater the number of cleanup steps included in a method, the greater the losses of analytes and the longer it takes to carry out the analysis. Most of the water must be removed from the extract to further concentrate the desired analytes. Much water is quickly removed by partitioning the organic solvent with sodium chloride-saturated water. These techniques remove large amounts of the water, but the remaining traces of water must be removed by filtering or adding dehydrated hygroscopic salts (e. g., Na_2SO_4) to the organic phase of the extract. Recoveries of extremely water-soluble pesticides such as acephate and methamidophos can vary depending on the concentration of other solutes and the mechanism of water removal.

Even after the removal of many water-soluble coextractives, extracts still contain large amounts of interfering compounds and only trace levels of pesticide residues. Buffering the aqueous phase close to a pH of 7 prior to removing water causes more ionic and polar biochemicals to partition into the water and results in removing large quantities of organic acids from the organic phase, which contains neutral and nonpolar chemicals, including pesticide residues. It is also possible to remove large amounts of nonpolar plant constituents before removing the water by filtering the aqueous/organic extract through reversed-phase solid-phase extraction (SPE) material or an activated carbon sorbent.

Because extracts must be concentrated 100-fold or more, a requirement for trace residue analyses, rapid and efficient concentration techniques are preferred. Rotary evaporators and Kuderna-Danish sample concentrators are good for concentrating thermally labile and highly volatile pesticides. Heating the extract in an open beaker with a stream of gas is also an efficient and inexpensive way to achieve concentration. Concentrated extracts, even after being subjected to the purification process, often contain quantities of interference chemicals that can easily interfere with analysis by overwhelming a chromatographic system or saturating a detector. Additional cleanup of extracts for MRMs maximizes the difference between physical and chemical properties of pesticide residues and those of interference chemicals. Two common techniques for cleanup are solid-phase extraction (SPE) and size-exclusion chromatography.

Chromatography and Detection

Multiresidue methods rely on chromatographic techniques to separate pesticide residues, to determine an analyte's identity on the basis of elution time (retention time), and to quantify responses from a specific detector. To this end, two chromatographic techniques are most common: gas chromatography (GC) and high performance liquid chromatography (HPLC).

Identification of pesticides using chromatography is based on the characteristic retention time of the pesticide on a particular chromatographic column under a given elution condition used in the separation. Retention times (R_t) of pesticides are often listed as a part of

MRMs. Two factors are major influences of R_t: types of columns used (liquid phases) and separation conditions. Gas chromatographic MRMs rely on multiple temperature gradient programs to enhance separation of pesticides and to reduce overall chromatographic time. Most HPLC MRMs rely on reversed-phase separation because of its reliability and cost-effectiveness. There are many different bonded liquid phases with different carbon loads and end capping that give different performance characteristics. The most commonly used HPLC column phases are octyl (C8) and octadecyl (C18) bonded to silica stationary phase. Methanol and acetonitrile are the two organic modifiers most commonly used with water in MRMs. As is the situation with the GC technique, HPLC MRMs rely on mobile-phase gradient schemes to vary the composition of the organic modifier to achieve the same goal of speed and cost efficiency.

Quantification of pesticide levels is as important as identification of pesticides for the regulatory laboratory, because the regulation of pesticide use is based on the maximum residue level (MRL or tolerance) that may be present. Thus, the correct estimation of pesticide residue concentration in a given matrix is critical, because levels that exceed the MRL are illegal. Most MRMs rely on external calibration techniques to quantify residues. Three to five concentrations of given pesticides are used to generate the GC, HPLC, or other calibration curve. The concentration of an incurred residue is quantified by comparison to the concentration-response curve.

There are several difficulties in correctly estimating or quantifying residue concentration. First, it is impractical to generate daily calibration curves for all analytes of interest. There are over 200 pesticides of interest in GC MRMs. Second, external calibration curves are often generated with standard pesticides in neat solvent (acetone or hexane) and not in a matrix blank. The so-called matrix effect on quantification of analytes is well known to analytical chemists. Sample matrix components (or coextractives) significantly influence the response of analytical instruments to pesticide residues. It would be ideal to use external calibration standards made in a matrix blank.

Different laboratories and organizations use different procedures for ensuring the best estimate of residue levels. The following is an example used in the California Department of Food and Agriculture laboratory. MRMs are validated initially by using a handful of representative pesticides. The external calibration curves for these pesticides are created by using standards in solvent on a daily basis. Over long periods of time, laboratories establish external calibration curves for all pesticides of interest and demonstrate the range of detection and linearity of detector response to the concentration of pesticides. When a pesticide residue is detected in a sample during a routine screening process, the estimation is made by using the external calibration curve. A pair of bracketing concentrations of the specific pesticide is chosen, and new external calibrations are then made using the same pesticide in a previously saved matrix blank. These calibration solutions are used to determine the residue concentration. In some cases only a single level of calibration might be used to reduce the time of analysis. This quantification scheme is a practical solution to what could otherwise be an unmanageable workload.

Confirmation

Unambiguous determination of pesticide residues is not always necessary, especially for initial screening. However, most pesticide regulatory surveillance and monitoring programs have established standard operating procedures (SOPs) to address the confirmation of initial findings of pesticide residues by addressing regulatory implications. A common approach and the most practical one for confirming a positive chromatographic response has been "the dual column confirmation", a technique that correlates two different retention times of a pesticide under two different chromatographic conditions. This technique is applicable in most situations, especially when differing retention times can be acquired simultaneously using a single chromatographic instrument. This can work well with a dual-column GC. However, it falls short when background matrix interferences become too great and the suspected pesticide residue response cannot be resolved sufficiently from them.

For the unambiguous identification of pesticide residues, MRMs rely on mass spectrometry (MS), another determinative technique that is different from GC. Mass spectrometry is a common choice because it gives direct physical and chemical information about the analyte and is easily coupled to chromatographic techniques. Various criteria for MS confirmation have been proposed for pesticide analysis. As GC/MS and LC/MS become more affordable, MRMs are being developed that are based on the use of MS for both initial and confirmatory determinations.

Selected from: Wheeler W B. Pesticides in Agriculture and the Environment. New York: Marcel Dekker, Inc., 2002: 160-181.

Words and Expressions

assay /ə'seɪ/ *n*. 化验；*v*. 化验
perishable /'perɪʃəbl/ *adj*. 容易腐烂的
lettuce /'letɪs/ *n*. [植] 莴苣，生菜，苦菜类
maceration /ˌmæsə'reiʃən/ *n*. 泡软，因绝食而衰弱
miscible /'mɪsəbl/ *adj*. 易混合的
acetonitrile /əˌsiːtə'naɪtraɪl/ *n*. [化] 乙腈
acetone /'æsɪtəun/ *n*. [化] 丙酮
ethyl acetate /'eθɪleɪsɪteɪt/ *n*. [化] 乙酸乙酯
compensate /'kɒmpenseɪt/ *v*. 偿还，补偿，付报酬
homogenate /hə'mɒdʒɪneɪt/ *n*. 均匀混合物，[生]（组织）匀浆
cellrupture 细胞破裂，细胞破碎
nonextractable /'nɒnɪks'træktəbl/ *adj*. 不可萃取的
filtration /fɪl'treɪʃn/ *n*. 过滤，筛选
matrix /'meɪtrɪks/ *n*. 矩阵
analyte /'ænəlait/ *n*.（被）分析物

hygroscopic /haɪgrə(ʊ)'skɒpɪk/ *adj*. 吸湿的
acephate /ə'sefət/ 乙酰甲胺磷
methamidophos /mɪθəmɪdəʊ'fəuz/ 甲胺磷
dehydrated /ˌdiːhaɪ'dreɪtɪd/ *adj*. 脱水的
hygroscopic /haɪgrə'skɒpɪk/ *adj*. 吸湿的；湿度计的；易潮湿的
coextract /kəʊɪks'trækt/ 同时萃取，共萃取
buffering /'bʌfərɪŋ/ 缓冲（作用），减震，阻尼，隔离
sorbent /'sɔːbənt/ *n*.［化］吸附剂
beaker /'biːkər/ *n*. 大口杯，有倾口的烧杯
chromatographic /krəʊˌmætə'græfɪk/ *adj*. 色谱法的
saturate /'sætʃəreɪt/ *v*. 使饱和，浸透，使充满
elution time 洗脱时间
retention time［分化］保留时间，［电子］保持时间
reversed /rɪ'vɜːst/ *v*. 翻转；*adj*. 颠倒的
octyl /'ɔktil/ *n*.［化］辛基
octadecyl /ɒk'tædəsɪl/ *n*.［化］十八（烷）基
hexane /'heksein/ *n*.［化］（正）己烷
surveillance /sɜː'veɪləns/ *n*. 监视，监督

Notes

［1］size-exclusion chromatography：凝胶过滤色谱，也叫做分子排阻色谱。一种利用带孔凝胶珠作基质，按照分子大小分离蛋白质或其他分子混合物的色谱技术。

［2］solid-phase extraction（SPE）：固相萃取，它利用固体吸附剂将液体样品中的目标化合物吸附，与样品的基体和干扰化合物分离，然后再用洗脱液洗脱或加热解吸附，达到分离和富集目标化合物的目的。

［3］the Code of Federal Regulations（CFR-40）：美国联邦法规。

［4］standard operating procedures（SOPs）：标准作业程序，指将某一事件的标准操作步骤和要求以统一的格式描述出来，用于指导和规范日常的工作。SOPs 的精髓是将细节进行量化。通俗来讲，SOPs 就是对某一程序中的关键控制点进行细化和量化。

Exercises

Ⅰ Answer the following questions according to the text.

1. Why are MRMs chosen for regulatory pesticide residue analysis although they sometimes provide less method sensitivity or analytical precision than SRMs?

2. How many fundamental steps do MRMs consist of and what are they?

3. What are the three different liquid phases commonly used in MRMs?

4. What are the difficulties in correctly estimating or quantifying residue concentration?

Ⅱ Translate the following English phrases into Chinese.

method sensitivity　gas chromatography　single-residue method　rotary evaporator

analytical precision　multiresidue method　analytical method　trace residue analyses
matrix interference　water-miscible solvent　reversed-phase separation
high performanceliquid chromatography　sodium chloride-saturated water

Ⅲ **Translate the following Chinese phrases into English.**

滤纸　流动相　色谱柱　样品处理　发育速度　质谱分析法
活性炭　有机酸　样品分析　保留时间　标准操作规程　反相固相萃取

Ⅳ **hoose the best answer for each of the following questions according to the text.**

1. According to the text, which of the following statement is not true?

A. Rotations of different pesticides on a crop are recommended to reduce the buildup of resistance by pests.

B. Mixtures of pesticides are often used for more effective control of pests.

C. The single-residue methods (SRMs) can be used for assaying a wide range of pesticides in many different types of samples.

D. rapid analysis is essential for assaying perishable samples.

2. Which of the followings is not the advantage of MRMs?

A. ability to detect a large number of pesticides

B. applicability to a wide range of matrices

C. ability to provide more method sensitivity or analytical than SRMs

D. relative ease and speed of sample analysis

3. Which of the followings is not the perishable sample?

A. cucumber　B. strawberry　C. lettuce　D. rice

4. In the step of extraction, additional water is often added to some low-moisture samples in order to ________.

A. dilute the extraction solvent

B. increase the water proportion of the extract solvent

C. get more extraction solvent

D. homogenize the extraction solvent

5. In the step of purification, the following techniques are used except ________.

A. concentration techniques　B. centrifugation

C. solid-phase extraction　D. size-exclusion chromatography

6. What are the two commonly used techniques among MRMs to separate pesticide residues, to determine an analyte's identity on the basis of elution time (retention time), and to quantify responses from a specific detector?

A. SPE and GC　B. HPLC and SPE　C. MS and GPC　D. GC and HPLC

Ⅴ **Translate the following short passages into Chinese.**

1. The single-residue methods (SRMs) describe analysis of a single pesticide (or a group of related compounds derived from it) in a specific crop because they have been developed to register particular pesticides for particular applications or crops. The multiresidue methods (MRMs) can be used for assaying a wide range of pesticides in many different types of samples.

2. Most laboratories chop and homogenize entire samples unless the applicable

government regulation requires the preservation of an unaltered portion of the submitted sample. Samples are often homogenized by using common commercial food processors, providing both maceration and mixing at the same time.

3. It is a challenging task to isolate and detect pesticide residues of interest in the presence of high levels of background chemicals, often called matrix interferences. Most crude extracts require some purification prior to analysis.

4. Quantification of pesticide levels is as important as identification of pesticides for the regulatory laboratory, because the regulation of pesticide use is based on the maximum residue level (MRL or tolerance) that may be present.

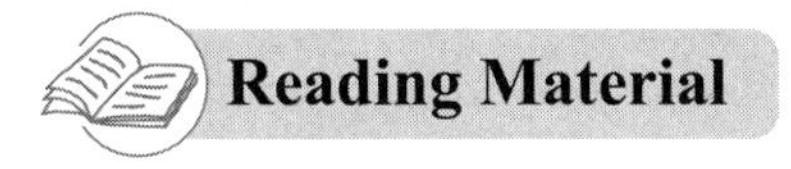

Analytical Methods

The aim of analyzing all of compounds both qualitatively and quantitatively in a single operation is unrealistic, and cannot be achieved by any single analytical method, as each method has specific limitations.

Chromatography

Chromatographic methods have the advantage of sensitivity and specificity as they can be combined with selective detectors (MS for GC, and photodiode array detection, MS and NMR in case of HPLC).

1. Gas Chromatography (GC)

Although GC requires that compounds be volatile, many are neither volatile nor thermostable, and therefore require derivatization (e. g. , acetylation, methylation, trimethylsilylation) prior to GC-analysis. The major advantage of GC is the large number of compounds that can be separated in a single analysis, and the large dynamic range. Coupling GC with MS renders the method highly selective. GC-MS has been used for more than 40 years for targeted approaches, and consequently extensive databases of MS data in combination with GC-retention time data exist. These targeted approaches may relate to terpenoids and fatty acids. The main drawbacks of GC analysis are that the sample preparation methods are quite elaborate, and the average time required for a single analysis is over 1 h.

2. High-Performance Liquid Chromatography (HPLC)

Most often, HPLC is combined with a PDA detector, which allows recording of the ultraviolet (UV) spectra of all compounds, thus adding selectivity to the system. Unfortunately, many metabolites have no UV-absorption. However, by combining HPLC with MS these compounds can be identified, simultaneously adding further selectivity to the system. The sensitivity of HPLC-MS varies for all compounds, with sugars notably difficult to detect. For all HPLC-applications, quantitative analysis can only be achieved by creating calibration curves for each individual compound.

As with all chromatographic methods, a major problem for HPLC is the reproducibility of separations. This occurs for several reasons, including small differences in parameters such as temperature and solvent quality, while matrix effects of the complex sample extract may influence the retention behavior of the compounds.

3. Capillary Electrophoresis (CE)

CE, which has also been applied to the analysis of metabolomes, has a very high resolution power, particularly for ionic compounds, when compared to other chromatographic methods. Moreover, its direct coupling to MS further contributes to the selectivity and provides a sensitive detection method. However, it is inherently limited to ionic compounds such as alkaloids, amino acids, or organic acids.

4. Thin-Layer Chromatography (TLC)

TLC has a large potential to detect diverse group of compounds using a wide array of detection methods, including UV and coloring reagents. Although high-performance thin-layer chromatography (HPTLC) has a higher efficiency, the low resolution and reproducibility remain the limiting factors in its use for metabolomics. For metabolic fingerprinting, TLC has been used for more than 45 years in the quality control of botanicals.

Spectroscopy

1. Mass Spectrometry (MS)

MS detects the molecular ion of a compound after ionization and, if desired, also the fragmentation of that molecular ion. MS analysis in metabolomics uses separation of the compounds based on their molecular weight. By using high-resolution MS, compounds with different elemental formula but similar mass can be separated. Assuming that the molecular weights of the metabolites which together form the metabolome are in the range of about 50 to 2000, MS can separate about 2000 compounds. In complex mixtures, individual compounds will be detected on the basis of their molecular weight, but clearly there will be an extensive overlap of compounds with the same molecular weight. However, by using tandem MS-MS, greater selectivity can be obtained through the specific fragmentation of each compound.

2. Nuclear Magnetic Resonance (NMR) Spectrometry

NMR spectrometry is a physical measurement of the resonance of magnetic nuclei such as ^{1}H and ^{13}C in a strong magnetic field. Because of small local differences in the magnetic field in a molecule, each proton and carbon will show a different resonance, resulting in a highly specific spectrum for each compound. Such a spectrum is highly reproducible, and in ^{1}H NMR the intensity of a signal of a proton is directly correlated with the molar concentration-that is, all compounds in a mixture can be compared, without the need for a calibration curve. One single internal standard can be used to make the quantitative analysis of all compounds in a mixture.

The main disadvantage of NMR is that it is less sensitive than the other methods mentioned. Typically, the amount of material needed for an NMR analysis is about 50 mg biomass (dry weight), though in practice this is similar to what is required for the other methods. For NMR the complete extract is needed for non-destructive analysis, after which the sample can be kept for further analysis. In contrast, in other methods only part of the extract is used for the analysis, though this is destructive and requires back-up material for future use. The sensitivity of NMR can be improved by increasing the measuring time, and also by improving the spectrometer, and this has been achieved in recent years. In NMR a higher field strength of the magnet improves sensitivity, and also influences the spectrum of a compound; the stronger the field the better the separation of the signals, and second-order spectra become first-order. This may hamper direct comparison of spectra recorded at different field strengths, but the problem can be resolved by using two-dimensional (2D)-NMR spectrometry. A 2D J-resolved spectrum shows in the second dimension the coupling constants of each proton signal as a set of signals at the chemical shift of the proton. By projection of these signals onto the chemical shift axis, all proton signals become singlets.

Selected from: Kayser O, Quax Wim J. Medicinal Plant Biotechnology. Weinheim: WILEY-VCH Verlag GmbH, 2007: 11-16.

Words and Expressions

photodiode /ˌfəutəu'daɪəud/ *n*.[电子] 光敏二极管，光电二极管
thermostable /θɜ:məʊ'steɪbəl/ *adj*.[化] 耐热的，热稳定的
derivatization /də'rɪvəˌtɪzeɪʃən/ 衍生化
acetylation /əˌseti'leiʃən/ 乙酰化
methylation /ˌmeθɪ'leɪʃən/ *n*.[有化] 甲基化，甲基化作用
trimethylsilylation /traɪmeθəlsɪlɪ'leɪʃən/ 三甲硅烷基化
terpenoid /'tə:pinɔid/ 萜类化合物，类萜
metabolite /mɪ'tæbəlaɪt/ *n*. 代谢物
calibration /ˌkælɪ'breɪʃn/ *n*. 标度，刻度，校准
capillary /kə'pɪləri/ *n*. 毛细管，*adj*. 毛状的，毛细作用的
electrophoresis /ɪˌlektrəfə'ri:sɪs/ *n*. 电泳
fingerprinting /'fɪŋgəprɪntɪŋ/ *n*. 指纹识别；*v*. 取……的指印，辨出
ionization /ˌaɪənaɪ'zeɪʃn/ *n*. 离子化，电离
metabolome /metæbəʊ'lɒm/ 代谢物组
resonance /'rezənəns/ *n*. 共鸣，谐振，共振，共振子
magnetic /mæg'netɪk/ *adj*. 磁的，有磁性的，有吸引力的
nucleus /'nju:kliəs/ *n*. 核，原子核，细胞核
proton /'prəutɒn/ *n*.[核] 质子
coupling constant 耦合常数 [恒量]
chemical shift 化学位移，异构位移

PART XI

PESTICIDE DEGRADATION

Unit 21 Environmental Fate

Assessing the transport and fate of pesticides in the environment is complicated. There are a myriad of transport and fate pathways at the local, regional, and global levels. Pesticides represent a diverse group of chemicals of widely varying properties and use patterns, and the environment is, of course, diverse in makeup and ever-changing, from one location to another and from one time to another.

In the past, particularly from roughly the 1940s to 1970s, knowledge of how pesticides and other chemicals behaved in the environment was obtained by retrospective analysis for these chemicals after they had been used for many years. By analyzing soil, water, sediment, air, plants, and animals, environmental scientists were able to piece together profiles of behavior. Dibromochloropropane (DBCP), ethylene dibromide (EDB), and chemicals with similar uses as soil nematicides and similar properties came to be recognized as threats to groundwater in general use areas. DDT and other chlorinated insecticides and organic compounds of similar low polarity, low water solubility, and exceptional stability threatened some aquatic and terrestrial animals because of their potential for undergoing bioaccumulation and their chronic toxicities. The chlorofluorocarbons (CFCs) and methyl bromide were found to be exceptionally stable in the atmosphere and able to diffuse to the stratosphere, where they entered into reactions that result in destruction of the ozone layer.

But as large a testimony as these examples and others were to the skill of environmental analytical chemists, environmental toxicologists, ecologists, and other environmental scientists in detecting small concentrations and subtle effects of chemicals, the retrospective approach is fraught with difficulty.

The trend from roughly the 1970s to the present has thus focused on ways to predict environmental behavior before the chemical is released. For pesticides, premarket testing of environmental fate and effects is now built into the regulatory processes leading to regulatory approval. The Environmental Fate Guidelines of the U.S. Environmental Protection Agency, for example, specify the tests and acceptable behavior required for registration of candidate pesticides in the United States. Europe, Canada, Australia, and other nations and economic organizations produce similar guidelines and test protocols to screen for potential adverse environmental behavior characteristics.

The Dissipation Process

Once a substrate (agriculture commodity, body of water, wildlife, soil, etc.) has been exposed to a chemical, dissipation processes begin immediately. The initial residue dissipates at an overall rate that is a composite of the rates of individual processes (volatilization, washing off, leaching, hydrolysis, microbial degradation, etc.). When low-level exposure results in the accumulation of residues over time, as in the case of bioconcentration of residues from water by aquatic organisms, the overall environmental process includes both the accumulation and dissipation phases. However, for simple dissipation, such as occurs in the application of pesticides and resulting exposure from residues in food or water or air, the typical result is that concentrations of overall residue decrease with time after end of exposure or treatment.

Environmental Compartments

Once a pesticide gains entry to the environment by purposeful application, accidental release, or waste disposal, it may enter one or more compartments. The initial compartment contacted by the bulk of the pesticide will be governed largely by the process of use or release. In time, however, residues will tend to redistribute and favor one or more compartments or media over others, in accordance with the chemicals physical properties, chemical reactivity, and stability characteristics and the availability and quality of compartments in the environmental setting where the use or release has occurred.

Some chemicals inherently favor water and thus will migrate to it when the opportunity arises. These are primarily chemicals of high water solubility and high stability in water, such as salts of carboxylic acid herbicides (2,4-D, MCPA, TCA). Others favor the soil or sediment compartment because they are preferentially sorbed to soil and they may lack other characteristics (volatility, water solubility) that lead to removal from soil. Examples include paraquat, which is strongly sorbed to the clay mineral fraction of soil, and highly halogenated pesticides such as DDT, toxaphene, and the cyclodienes, which sorb to and are stabilized in soil organic matter. Others, such as the fat-soluble organochlorines, favor storage in fatty animal tissue when the opportunity arises. Volatile chemicals such as methylbromide and telone (1, 3-dichloropropene) migrate to the air compartment. The elements of predicting environmental behavior, based on properties of the chemical of interest, become apparent through these well-established "benchmark" chemicals.

Activation-Deactivation

Most environmental transformations lead to products that are less of a threat to biota and the environment in general. The products may be less toxic than the parent or of lower mobility and persistence relative to the parent. They may, in short, be simply transient intermediates on the path to complete breakdown, that is, mineralization of the parent chemical. Thus, 2,4-D may degrade to oxalic acid and 2,4-dichlorophenol. The latter is of some concern, but it lacks the herbicidal toxicity of 2, 4-D and appears to be further degraded in most environments by sunlight, microbes, etc. Organophosphates can be hydrolyzed in the environment to phosphoric or thiophosphoric acid derivatives and a substituted phenol or alcohol. These products, in the case of most organophosphates, are not serious threats to humans or the environment.

In part because of the concern over environmental activation, the US EPA requires extensive information on the occurrence and toxicity of environmental and metabolic transformation products of pesticides submitted for registration. The tests include products of hydrolysis, photolysis, oxidation, and microbial metabolism in both laboratory and field tests. But, increasingly, regulations are also geared to products that might be formed during illegal use or during fires, explosions, spills, disinfection, and other situations that expose chemicals to conditions for which they were not intended. Unfortunately, not all such situations can be anticipated, requiring continual vigilance by the registrant and regulatory agencies as a part of product stewardship and environmental protection.

Chemical Reactions

By far the greatest complication in fully defining, or predicting, environmental fate processes arises with chemical degradation of the parent chemical into an array of degradation products. Abiotic reactions include hydrolyses and oxidations that occur in air, in water, and at the surface of soils, with or without light activation, but without intervention by microorganisms, plants, or animals. Biotic reactions are under enzymatic control, but both kinetics and degree of degradation vary considerably depending on whether plants, animals, or microorganisms are involved and, for microorganisms, the population density. The pathways of biotic and abiotic degradation are often the same, so that analysis for product profiles does not always help in detecting which type of process operates or predominates in a given setting. However, there are experimental techniques for differentiating biotic and abiotic reactions, just as there are for separately determining the operation of chemical and physical dissipaton processes and the type of process.

Any attempt at in-depth coverage of reaction pathways for pesticides here would be superficial and incomplete because of the variety of pesticides and, consequently, reaction products. A few generalizations, however, will be offered, followed by a discussion of reaction rates emphasizing microbial degradation and citation of relatively recent references to the subject.

Environmental reactions fall into just a relatively few reaction types: oxidations in air and on surfaces exposed to air, reductions in anaerobic sediments and soils, hydrolyses in water and moist soil, and conjugations in plants and animals.

Microbial Degradation

The important role played by microorganisms in degrading pesticides has been studied in great detail during the past 25 years or so. It is believed that degradation by microbes (bacteria, fungi, algae) accounts for over 90% of all degradation reactions in the environment and is the nearly exclusive breakdown pathway in most surface soils, near plant root zones, and in nutrient-rich waters including sewage ponds and sewage treatment systems. There are three types of bacterial chemical degradation possibilities, differentiated by the kinetics of breakdown of the chemical substrate.

Type a. Substrate degradation begins immediately upon contact. This indicates that the substrate can be used immediately as an energy source by the bacterial community, resulting in consumption of the substrate and a population increase among the degraders. Substrates that most resemble natural energy sources for bacteria—sugars and other simple carbohydrates, amino acids and simple proteins, aliphatic alcohols and acids, etc. —are the favored substrates in this class. Pesticides such as methomyl, glyphosate, sulfonylureas, and some prethyroids are included in this group because of their similarity to natural substrates in physicochemical properties and ability to act as nutrient sources.

Type b. Substrate degradation occurs after a lag period of bacterial acclimation. Group b includes substrate-bacteria combinations that require adaptation or acclimatization of the bacteria before the substrate can be used as an energy source. Adaptation may involve an induction of latent enzymes in the microbial community or a population shift to favor degrading species in a mixed microbial population, or some combination of the two. Once adaptation has occurred, the degradation rate increases until the substrate source is depleted, the same as in Type a systems. Type b systems predominate for most pesticide-microbe combinations, perhaps because most pesticides have structures different enough from natural food sources that enzyme systems are not immediately present in the natural microflora for deriving energy from them.

Type c. The substrate is not significantly usable as an energy source by microbial populations. If a chemical cannot be used as an energy source, even after prolonged periods of adaptation and addition of nutrients, water, air, etc. , it is regarded as "recalcitrant" to microbial degradation. Several chlorinated hydrocarbon insecticides, chlorodibenzodioxins, and polychlorinated biphenyls (PCBs) fall into this group, along with synthetic polymers and certain other organic chemicals. Recalcitrant chemicals can be transformed by microbes, but the transformation is incidental to the normal metabolism of acceptable substrates by the microorganisms.

Selected from: Wheeler W B. Pesticides in Agriculture and the Environment. New York: Marcel Dekker, Inc. , 2002: 125-148.

Words and Expressions

myriad /'mɪriəd/ *n*. 无数; *adj*. 无数的，种种的

retrospective /ˌretrə'spektɪv/ *adj*. 回顾的

sediment /'sedɪmənt/ *n*. 沉淀物，沉积

dibromochloropropane /dai'brəuməˌklɔːrə'prəupein/ [化] 二溴氯丙烷

ethylene dibromide [化] 二溴化乙烯
chlorinate /'klɔːrɪneɪt/ *vt*. [化] 使氯发生作用
polarity /pə'lærəti/ *n*. 极性
terrestrial /tə'restriəl/ *adj*. 陆地的，地球的，陆生的；*n*. 陆地生物，地球人
bioaccumulation /'baiəuˌkjuːmju'leiʃən/ *n*. 生物体内积累
stratosphere /'strætəsfɪə(r)/ *n*. [气] 同温层，最上层，最高阶段
chlorofluorocarbon /'klɔːrəuˌfluərəu'kɑːbən/ 氯氟烃
methyl bromide 甲基溴，溴化甲烷
testimony /'testɪməni/ *n*. 证词（尤指在法庭所作的），宣言，陈述
toxicologist /ˌtɒksɪ'kɒlədʒɪst/ *n*. 毒物学者
fraught /frɔːt/ *adj*. 充满……的
premarket /priː'mɑːkɪt/ *adj*. 上市前的
dissipation /ˌdɪsɪ'peɪʃn/ *n*. 消散，挥霍，浪费，狂饮
composite /'kɒmpəzɪt/ *adj*. 合成的，复合的；*n*. 合成物
volatilization /vɔˌlætilai'zeiʃən/ *n*. 挥发，发散
leaching /'liːtʃɪŋ/ *n*. 滤取，滤去
bioconcentration /ˌbaiəuˌkɔnsən'treiʃən/ *n*. 生物富集，生物浓缩
kinetics /kɪ'netɪks/ *n*. 动力学
sorb /sɔːb/ *n*. [植] 山梨树；*vt*. 吸附，吸收
compartment /kəm'pɑːtmənt/ *n*. 区划，间隔，舱，分立而不相属的机能、作用等
halogenate /hə'lɒdʒɪneɪt/ *v*. [化] 卤化
toxaphene /'tɒksəfiːn/ *n*. [化] 毒杀芬，八氯莰烯（用作杀虫剂）
methylbromide /meθɪlb'rɒmaɪd/ 溴甲烷
telone 1,3-二氯丙烯
benchmark /'bentʃmɑːk/ [计] *n*. 基准
transient /'trænziənt/ *adj*. 短暂的，瞬时的；*n*. 瞬时现象
biota /baɪ'əʊtə/ *n*. [生态] 生物区
mineralization /ˌminərəlai'zeiʃən/ *n*. 矿化，使含矿物
oxalic acid *n*. [化] 草酸
2,4-dichlorophenol 2,4-二氯苯酚
phosphoric /fɒs'fɒrɪk/ *adj*. [化] 磷的（尤指含五价磷的），含磷的
thiophosphoric *adj*. 硫代磷酸的
vigilance /'vɪdʒɪləns/ *n*. 警戒，警惕，失眠症，警惕性
abiotic /ˌeɪbaɪ'ɒtɪk/ *adj*. 无生命的，非生物的
anaerobic /ˌænə'rəubɪk/ *adj*. [微] 没有空气而能生活的，厌氧性的
superficial /ˌsuːpə'fɪʃl/ *adj*. 表面的，肤浅的，浅薄的
proficiency /prə'fɪʃnsi/ *n*. 熟练，精通，熟练程度
algae /'ældʒiː/ *n*. 藻类，海藻
exogenous /ek'sɒdʒənəs/ *adj*. [生] 外生的，[医] 外因的
carbohydrate /ˌkɑːbəu'haɪdreɪt/ *n*. [化] 碳水化合物，糖类
aliphatic /ˌælɪ'fætɪk/ *adj*. 脂肪族的，脂肪质的

methomyl /'metəmɪl/ 灭多威

sulphonylurea /ˌsʌlfənil'juəriə/ *n.*［药］磺脲（一种治糖尿病的口服药）

acclimatization /əˌklaɪmətaɪ'zeɪʃn/ *n.* 环境适应性

recalcitrant /rɪ'kælsɪtrənt/ *adj.* 反抗的；*n.* 反抗的人，顽抗者

chlorodibenzodioxin 氯苯氧化芑

polychlorinated 多氯化的

Notes

［1］2,4-D（2,4-dichlorophenoxyacetic acid）：氯化苯氧乙酸类除草剂，也可用作植物生长调节剂。

［2］MCPA（2-甲基-4-氯苯氧基乙酸）：农业上用作植物生长调节剂，也可作除草剂，易被根和叶部吸收和传导，用于小粒谷物、水稻、豌豆、草坪和非耕作区芽后防除多种一年生和多年生阔叶杂草。

Exercises

Ⅰ Answer the following questions according to the text.

1. Please give some examples to illustrate that with their inherent characteristics, the chemicals would respectively migrate to different compartments, such as air, water, soil or sediment, and other kind of compartments.

2. How many reaction types do environmental reactions fall into?

3. What are the three types of bacterial chemical degradation possibilities? Give some examples to state briefly each of them.

4. Why are some chemicals considered as "recalcitrant" to mircrobial degradation? Give some examples of such chemicals.

Ⅱ Translate the following English phrases into Chinese.

ozone layer	waste disposal	photolysis reaction	substrate degradation
low polarity	persistent pesticide	low water solubility	transient intermediate
piece together	dissipation process	degradation product	microbial metabolism

Ⅲ Translate the following Chinese phrases into English.

水生生物　高水溶性　氧化反应　非生物反应　生物体内积累

化学降解　生物降解　环境转移　环境毒理学　环境分析化学

Ⅳ Choose the best answer for each of the following questions according to the text.

1. Which of the following chemicals are exceptionally stable in the atmosphere and able to diffuse to the stratosphere where their reactions would destroy the ozone layer?

A. dibromochloropropane and ethylene dibromide

B. soil nematicides

C. chlorofluorocarbons and methyl bromide

D. DDT and other chlorinated insecticides

2. Which of the following is not a substrate?

A. body of life

B. agriculture commodity

C. soil

D. pesticide

3. Which of the following statements about the residue in the dissipation processes is wrong?

A. A plot of remaining residue concentration versus time is asymptotic to the time axis.

B. Theoretically, all environmental exposures lead to residues that have endless lifetimes.

C. Residue will approach zero but never be zero.

D. Residue will disappear entirely with time going by.

4. Which of the following chemicals will migrate to the soil or sediment compartment?

A. salts of carboxylic acid herbicides

B. paraquat, toxaphene and the cyclodienes

C. methylbromide and telone

D. fat-soluble organochlorines

5. All of the following chemicals present less serious threats to humans or the environment except ________.

A. oxalic acid

B. phosphoric or thiophosphoric acid derivatives

C. organophosphates

D. a substituted phenol or alcohol

6. Which of the following statements about biotic and abiotic reaction/degradation is true?

A. Pathways of biotic and abiotic degradation are usually different.

B. Abiotic reactions include hydrolyses and oxidations with intervention by microorganisms, plants, or animals.

C. Biotic reactions are under enzymatic control.

D. It is always helpful to analyze product profiles to detect which type of process (biotic and abiotic reaction/degradation) operates or predominates in a given setting.

7. For which group of chemicals substrate degradation will begin immediately upon contact?

A. chlorodibenzodioxins and polychlorinated biphenyls

B. methomyl, glyphosate and sulfonylureas

C. some chlorinated hydrocarbon insecticides

D. None of the above

Ⅴ Translate the following short paragraphs into Chinese.

1. The initial residue dissipates at an overall rate that is a composite of the rates of individual processes, such as volatilization, washing off, leaching, hydrolysis, microbial degradation, etc.

2. Residues will tend to redistribute and favor one or more compartments or media over others, in accordance with the chemicals physical properties, chemical reactivity, and stability characteristics and the availability and quality of compartments in the environmental setting where the use or release has occurred.

3. Organophosphates can be hydrolyzed in the environment to phosphoric or

thiophosphoric acid derivatives and a substituted phenol or alcohol.

4. It is believed that degradation by microbes (bacteria, fungi, algae) accounts for over 90% of all degradation reactions in the environment and is the nearly exclusive breakdown pathway in most surface soils, near plant root zones, and in nutrient-rich waters including sewage ponds and sewage treatment systems.

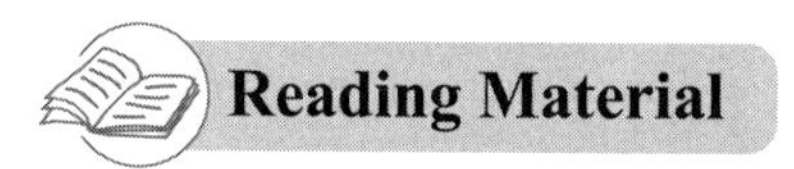

Effects of Soil Parameters on Pesticides Fate

Soil Humidity

Higher soil humidity has been associated with faster pesticide degradation and volatilization. When soils become wet after a period of drought, pesticides are lost from the soils through rapid volatilization due to the replacement of the adsorbed pesticides with water molecules. Shelton and Parkin (1991) reported that decreasing levels of soil moisture have consistently been observed to retard or inhibit rates of microbial metabolism, although the pattern or degree of inhibition is dependent upon the substrate, microbial process, and the nature of the soil microflora. These authors further discussed that fluctuations in soil moisture content may affect rates of biodegradation indirectly by affecting the bioavailability of the substrate compound. For example, they noted that concentrations of soluble, and hence bioavailable, carbofuran increased from 15.5 to 25.5 pg/mL with decreasing soil moisture content. Similarly, Chelinho et al. (2014) noted a lower toxicity of carbofuran in a soil with a higher moisture content and attributed this to increased degradation and greater dilution of the pesticide in the soil pore water.

Clay Content

In line with the above, slower pesticide degradation of pesticides in wet soils is usually explained from inactivation of aerobic soil microbes responsible for their degradation. In air-dry soil, an accelerated degradation by clay catalyzed hydrolysis has been reported for chlorpyrifos and acephate. However, Laabs et al. (2002b) did not find any substantial influence of contrasting clay contents of different tropical soils (13% versus 47% clay) on the dissipation dynamics of the studied pesticides.

pH

Acidic soils contribute to a greater desorption and hence exposure and persistence of polar herbicides as compared to alkaline or neutral soils. Degradation rates are also known to be often affected by prevailing pH levels. Laabs et al., for example, related a greater and lower persistence of endosulfan and triazines to an acidic reaction of the studied tropical soils (pH 4~5, 1 mol/L KCl), leading to an enhanced and slower degradation, respectively. However, after reviewing tropical and non-tropical studies on the environmental risks of pesticides, Sanchez-Bayo and Hyne noted that tropical soil pH in these studies were acidic in 59% of the cases studied, with the remainder being alkaline (23%) or neutral (18%). They further concluded that acidic soils are common in Brazil and Southeast Asian countries; soils from some African countries and Caribbean islands tend to be neutral, whereas those from the Indian subcontinent are alkaline in most cases. Subsequently, a possible effect of pH on pesticide fate in tropical soils depends on the local conditions and may differ between in-field and off-field areas.

Soil Organic Carbon Content and Pesticide K_{oc}

Tropical soils have been considered to be rather deficient in organic carbon (OC) because of the intense microbial degradation that takes place uninterrupted throughout the entire year, whereas soils in temperate regions are rich in OC because their microbial activity is restricted only to the warm seasons. This relatively low OC content would hence imply a limited binding capacity of most tropical soils for pesticides. This would suggest a higher bioavailability of pesticides in tropical soils. Since lower soil OC contents also imply a greater pesticide mobility through runoff and leaching, available (i. e. dissolved) pesticide residue levels may, however, more rapidly decline leading to reduced exposure of tropical soil organisms. In addition, this would also influence the pesticide exposure via ingestion of dead organic material on which many edaphic organisms feed, although this has been suggested already.

It is questionable, however, whether tropical agricultural soils contain consistently lower OC contents than their temperate counterparts. Although the rate of organic decomposition is five times greater in the tropics, annual addition of organic carbon to the soil is also five times greater in tropical udic environments than in temperate udic environments. But the type of organic matter (OM) and its carbon content (typically about 58% for temperate soils) may differ between tropical and temperate regions. Sanchez-Bayo and Hyne discussed that a large proportion (58%) of the tropical soils tested in the studies they reviewed had normal OC contents (i. e., between 1.5% and 3%). In addition, these authors indicated that about 45% of the herbicides and fungicides examined showed higher K_{oc} (sorption constant for binding to organic carbon) values than they normally have in soils from temperate regions. Especially in the case of insecticides, this would indicate a trend towards stronger adsorption onto tropical soils, with 77% of the compounds showing slightly higher K_{oc} values, although none of them to a significant extent.

Anion Exchange Capacity

Anion exchange is more likely to occur in tropical soils since they contain significant higher quantities of positively charged adsorption surfaces in the form of aluminium and iron (hydr) oxides than in temperate soils. Anion exchange is known to play a significant role in the adsorption of several pesticides. Hyun and Lee, for example, demonstrated that anion exchange of prosulfuron accounted for up to 82% of the overall sorption in the pH range 3~7. These authors further concluded that since both hydrophilic and hydrophobic sorption of prosulfuron decreased with increasing pH, addition of fertilizer and lime amendments may enhance the potential for off-site leaching of recently applied acidic pesticides.

Conclusions

From the above, it appears that there are no inherent differences in pesticide fate due to soil properties uniquely possessed by tropical soils, as it has been concluded by several other authors. Tropical soils themselves defy easy categorization, and their properties are as varied in nature as those from temperate zones. In the literature there is also consensus that DT_{50} values can easily differ by a factor of 10 for the same compound applied in the same climatic zone but on different soils. The exposure of soil organisms to pesticides depends on the regional characteristics of the soils, rather than on climatic characteristics of the tropics. Such influence evidently also depends on pesticide properties and may for example be greater for those depending on aerobic microbial degradation.

Selected from: Daam M A, Chelinho S, Niemeyer J C, *et al*. Environmental risk assessment of pesticides in tropical terrestrial ecosystems: Test procedures, current status and future perspectives. Ecotoxicology and Environmental Safety, 2019: 181, 534-547.

Words and Expressions

bioavailability /'baiəuə,veilə'biləti/ *n*. 生物利用度，生物药效率
carbofuran /,kɑːbəu'fjuərən/ *n*. 克百威
chlorpyrifos /k'lɔːpɪrɪfəuz/ *n*. [农药] 毒死蜱
acephate /ə'sefət/ 乙酰甲胺磷
glaciate /'gleɪsɪeɪt/ *v*. 使冰冻，被冰覆盖，使受冰川作用
endosulfan /,endəu'sʌlfən/ *n*. [农药] 硫丹（烈性杀虫剂）
triazine /'traɪəziːn/ *n*. [有化] 三嗪
prosulfuron 氟磺隆，三氟丙磺隆，氟丙磺隆
hydrophilic /,haɪdrə(ʊ)'fɪlɪk/ *adj*. [化学] 亲水的（等于 hydrophilous）
hydrophobic /,haɪdrə'fəubɪk/ *adj*. 疏水的，疏水性，斥水性

Unit 22 Microbial and Enzymatic Degradation

Microbes

Soil microbes perform substantial role in litter degradation, promotion of plant growth, nutrient cycling and the degradation of pollutants and pesticides. Such functions are significant to the farmers and society; therefore, the applications of pesticide which hamper the ecosystem services need to be minimized. The effect of pesticide on the activity of soil microbes has been observed by testing the efficiency of carbon utilization and nitrification. The success of the nitrification test against the pesticide side effect is likely due to the fact that no soil fungi are known to be involved in the nitrification process and that only few species of bacteria are able to perform the process. Recently, the expressed functional genes (*amoA* and *amoB*) involved in the nitrification process have been used to quantify the effect of pesticides on nitrification. One other microbial-driven ecosystem function that has been found to be affected by pesticide use is the degradation of pesticides by increasing the pesticide degrading microbial populations.

For minimization of soil pollution and soil toxicant in agricultural fields, various treatment methods (i. e., filling of contaminated land sites, recycling, pyrolysis and incineration) have been used. The physico-chemical remedial strategies to clean up sites containing these compounds are neither cost effective nor adequate enough. Therefore, biological method would be one of the promising methods to exploit the ability of microorganisms for removal of pollutants from the contaminated sites.

This alternative treatment strategy (bioremediation) could be effective, minimally hazardous, economical, environment-friendly, and versatile. The microbial based (with the help of fungi, bacteria, actinomycetes and viruses) degradation may ease the pesticide problem which is a natural process, does not produce toxic intermediates more effective technology, less hazardous and environment-friendly. Soil microorganisms collectively decompose various xenobiotic compounds and return the item to the mineral salts which are utilized by plants. Hence, they play an important role in the dissipation of xenobiotic pesticides in the soil. Microorganisms especially bacteria, due to their continuous exposure to such environmental stress, develop a genetically determined system against toxicant. The highly diverse microbial communities present in fresh and marine water, sewage and soils are able to transform a wide range of organic chemicals. Compared to chemical methods, the microorganisms that survive in the environment, utilize the pesticides for obtaining the energy seem to be the best candidates for bioremediation of pesticide present in soil or water.

Pesticide degradation may be carried out by one or more ways described as biotransformation, biomineralization, bioaccumulation, biodegradation, bioremediation and cometabolism. Microbial degradation involves the use of effective microorganism for the enzymatic breakdown of toxic pesticides (organochlorines, organophosphates and carbamates) into a nontoxic compounds or it may be defined as; microbial mediated degradation of complex organic compound (pesticide) into simple inorganic chemicals on the contaminated sites; soil, ground water, sludge's, industrial water system and gas. Thus, the microorganisms which play substantial role in the in situ removal and detoxification of these toxicants from the concerned environment are of special significance.

Large number of bacterial strains; viz. *Arthobacter*, *Pseudomonas*, *Ralstonia*, *Rhodococcus*, *Bacillus*, *Nocardiopsis*, *Cryptococcus* and *Acetobacter* have been isolated from different parts of the world with amazing property to degrade xenobiotic contaminants. These microorganisms are appropriately termed as nature's biodegraders or scavengers because they recycle most natural waste materials into harmless compounds. The endogenous microbes remove pollutants from soils by their extraordinary ability to use a wide variety of xenobiotics as sole energy and carbon source. Such microorganisms are highly adaptable and have the capability to degrade the recalcitrant compounds through evolution of new genes, which encode enzymes that can use these compounds as their primary substrates.

The pesticide degrading genes of soil bacteria reside on plasmids (anti-catabolic plasmid) and encode the pollutant-degrading enzymes. These catabolic plasmids have been reported from different species of *Pseudomonas*, *Alcaligenes*, *Actinobacter*, *Cytophaga*, *Moraxella*, *Klebsiella* and *Arthrobacter*. *Pseudomonas aeruginosa* MRM6, *Pseudomonas aeruginosa* SFM4 and *Acromobacter xylosoxidans* ES9 degrade the endosulfan (a broad range of pesticide). The bacteria *Clostridium sporogenes* and *Bacillus coli* produce trace amounts of benzene and mono-chlorobenzene from lindane and can further metabolize it by producing the CO_2 from sub-merged soils. *Pseudomonas putida* and *Acinetobacter* have ability for dehalogenate the chloro-aromatic compounds. A group of brown-rot fungi (*Gloeophyllum trabeum*, *Fomitopsis pinicola*, and *Daedalea dickinsii*) degrades the DDT. *Mucor circinelloides* and *Galactomyces geotrichum* isolated from cattle manure compost (CMC) and *Pleurotus ostreatus*, from spent mushroom waste (SMW), degrade the DDT.

Enzyme

The microbial degradation of pesticide involves various types of enzyme. These methods are based on various types of genes coded with plasmid or chromosomal DNA. Pesticidal degradative genes in microbial strains have been found to be located on plasmids, transposons, and chromosomes. Genetically engineered microorganisms have substantial ability to degrade the pesticide because they secrete various specific enzymes which have a specific catabolic gene into plasmids. The recombinant DNA studies have made it possible to develop DNA and RNA probes that are being used to identify microbes from diverse environmental communities with a unique ability to degrade pesticides.

Pesticide degradation is a process that involves the complete rupture of an organic compound into inorganic constituents of pesticide by pesticide degrading microbes. The biodegradation involves transferring the substrates and products within a well-coordinated microbial community, a process referred to as metabolic cooperation under natural environments. Enzyme based pesticide degradation process is an innovative treatment technique for removal of pesticide from polluted environments. Fungi and bacteria are considered as the extracellular enzyme-producing microorganisms for degradation of pollutant. White rot fungi have been used as promising bioremediation agents, especially for compounds not readily degraded by bacteria. The production and secretion of extracellular enzymes in microbes by environmental effects could be of interest. Some of these extracellular enzymes are involved in lignin degradation, such as lignin peroxidase, manganese, peroxidase, laccase and oxidases. Transferases, isomerases, hydrolases and ligases enzymes are responsible for the biodegradation of the pesticides. These enzymes catalyze metabolic reactions including hydrolysis, oxidation, addition of an oxygen to a double bound, oxidation of an amino group (NH_2) to a nitro group, addition of a hydroxyl group to a benzene ring, dehalogenation, reduction of a nitro group (NO_2) to an amino group, replacement of a sulfur with an oxygen, metabolism of side chains, ring cleavage.

The metabolism of pesticides involves three phase process; in first phase process the initial properties of a parent compound are transformed through oxidation, reduction, or hydrolysis to generally produce a more water-soluble and usually a less toxic product than the parent, in second phase process involves conjugation of a pesticide to a sugar or amino acid, which increases the water solubility and reduces toxicity compared with the parent pesticide, and in third phase process helps in conversion of second phase metabolites into secondary conjugates, which are also non-toxic. In these processes fungi and bacteria are involved producing intracellular or extra cellular enzymes including hydrolytic enzymes, peroxidases, oxygenases, etc.

Hydrolases are a broad group of enzymes which helps in pesticide biodegradation. It catalyzes the hydrolysis of several major biochemical classes of pesticide (esters, peptide bonds, carbon-halide bonds, ureas, thioesters, etc.). Organophosphate hydrolases enzyme isolated from *Pseudomonas diminuta* and *Flavobacterium* sp.. The first isolated phosphotriesterase belongs to the *Pseudomonas diminuta* strain MG, this enzyme encoded by a gene called *opd* (organophosphate-degrading) and shows a highly catalytic activity towards organophosphate pesticides. These enzymes specifically hydrolyze phosphoester bonds, such as P—O, P—F, P—NC, and P—S. Other microbial enzymes such as organophosphorus hydrolase (OPH; encoded by the *opd* gene), methyl-parathion hydrolase (MPH; encoded by the *mpd* gene), and hydrolysis of coroxon (HOCA; encoded by the *hocA* gene), were isolated from *Flavobacterium* sp., *Plesimonas* sp. strain M6 and *Pseudomonas moteilli*, respectively.

Esterase have been cloned and proteins have sequenced from several microorganism, e. g. two ferulic acid esterases from *Aspergillus tubingensis*, a cephalosporin esterase from the yeast *Rhodosporidium toruloides*, a chrysanthemic acid esterase from *Arthrobacter globiformis*, and several other esterase from *Pseudomonas fluorescens* strain. These enzymes used for biodegradation of various types of organophosphate, and other herbicide and insecticide. Cytochrome P450 oxidoreductases helps in degradation of atrazine, norflurazon and chlortoluron from soils.

Cytochrome CYP76B1 isolated from *Helianthus tuberosus* was cloned into tobacco and Arabidopsis. This enzyme has capability to catalyse the oxidative dealkylation of phenylurea herbicides such as linuron, chlortoluron and isoproturon. Prokaryotic cytochrome P450 s isolated from *Pseudomonas putida* and *Sphingobium chlorophenolicum* have great potential to degrade chlorinated pentachlorobenzene and hexachlorobenzene. Toluene dioxygenases (TOD) enzyme isolated from *P. putida* F_1 that is highly used as degradation of toluene, polychlorinated hydrocarbons, ethylbenzene and *p*-xylene.

Haloalkane dehalogenases (Dh1A) enzyme incoded gene *LinB* isolated from *Xanthomonas thobacter autotrophicus* GJ10; haloacetate dehalogenase (DehH1) from *Moraxella* sp. B; and 2-hydroxymuconic semialdehyde hydrolase (DmpD) from *Pseudomonas* sp. CF600 which is responsible for degradation of hexachlorocyclohexane. Oxidoreductase is an enzyme that catalyses oxidation and reduction reaction. It is important enzyme like glyphosate oxidase (GOX) for glyphosate degradation. This enzyme is a flavoprotein amine oxidase from *Pseudomonas* sp. LBr that catalyses the oxidation of glyphosate to form aminomethylphosphonate and releases the keto acid glyoxylate.

Mycobacterium tuberculosis ESD strains secrete endosulfan monooxygenase II enzyme which degrade the β-endosulfan to the monoaldehyde and hydroxyether but transforms α-endosulfan to the more toxic endosulfan sulfate. Alternatively, hydrolysis of endosulfan in some bacteria (*Pseudomonas aeruginosa*, *Burkholderia cepaeia*) yields the less toxic metabolite endosulfan diol. Endosulfan can spontaneously hydrolyze to the diol in alkaline conditions, so it is difficult to separate bacterial from abiotic hydrolysis. *Phanerochaete chrysosporium* and *T. versicolor* have ability to degrade simazine, dieldrin and trifluralin pesticides independently by laccase activity.

Selected from: Verma J P, Jaiswal D K, Sagar R. Pesticide relevance and their microbial degradation: a-state-of-art. Rev Environ Sci Biotechnol, 2014, 13: 429-466.

Words and Expressions

nitrification /ˌnaitrifiˈkeiʃən/ *n*.［化学］硝化作用，氮饱和，氮化合

pyrolysis /paɪˈrɒlɪsɪs/ *n*.［化学］热解，［化学］高温分解

incineration /ɪnˌsɪnəˈreɪʃn/ *n*. 焚化，烧成灰

bioremediation /ˈbaiəuriˌmiːdiˈeiʃən/ *n*. 生物修复，生物降解，生物除污

environment-friendly *adj*. 环保的，环境友好的

actinomycete /ˌæktɪnəˈmaɪsiːt/ *n*. 放射菌类

biotransformation /ˌbaɪəutrænsfəˈmeɪʃən/ *n*.［环境］生物转化

biomineralization /ˈbaiəuˌminərəlaiˈzeiʃən/ *n*.［地质］生物矿化

bioaccumulation /ˈbaiəuˌkjuːmjuˈleiʃən/ *n*. 生物体内积累

biodegradation /ˈbaɪəuˌdegrəˈdeɪʃən/ *n*.［生物］生物降解，生物降解作用

cometabolism /ˈkɒmɪtæbəlɪzəm/ *n*. 辅助代谢，次级代谢

viz /viz/ *abbr*. 即，也就是（videlicet）

Arthobacter 节杆菌

Pseudomonas /ˌsjuːdə(ʊ)ˈməʊnəs/ *n.* 假单胞菌

Ralstonia 劳尔氏菌属，青枯菌属

Rhodococcus /ˌrəudəuˈkɔkəs/ *n.* 红球菌属

Bacillus /bəˈsɪləs/ *n.* 杆菌，芽孢杆菌

Nocardiopsis 拟诺卡氏菌属

Cryptococcus /ˌkriptəˈkɔkəs/ *n.* 隐球菌，隐球酵母

Acetobacter /əˌsiːtəˈbæktə/ *n.* 醋菌属，醋杆菌属

scavenger /ˈskævɪndʒər/ *n.* 食腐动物，[助剂] 清除剂，拾荒者

recalcitrant /rɪˈkælsɪtrənt/ *n.* 顽抗者，不服从的人；*adj.* 反抗的，反对的

catabolic /ˌkætəˈbɔlik/ *adj.* 分解代谢的，异化的

Alcaligenes /ˌælkəˈlidʒəˌniːz/ *n.* [微] 产碱杆菌属

Actinobacter 杆菌，不动杆菌

Cytophaga /saiˈtɔfəgə/ *n.* [微] 噬细胞菌属

Moraxella /ˌmɔːrækˈselə/ *n.* 莫拉克斯氏菌属

Klebsiella /ˌklebziˈelə/ *n.* 克雷伯氏菌属

Arthrobacter /ˈæθrəˌbæktə/ *n.* 节细菌属

Pseudomonas aeruginosa 铜绿假单胞菌

Acromobacter xylosoxidans 木糖氧化无色杆菌，氧化无色杆菌

Clostridium sporogenes 生孢梭菌，梭状芽孢杆菌

lindane /ˈlɪndeɪn/ *n.* 林丹（一种农用杀虫剂）

Pseudomonas putida 恶臭假单胞菌，假单胞菌

Acinetobacter /eiˌsinitəuˈbæktə/ *n.* 不动杆菌属

dehalogenate 脱卤素

Gloeophyllum trabeum 密粘褶菌，密褐褶孔菌

Fomitopsis pinicola 松生拟层孔菌，红缘层孔菌

Daedalea dickinsii 白肉迷孔菌

Mucor circinelloides 枝毛霉，毛霉菌

Galactomyces geotrichum 半乳糖霉菌，地霉属

transposon /trænˈspəuzɔn/ *n.* 转位子

chromosome /ˈkrəuməsəum/ *n.* 染色体

Genetically engineered microorganisms 遗传工程微生物，基因工程微生物

lignin /ˈlɪgnɪn/ *n.* [木] 木质素

lignin peroxidase 木质素过氧化物酶

manganese /ˈmæŋgəniːz/ *n.* [化学] 锰

laccase /ˈlækeis/ *n.* [生化] 虫漆酶，[生化] 漆酶

peptide bond 肽键

thioester /ˌθaiəuˈestə/ *n.* 硫酯，[有化] 硫代酸酯

Pseudomonas diminuta 缺陷假单胞菌

Flavobacterium sp. 黄杆菌属

methyl-parathion 甲基对硫磷

Plesimonas sp.　邻单胞菌
Pseudomonas moteilli　蒙特氏假单胞菌
ferulic acid　阿魏酸
Aspergillus tubingensis　塔宾曲霉
Rhodosporidium toruloides　圆红冬孢酵母
chrysanthemic acid　菊酸
Arthrobacter globiformis　球形节杆菌
Pseudomonas fluorescens　荧光假单胞菌
norflurazon /nɔːrf'luəreɪzəʊn/ 氟草敏
Helianthus tuberosus　菊芋
dealkylation /diːælki'leiʃən/ *n*. 脱烷基化作用
phenylurea /fenɪluː'riə/ 苯基脲，苯基尿素
linuron /'lɪnjuərɒn/ *n*. [农药] 利谷隆
chlortoluron /k'lɔːtəluərɒn/ *n*. [农药] 绿麦隆
isoproturon /aɪsəʊprəʊ'tuərɒn/ *n*. [农药] 异丙隆
prokaryotic /prəʊˌkæriː'ɒtɪk/ *adj*. 原核的
Sphingobium chlorophenolicum　氯酚鞘氨醇杆菌
pentachlorobenzene /pen'tækləɾəbenziːn/ 五氯苯
hexachlorobenzene /'heksəˌklɔːrə'benziːn/ *n*. [农药] 六氯苯，六氯代苯
toluene dioxygenases　甲苯加双氧酶
polychlorinated hydrocarbon　多氯烃，多氯化合物
ethylbenzene /ˌeθil'benziːn/ *n*. [有化] 乙苯
p-xylene /'piːz'aɪliːn/ *n*. 对二甲苯
haloalkane dehalogenases　卤代烷脱卤酶
Moraxella sp.　莫拉克斯氏菌属
flavoprotein /ˌfleɪvə'prəutiːn/ *n*. [生化] 黄素蛋白
Pseudomonas sp.　假单胞菌
hexachlorocyclohexane /'heksəˌklɔːrəˌsaikləu'h-eksein/ *n*. [有化] 六氯环己烷，六六六
aminomethylphosphonate　氨甲基磷酸盐
Mycobacterium tuberculosis　结核杆菌，结核分枝杆菌
Burkholderia cepaeia　洋葱伯克霍尔德菌
Phanerochaete chrysosporium　黄孢原毛平革菌，黄孢平革菌，白腐病病菌
simazine /sɪ'mɑːzɪn/ *n*. [农药] 西玛津
dieldrin /'diːldrɪn/ *n*. [农药] 狄氏剂
trifluralin /traɪ'fluərəlɪn/ *n*. [农药] 氟乐灵

Notes

[1] coroxon：蝇毒磷，分子式为 $C_{14}H_{16}ClO_5PS$，对人畜高毒，也称香豆磷（反刍动物驱虫剂）。蝇毒磷系非内吸性杀虫剂，对双翅目害虫特别有效，还用于防治体外寄生虫，防治皮蝇效果显著。

Exercises

Ⅰ Answer the following questions according to the text.

1. Why could bioremediation be effective? not the chemical methods.
2. What is the advantage of microbial degradation?
3. What are the methods of microbial degradation?
4. Which bacteria can degrade pesticides? Please give examples.
5. Which three processes do involved the metabolism of pesticides?

Ⅱ Translate the following English phrases into Chinese.

litter degradation　environmental stress　soil pollution　recombinant DNA
carbon utilization　nontoxic compound　carbon source　xenobiotic compound

Ⅲ Translate the following Chinese phrases into English.

微生物群落　无机盐　内源微生物　硝化作用　养分循环
土壤有毒物　酶变败　母体化合物　生物修复　染色体 DNA

Ⅳ Choose the best answer for each of the following questions according to the text.

1. What treatment methods have been adopted to minimize soil poisons in agricultural fields?

A. filling of contaminated land sites　B. recycling
C. pyrolysis and incineration　D. all of the above

2. What method would be one of the promising methods for removal of pollutants?

A. microorganisms　B. physical methods
C. chemical methods　D. incineration

3. Which microbes are most effective at degrading pesticides?

A. fungi　B. bacteria　C. actinomycetes　D. viruses

4. Where were pesticide degrading genes found in microbial strains?

A. plasmids　B. transposons　C. chromosomes　D. all of the above

5. Which of the following reactions are involved in primary metabolism of pesticides?

A. oxidation　B. reduction　C. hydrolysis　D. all of the above

Ⅴ Translate the following short paragraphs into Chinese.

1. Soil microbes perform substantial role in litter degradation, promotion of plant growth, nutrient cycling and the degradation of pollutants and pesticides.

2. For minimization of soil pollution and soil toxicant in agricultural fields, various treatment methods (i. e. , filling of contaminated land sites, recycling, pyrolysis and incineration) have been used.

3. The physico-chemical remedial strategies to clean up sites containing these compounds are neither cost effective nor adequate enough. Therefore, biological method would be one of the promising methods to exploit the ability of microorganisms for removal of pollutants from the contaminated sites.

4. Pesticide degradation may be carried out by one or more ways described as biotransformation, biomineralization, bioaccumulation, biodegradation, bioremediation and cometabolism.

5. Microbial mediated degradation of pesticide into simple inorganic chemicals on the contaminated sites; soil, ground water, sludge's, industrial water system and gas.

6. These enzymes catalyze metabolic reactions including hydrolysis, oxidation, addition of an oxygen to a double bound, oxidation of an amino group (NH_2) to a nitro group, addition of a hydroxyl group to a benzene ring, dehalogenation, reduction of a nitro group (NO_2) to an amino group, replacement of a sulfur with an oxygen, metabolism of side chains, ring cleavage.

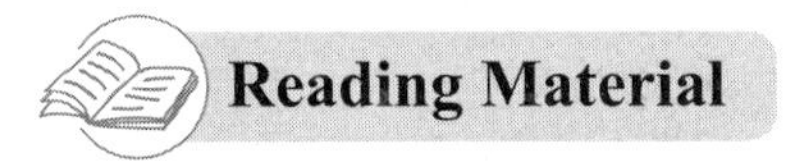

Microorganisms Involved in Degradating Pesticides

Microorganisms such as bacteria, fungi, or algae can be applied in removal of organic or inorganic chemicals from soil. These potential living organisms are able to degrade toxic chemical, belonging either to wild type or is genetically modified. Microbes can completely metabolize the pollutants or alter the chemical structure that leads to facilitate its degradation and make local environment effectively contaminant-free. The diverse range of chemicals and chemically derived materials such as organic complexes, heavy metals, and chlorinated and nonchlorinated pesticides has been metabolized by the microbes. Microbes utilized pesticides as carbon and nutrient source by their own diverse enzymatic machinery. A general estimation states that 1 g of soil sample contains more than one hundred million bacteria, which includes 5000～7000 inimitable strains along with more than 10,000 colonies of fungi. Application of indigenous microbes or natural attenuation is effective technics applied for the elimination of toxic contaminants from the natural environmen.

Bacterial-assisted Biodegradation

Bacteria belonging to genera *Flavobacterium*, *Arthrobacter*, *Azotobacter*, *Burkholderia*, and *Pseudomonas* are known to degrade the pesticides. The pesticide's biodegradation includes oxidizing the parent compound into carbon dioxide, water, or less toxic compounds. Pesticide degradation by microbes relies not only on the enzyme system but also on conditions such as temperature, pH, and nutrients. Some of the pesticides are readily degraded, but the presence of anionic species in some compounds hinders from their degradation. Degradation mode of pesticides and process differs between the compound to be degraded and bacterial species involved in the process.

Pseudomonas sp. and *Klebsiella pneumoniae* possess hydrolytic enzymes that degrade organophosphorus compounds and neonicotinoids. Limiting environmental conditions sometimes poses partial degradation of pesticides leading to the accumulation of metabolites in soil that further inhibits the microbial population in soil. Like dichlorodiphenyltrichloroethane (DDT) that is partially degraded to dichlorodiphenyldichloroethane (DDD) and dichlorodiphenyldichloroethylene (DDE) metabolites that are even more toxic than the parent compound. By optimizing the environmental condition, degradation of pesticides and release of metabolites in soil can be controlled by the introduction of microbes in soil that are capable of degrading pesticides.

Fungi-assisted Biodegradation

Pesticides biodegradation follows distinct pathways depending on its nature, environmental condition and primarily depends on microbe type. Bacteria and fungi play an important role in the degradation process, but fungi basically biotransform pesticides into nontoxic substances by introducing structural changes and then releasing them into soil where it is receptive of further degradation by bacteria. Among them, *Phanerochaete chrysosporium* degrades a broad range of pesticides. Lindens, atrazine, diuron, terbuthylazine, metalaxyl, DDT, gamma-hexachlorocyclohexane, dieldrin, aldrin, heptachlor, chlordane, mirex are several classes of pesticides that are degraded by white-rot fungi to different extent. Similarly, several other fungi such as *Agrocybe semiorbicularis*, *Auricularia auricula*, *Coriolus versicolor*, *Dichomitus squalens*, *Flammulina velutipes*, *Hypholoma fasciculare*, *Pleurotus ostreatus*, *Stereum hirsutum*, and *Avatha discolor* have demonstrated their capacity to degrade different groups of pesticides such as phenylamide, triazine, phenylurea, dicarboximide, chlorinated, and organophosphorus compounds.

Enzymatic-assisted Biodegradation

Enzymes generated in crops and microbes present in soil during various metabolic processes are the key to the bioremediation of pesticides. Enzymes play a main role in the biodegradation of any xenobiotic and can refurbish pollutants at a noticeable pace and can help in restoring the polluted environment in future. Enzymes also contribute to the degradation of pesticides, both in the target organism, through inherent detoxification processes and developed metabolic resistance, and in the wider setting, through soil and water microorganism biodegradation. In particular, fungal enzymes oxidoreductase, laccase, and peroxidases are used prominently to remove polyaromatic hydrocarbons contaminants in fresh, marine, or terrestrial water. Organophosphorus degrading enzymes have been studied in great detail over the years, and several bacteria, fungi, and cyanobacteria have also been isolated that utilizes organophosphorus compounds as carbon and nutrients source.

Selected from: Sun Y, Kumar M, Wang L, *et al*. Biotechnology for soil decontamination: opportunity, challenges, and prospects for pesticide biodegradation. Bio-Based Materials and Biotechnologies for Eco-Efficient Construction, 2020: 261-283.

Words and Expressions

Flavobacterium /ˌfleivəu'bæktiəriəm/ *n*. 黄杆菌属，黄质菌属，产黄菌属
Arthrobacter /'æθrəˌbæktə/ *n*. 节细菌属
Azotobacter /ə'zəutəuˌbæktə/ *n*. 固氮菌
Burkholderia 伯克霍尔德菌

Klebsiella pneumoniae 克雷伯氏肺炎菌，肺炎克雷伯菌
dichlorodiphenyltrichloroethane *n*.［有化］二氯二苯三氯乙烷
dichlorodiphenyldichloroethane *n*.［有化］二氯二苯二氯乙烷
dichlorodiphenyldichloroethylene *n*.［有化］二氯二苯二氯乙烯
atrazine /ˈætrəziːn/ *n*.［农药］莠去津
diuron /ˈdaɪjʊrɒn/ *n*.［农药］敌草隆（一种剧毒性除草剂）
terbuthylazine *n*.［农药］特丁津
metalaxyl *n*.［农药］甲霜灵
gamma-hexachlorocyclohexane 丙体-六六六
aldrin /ˈɔːldrɪn/ *n*.［农药］艾氏剂
heptachlor /ˈheptəklɔː/ *n*.［有化］七氯（一种杀虫剂）
chlordane /ˈklɔːdeɪn/ *n*.［农药］氯丹（一种强力杀虫剂）
mirex /ˈmaɪreks/ *n*. 灭蚁灵
Agrocybe semiorbicularis 半球盖田头菇
Auricularia auricula 黑木耳，木耳
Coriolus versicolor 彩绒革盖菌，云芝
Dichomitus squalens 污叉丝孔菌
Flammulina velutipes 金针菇，毛柄金钱菌
Hypholoma fasciculare 簇生黄韧伞
Pleurotus ostreatus 糙皮侧耳，糙皮侧耳菌
Stereum hirsutum 毛韧革菌，粗毛硬革菌
Avatha discolor 宇夜蛾
dicarboximide 二甲酰亚胺

PART XII
TOXICOLOGY

Unit 23 Toxicology

Terminology

The Greek word τοξιΧον (toxicon) was used for poisonous liquids in which arrowheads were dipped. The word toxicology, derived from this word, has been used as the name of the science within human medicine that describes the effect of poisons on humans. The definition includes uptake, excretion, and metabolism of poisons (toxicokinetics), as well as the symptoms and how they develop (toxicodynamics). We can say that the toxicodynamics tell us what the toxicants do to the organisms; and toxicokinetics, what the organism does with the substance. Toxicology also includes the legislation enforced to protect the environment and human health, and the risk assessments necessary for this purpose. Today a toxicologist is not exclusively working with the species Homo sapiens or model organisms like rats, but all kinds of organisms.

The term ecotoxicology is defined as "the science occupied with the action of chemicals and physical agents on organisms, populations, and societies within defined ecosystems. It includes transfer of substances and interactions with the environment". Ecotoxicology is sometimes used synonymously with environmental toxicology; however, the latter also encompasses the effects of environmental chemicals and other agents on humans. Because the basic chemical and physical processes behind the interaction between biomolecules and chemicals are independent of the type of organism, it is not necessary to have a too rigid division between the various branches of toxicology.

Dose-response Curves

The law of mass action tells us that the amount of reaction products and the velocity of a chemical reaction increase with the concentrations of the reactants. This means that there is always a positive relation between dose and the degree of poisoning. A greater dose gives a greater concentration of the toxicant around the biomolecules and therefore more serious symptoms because more biomolecules react with the toxicant and at a higher speed. This simple and fundamental law of mass action is one of the reasons why a chemist does not believe in homeopathy. Paracelsus (1493—1541) said "All substances are poisons; there is none which is not a poison. The right dose differentiates a poison from a remedy". The connection between dose or concentration of the toxicant and the severity of the symptoms is fundamental in toxicology.

If we have many groups with a high number of individuals and then plot the relative response against the dose, we very often get an oblique S-shaped graph, with an inflection point at 50% response. The graph may be made symmetrical by plotting log dose instead of dose. Furthermore, the S-shaped graphs can be changed into straight lines by transforming the responses to probit response. We then presuppose that the sensitivity of the organisms has a normal distribution, which predicts that most individuals have average sensitivity, a few are very robust, very few are almost resistant, and some have high sensitivity.

The log transformation of dose or concentration is easily done with a calculator. Using the formula for the inverse normal distribution in the data sheet Excel, one can easily do the calculation of the probit values. The mean or median is set to 5 and the standard deviation to 1. By writing the relative response into the formula, Excel will return the probit value. Note that the probit of 0.5 (50% response) is 5, and the probit of 0.9 (90% response) is 6.282.

It cannot be expressed as a simple function, and some mathematical skill is necessary to interpret its meaning. Therefore, the much simpler logit transformation $L=\ln[P/(1-P)]$ is often used. The logit values (L) can be calculated from the relative response values (P). The logit transformation also gives almost straight lines if the sensitivity is normally distributed. The most serious problem with dose-response graphs, however, is not this mathematical inconvenience. The low reproducibility is a more serious problem. As an example, if you know exactly the LD_{50} (lethal dose in 50% of the population) and give this to 10 animals, the probability that 5 die is only 0.246. The confidence intervals of the responses for the same dose, or for the doses calculated to give a specified response (e.g., LD_{50}), will be large and are not easily calculated without special data programs.

LD_{50} and Related Parameters

The statistical problems in making good dose-response curves can only be overcome by using many organisms in the experiment. Abetter method may be to determine one dose, for instance, the dose that is expected to kill or harm 50% of the individuals, and not to construct a graph. This can be done satisfactorily with much fewer individuals. The latter method is definitely better when studying vertebrates. Most countries have strict legislation concerning the use of vertebrates in research, and it is difficult to get permission to do experiments involving hundreds of animals. Furthermore, most vertebrates suitable for research are expensive. Therefore, we seldom find graphs of dose-response relationships on mammals in the more recent scientific literature. More often, we find a value called LD_{50} that can be determined with reasonable accuracy by using few individuals. LD_{50} is the dose expected to kill half of the exposed individuals. Sometimes we are interested in determining the doses that kill 90% or 10%, etc., and these doses are called LD_{90} and LD_{10}, respectively. They can easily be determined from a dose-response curve, but these values are less accurate than LD_{50}. If we study endpoints other than death, we use the term ED_{50} (effective dose in 50% of the population), and if we study concentrations and not doses, we use the terms LC_{50} (lethal concentration in 50% of the population) and EC_{50} (effective concentration in 50% of the population). Protocols for determination of LD_{50} for rodents are available in order to minimize the number of animals necessary for a satisfactory determination. According to Commission of the European Communities' Council Directive 83/467/EEC, 20 rats may be sufficient for an appropriate LD_{50} determination. LD_{50} values are often given as milligram of toxicant per kilogram of body weight of the test animals, assuming that twice as big a dose is necessary to kill an animal of double weight. It is therefore easier to compare toxicity data from different species, life stage, or sex. The LD_{50} values or the related values should not be taken as accurate figures owing to the intrinsic nature of these parameters, as well as the difficulties of determination. Even if you know the exact LD_{50} value, for example, of parathion to mice ($LD_{50}=12$ mg/kg), and give these doses to a group of animals, for instance, $n=10$, the probability that $r=5$ will die is only $P=0.246$. This can easily be calculated from the binomial formula. However, you can be confident that between 1 and 9 will die ($P=0.998$). LD_{50} values are therefore very useful if you do not need to know the exact number of fatalities, but merely want to describe the toxicity of a compound by one figure. Table 1 shows how toxicants are classified according to their LD_{50}.

Table 1 Common Classification of Substances

Toxicity Class	LD_{50}/(mg/kg)	Examples, LD_{50}/(mg/kg)
Extremely toxic	Less than 1.0	botulinum toxin: 0.00001
		aldicarb: 1.0
Very toxic	1～50	parathion: 10
Moderately toxic	50～500	DDT: 113～118
Weakly toxic	500～5000	NaCl: 4000
Practically nontoxic	5000～15,000	glyphosate: 5600
		ethanol: 10,000
Nontoxic	More than 15,000	water

Acute and Chronic Toxicity

An important distinction has to be made between acute and chronic toxicity. Substances that are eliminated very slowly and therefore accumulate if administered in several small doses over a long time may, when the total dose is large enough, cause symptoms. A good example is cadmium that accumulates in the kidneys. Another example is organophosphates that in repeated small doses eventually inhibit acetylcholinesterase more than 80%, which will produce neurotoxic symptoms. Because the inhibition is partly irreversible, many small doses may cause poisoning even though the poison itself does not accumulate. Other poisons (e.g., ethanol) may be given in large, but sublethal doses for years before any sign of chronic toxicity is observed (liver cirrhosis), whereas the acute toxicity results in well-known mental disturbances. In many cases, acute or subacute doses may give chronic symptoms or effects many years after poisoning (cigarette smoking and cancer) or effects in the following generation (stilbestrol may give vaginal cancer in female offspring at puberty).

We use the following terms:

Acute dose — The dose is given during a period shorter than 24 h.

Subacute dose — The doses are given between 24 h and 1 month.

Subchronical dose — The doses are given between 1 and 3 months.

Chronical dose — The doses are given for more than 3 months.

These terms apply to mammals, whereas the times are shorter for short-lived animals or plants used in tests. The dose of a pesticide toward a pest will usually be acute, whereas the dose that consumers of sprayed food will be exposed to is chronic.

Interactions

One toxicant may be less harmful when taken together with another chemical. If we use blindness as an endpoint for methanol poisoning, then whisky or other drinks that contain ethanol would reduce the toxicity of methanol considerably. When ethanol is present, methanol is metabolized more slowly to formaldehyde and formic acid, which are the real harmful substances. Ethanol is therefore an important antidote to methanol poisoning. Malathion is an organophosphorus insecticide with low mammalian toxicity, but if administered together with a small dose of parathion, its toxicity increases many times. This is because paraoxon, the toxic metabolite of parathion, inhibits carboxylesterases that would have transformed malathion into the harmless substance malathion acid. In another example, a smoker should not live in a house contaminated with radon. Although smoking and radon may both cause lung cancer on their own, smoke and radon gas interact and the incidence will increase 10 times or more when smokers are exposed to radon.

Two or more compounds may interact to influence the symptoms in an individual and change the number of individuals that get the symptoms in question. Interaction may be caused by simultaneous or successive administration.

Selected from: Stenersen J. Chemical Pesticides: Mode of Action and Toxicology. Florida: CRC Press LLC, 2004: 9-27.

Words and Expressions

toxicology /ˌtɒksɪˈkɒlədʒi/ *n*.［毒物］毒物学，［毒物］毒理学

toxicon *n*. 毒素

excretion /ɪkˈskriːʃn/ *n*. 排泄，排泄物，分泌，分泌物

toxicokinetics /tɒksɪkəʊkaɪˈnetɪks/ *n*.［医］毒物代谢动力学

toxicodynamics /tɒksɪkɒdaɪˈnæmɪks/ *n*.［医］毒物作用动力学

ecotoxicology /ˌiːkəʊtɒksɪˈkɒlədʒɪ/ *n*. 生态毒理学

toxicant /ˈtɒksɪkənt/ *n*. 有毒物，［毒物］毒药；*adj*. 有毒的

homeopathy /ˌhəʊmiˈɒpəθi/ *n*.［临床］顺势疗法，同种疗法

oblique /əˈbliːk/ *n*. 倾斜物；*adj*. 斜的，不光明正大的；*vi*. 倾斜

binomial /baɪ'nəʊmiəl/ *adj*. 二项式的；*n*. 二项式

cadmium /'kædmiəm/ *n*. [化] 镉

acetylcholinesterase /ə'siːtil,kəuli'nestəreis/ *n*. 乙酰胆碱酯酶

neurotoxic /,njuərəu'tɒksɪk/ *adj*. 毒害神经的

sublethal /,sʌb'liːθəl/ *adj*. 亚致死（量）的

cirrhosis /sə'rəusɪs/ *n*. [医] 硬化

subacute /,sʌbə'kjuːt/ *adj*. 稍尖的，亚急性的

stilbestrol /stɪl'biːstrɒl/ 己烯雌酚

malathion /,mælə'θaɪən/ 马拉硫磷

parathion /,pærə'θaɪən/ *n*. [农药] 对硫磷，硝苯硫磷酯

paraoxon /,pærə'ɔksɔn/ *n*. 对氧磷

Notes

[1] Homo sapiens：智人，是人属下的唯一现存物种，形态特征比直立人更为进步。分为早期智人和晚期智人。早期智人曾叫古人，生活在距今 25 万～4 万年前，主要特征是脑容量大，在 1300mL 以上；眉嵴发达，前额较倾斜，枕部突出，鼻部宽扁，颌部前突，一般认为是由直立人进化来的。晚期智人是解剖结构上的现代人，大约从距今四五万年前开始出现，与前者形态上的主要差别在于前部牙齿和面部减小，眉嵴减弱，颅高增大，到现代人则更加明显。晚期智人臂不过膝，体毛退化，有语言和劳动，有社会性和阶级性。

[2] radon：氡，是一种化学元素，符号 Rn。氡通常的单质形态是氡气，为无色、无嗅、无味的惰性气体，具有放射性。氡的化学性质不活泼，不易形成化合物。氡没有已知的生物作用。当人吸入体内后，氡发生衰变的阿尔法粒子可对人的呼吸系统造成辐射损伤，引发肺癌。而建筑材料是室内氡的最主要来源，如花岗岩、砖砂、水泥及石膏之类，特别是含放射性元素的天然石材，最容易释出氡。

Exercises

Ⅰ Answer the following questions according to the text.

1. what are the toxicology and ecotoxicology? and what are their differences and connections?

2. How do you plot a dose response curve?

3. What is LD_{50}, LD_{90} and LD_{10}?

4. What are the acute and chronic toxicity? and what are their differences and connections?

5. Learn to use toxicity classification standard to determine the toxicity of pesticides.

Ⅱ Translate the following English phrases into Chinese.

standard deviation　Botulinum toxin　extremely toxic　very toxic
lethal concentration　binomial formula　chronical dose　acute dose
dose-response curve　chronic toxicity　acute toxicity　lethal dose

Ⅲ Translate the following Chinese phrases into English.

低毒　亚慢性剂量　亚致死剂量　正态分布　置信区间　质量作用定律

中毒　环境毒物学　亚急性剂量　有效剂量　有效浓度　无毒

Ⅳ Choose the best answer for each of the following questions according to the text.

1. According to toxicity classification standard, which of the following is belong to very toxic?

A. less than 1.0 mg/kg　　B. 1.0～50.0 mg/kg

C. 50.0～500.0 mg/kg　　D. 500～5000 mg/kg

2. According to toxicity classification standard, which of the following is belong to weakly toxic?

A. less than 1.0 mg/kg　　B. 1.0～50.0 mg/kg

C. 50.0～500.0 mg/kg　　D. 500～5000 mg/kg

3. If the symptoms of poisoning are present in the offspring, what is the toxicity?

A. acute toxicity　　B. subacute toxicity

C. chronic toxicity　　D. environmental toxicity

4. What is the dose expected to kill a quater of the exposed individuals?

A. LD_{90}　　B. LD_{50}　　C. LD_{25}　　D. LD_{10}

5. if malathion is use together with a small dose of parathion, its toxicity increases many times. what is interactions?

A. synergism　　B. antagonism　　C. additive effect　　D. all of the above

Ⅴ Translate the following short passages into Chinese.

1. The word toxicology has been used as the name of the science within human medicine that describes the effect of poisons on humans. The definition includes uptake, excretion, and metabolism of poisons (toxicokinetics), as well as the symptoms and how they develop (toxicodynamics).

2. All substances are poisons; there is none which is not a poison. The right dose differentiates a poison from a remedy.

3. LD_{50} is the dose expected to kill half of the exposed individuals. Sometimes we are interested in determining the doses that kill 90% or 10%, etc., and these doses are called LD_{90} and LD_{10}, respectively.

4. In many cases, acute or subacute doses may give chronic symptoms or effects many years after poisoning (cigarette smoking and cancer) or effects in the following generation.

Classification of Toxicants

We will approach the toxicology of pesticides from the biochemist's perspective. Because the cells in all organisms are very similar, it is possible to classify toxicants into roughly seven categories according to the type of biomolecule they react with.

Enzyme Inhibitors

The toxicant may react with an enzyme or a transport protein and inhibit its normal function. Enzymes may be inhibited by a compound that has a similar, but not identical structure as the true substrate; instead of being processed, it blocks the enzyme. Typical toxicants of this kind are the carbamates and the organophosphorus insecticides that inhibit the enzyme acetyl cholinesterase. Some extremely efficient herbicides that inhibit enzymes important for amino acid synthesis in plants, e. g. , glyphosate and glufosinate, are other good examples in this category.

Enzyme inhibitors may or may not be very selective, and their effects depend on the importance of the enzyme in different organisms. Plants lack a nervous system and acetylcholinesterase does not play an important role in other processes, whereas essential amino acids are not produced in animals. Glyphosate and other inhibitors of amino acid synthesis are therefore much less toxic in animals than in plants, and the opposite is true for the organophosphorus and carbamate insecticides. Sulfhydryl groups are often found in the active site of enzymes. Substances such as the Hg^{2+} ion have a very strong affinity to sulfur and will therefore inhibit most enzymes with such groups, although the mercury ion does not resemble the substrate. In this case, the selectivity is low.

Disturbance of the Chemical Signal Systems

Organisms use chemicals to transmit messages at all levels of organization, and there are a variety of substances that interfere with the normal functioning of these systems. Toxicants, which disturb signal systems, are very often extremely potent, and often more selective than the other categories of poisons. These toxicants may act by imitating the true signal substances, and thus transmit a signal too strongly, too long lasting, or at a wrong time. Such poisons are called agonists. A typical agonist is nicotine, which gives signals similar to acetylcholine in the nervous system, but is not eliminated by acetylcholinesterase after having given the signal. Other quite different agonists are the herbicide 2,4-D and other aryloxyalkanoic acids that mimic the plant hormone auxin. They are used as herbicides. An antagonist blocks the receptor site for the true signal substance. A typical antagonist is succinylcholin, which blocks the contact between the nerve and the muscle fibers by reacting with the acetylcholine receptor, preventing acetylcholine from transmitting the signal. Some agonists act at intracellular signal systems. One of the strongest man-made toxicants, 2,3, 7,8-tetrachlorodibenzodioxin, or dioxin, is a good example. Organisms use a complicated chemical system for communication between individuals of the same species. These substances are called pheromones. Good examples are the complicated system of chemicals produced by bark beetles in order to attract other individuals to the same tree so that they can kill them and make them suitable as substrates. Man-made analogues of these pheromones placed in traps are examples of poisons of this category. The kairomons are chemical signals released by individuals of one species in order to attract or deter individuals of another. The plants' scents released to attract pollinators are good examples.

Signals given unintentionally by prey or a parasite host, which attract the praying or parasitizing animal, are important. A good example is CO_2 released by humans, which attracts mosquitoes. The mosquito repellent blocks the receptors in the scent organ of mosquitoes.

Toxicants that Generate very Reactive Molecules that Destroy Cellular Components

Most redox reactions involve exchange of two electrons. However, quite a few substances can be oxidized or reduced by one-electron transfer, and reactive intermediates can be formed. Oxygen is very often involved in such reactions. The classical example of a free radical-producing poison is the herbicide paraquat, which steals an electron from the electron transport chain in mitochondria or chloroplasts and delivers it to molecular oxygen. The superoxide anion produced may react with hydrogen superoxide in a reaction called the Fenton reaction, producing hydroxyl radicals. This radical is extremely aggressive, attacking the first molecule it meets, no matter what it is. A chain reaction is started and many biomolecules can be destroyed by just one hydroxyl radical. Because one paraquat molecule can produce many superoxide anions, it is not difficult to understand that this substance is toxic. Copper acts in a similar way because the cupric ion (Cu^{2+}) can take up one electron to make the cuprous cation (Cu^{+}) and give this electron to oxygen, producing the superoxide anion ($O_2^- \cdot$).

Free radical producers are seldom selective poisons. They work as an avalanche that destroys membranes, nucleic acids, and other cell structures. Fortunately, the organisms have a strong defense system developed during some billion years of aerobic life.

Weak Organic Bases or Acids that Degrade the pH Gradients across Membranes

Substances may be toxic because they dissolve in the mitochondrial membrane of the cell and are able to pick up an H^+ ion at the more acid outside, before delivering it at the more alkaline inside. The pH difference is very important for the energy production in mitochondria and chloroplasts, and this can be seriously disturbed. Substances like ammonia, phenols, and acetic acid owe their toxicity to this mechanism. Selectivity is obtained through different protective mechanisms. In plants, ammonia is detoxified by glutamine formation, whereas mammals make urea in the ornithine cycle. Acetic acid is metabolized through the citric acid cycle, whereas phenols can be conjugated to sulfate or glucuronic acid. Phenols are usually very toxic to invertebrates, and many plants use phenols as defense substances.

Toxicants that Dissolve in Lipophilic Membranes and Disturb Their Physical Structure

Lipophilic substances with low reactivity may dissolve in the cell membranes and change their physical characteristics. Alcohols, petrol, aromatics, chlorinated hydrocarbons, and many other substances show this kind of toxicity. Other, quite unrelated organic solvents like toluene give very similar toxic effects. Lipophilic substances may have additional mechanisms for their toxicity. Examples are hexane, which is metabolized to 2,5-hexandion, a nerve poison, and methanol, which is very toxic to primates.

Toxicants that Disturb the Electrolytic or Osmotic Balance or the pH

Sodium chloride and other salts are essential but may upset the ionic balance and osmotic pressure if consumed in too high doses. Babies, small birds, and small mammals are very sensitive. Too much or too little in the water will kill aquatic organisms.

Strong Electrophiles, Alkalis, Acids, Oxidants, or Reductants that Destroy Tissue, DNA, or Proteins

Caustic substances like strong acids, strong alkalis, bromine, chlorine gas, etc., are toxic because they dissolve and destroy tissue. Many accidents happen because of carelessness with such substances, but in ecotoxicology they are perhaps not so important. More interest is focused on electrophilic substances that may react with DNA and induce cancer. Such substances are very often formed by transformation of harmless substances within the body. Their production, occurrence, and protection mechanisms will be described in some detail later.

Selected from: Stenersen J. Chemical Pesticides: Mode of Action and Toxicology. Florida: CRC Press LLC, 2004: 14-17.

Words and Expressions

glufosinate /glʌfəuzɪ'neɪt/ 草铵膦
sulfhydryl /sʌl'haidril/ *n*. 巯氢基
agonist /'ægənɪst/ *n*. 收缩筋，兴奋剂
aryloxyalkanoic 芳氧基烷基
succinylcholin /sʌksɪnɪl'kəulɪn/ 丁二酰胆碱 琥珀酰胆碱
acetylcholine /ˌæsɪtaɪl'kəuliːn/ *n*. [有化] 乙酰胆碱
2,3,7,8-tetrachlorodibenzodioxin 2,3,7,8-四氯二苯并二噁英，二噁英
dioxin /dai'ɔksin/ *n*. 二噁英，二氧（杂）芑
pheromone /'ferəməun/ *n*. 信息素
pollinator /'pɑlɪnetə/ *n*. 传粉者，传粉媒介，传粉昆虫 授花粉器
redox /'riːdɒks/ *n*. 氧化还原反应，[化学] 氧化还原剂
superoxide anion [生物物理] 超氧阴离子，超氧化物阴离子
ornithine /'ɔːnɪθiːn/ *n*. [生化] 鸟氨酸
lipophilic /ˌlɪpə'fɪlɪk/ *adj*. 亲脂性的，亲脂的
chlorinated hydrocarbons 氯化烃类，氯代烃
2,5-hexandion 2,5-己二酮
primates /prai'meitiːz/ *n*. 灵长类
osmotic /ɒz'mɒtɪk/ *adj*. 渗透性的，渗透的

Unit 24 Toxicity of Pyrethroids

Pyrethroids are divided into type Ⅰ and type Ⅱ depending on whether or not they contain α-cyano-3-phenoxybenzyl. Type Ⅰ pyrethroids induce tremors or convulsions, while type Ⅱ pyrethroids exposure primarily causes choreoathetosis. Moreover, qualitative differences in sodium channel modifications often lead to different toxicological symptoms.

Cytotoxicity

Methods for assessing cytotoxicity include MTT assay, LDH assay, the MTT assay is living cells, the detection wavelength is 500～600 nm; the detection target of the LDH method is damaged or dead cells, and the detection wavelength is 490～620 nm. The major *in vitro* models of cytotoxicity tests of pyrethroids are SH-SY5Y, HepG2, Caco-2 human cells. The SH-SY5Y neuroblastoma cell line has been widely used in neurological research and has become a priority target for testing the cytotoxicity of pesticides. This is because human dopaminergic neuroblasts-SY5Y cells not only have the biochemical properties of many neurons, but are also more vulnerable to oxidative damage mediated by reactive oxygen species (ROS) than other neuronal cells. Besides, studies have also shown that human hepatoma HepG2 cell line may be used to research pyrethroid metabolism and toxicity. Lastly, human intestinal Caco-2 cells have been widely used as models of intestinal barriers and are commonly used for absorption researches.

In addition, *in vitro* assays that have been used extensively with pyrethroids also include the yeast expressing steroid receptor (YES) and chemical activated luciferase gene expression (CALUX) assays. The YES assay is defined by first integrating the DNA of the hormone receptor sequence into the yeast genome to express the corresponding hormone receptor, and then using the specific chromogenic substrate to generate color change when the active ligand binds to the receptor. Additionally, CALUX assay that responds to estrogen chemistry by inducing the expression of firefly luciferase has been stably transfected with an estrogen-responsive luciferase reporter plasmid. Particularly, CALUX assay expressing mRNA of estrogen receptor (ER) isoforms alpha and beta is superior to YES assay typically only transfected with ER alpha.

Toxicity to Mammals

Human exposure to pyrethroids is divided into dietary intake and non-dietary intake. In humans, exposure to pyrethroids can cause acute symptoms such as nausea and vomiting, dyspnea, cough, bronchospasm, dermal effects. The dermal effects mean that pyrethroids-exposure induces localized paresthesias and allergies through inhalation or direct dermal contact, but these symptoms are usually caused by type Ⅱ pyrethroids containing an α-cyano group. Significant evidence supported that pyrethroids exposure affected male reproductive system by reducing sperm quality, damaging sperm DNA and disrupting reproductive hormones.

It's reported that residues of permethrin, bifenthrin, cypermethrin in vegetable (spinach, cauliflower, beans, etc.) reached 4～1100, 2～300, 1～500 ng/g, respectively. And the frequent consumption of foods with pyrethroid residues increases the risk of developmental and neurological diseases in children aged 3～11 years. Dietary ingestion followed by non-dietary ingestion are the dominate exposure routes for children. However, there are not many studies on the chronic toxicity of pyrethroids. Nowadays, the most common chronic toxicity tests usually include only neurotoxicity, reproductive toxicity, and endocrine disruption.

Most importantly, assessing the toxicity of pesticides and their metabolites to mammals is critical to the development of agrochemicals. The environmental pollution of pyrethroids can be easily found by using invertebrates and fish as models. However, direct testing of novel candidate compounds for toxicity to large mammals such as humans is often unethical, so rodents are used instead. And studies have shown that reliable animal models of in vivo toxicity testing are mice/rats.

1. Acute oral Toxicity

"Acute Oral Toxicity-Acute Toxic Class Method" described in Organisation for Economic Co-operation and Development (OECD/OCDE) Guideline for Testing of Chemical (2001), "National Food Safety Standard-Acute Oral Toxicity Test (GB/T 15193.3—2014)" mentioned in China national standard document both are available methods for monitoring acute oral toxicity of pesticides. The acute oral toxicity test can provide health information from animals taking the analytes orally within a short period of time, serve as the basis for acute toxicity classification, provide the basis for dose selection and observation indicators for further toxicity tests, and initially estimate the target organs and possible toxicity mechanism.

Several commonly used acute toxicity test design methods mainly include Horn's method, limit test, up-down procedure (UDP), Korbor's method, logarithmic plotting method. Among them, the Korbor's method and logarithmic plotting method have less stringent requirements and are suitable for most samples. The Korbor's method is not only applicable to a wide range of applications, but also requires fewer animals than other methods. However, the later data processing process of Korbor's method is also cumbersome. Statistics on the dose, dose log, number of animals, number of animal deaths, and percentage of animal deaths are required and the LD_{50} should be calculated group by group according to the relevant formula.

The acute oral toxicity of pyrethroids in mice/rats is described. β-cyfluthrin (LD_{50} = 1.27～35.48 mg/kg weight) are very toxic (LD_{50} = 1.0～50.0 mg/kg weight) to mammals. In addition, deltamethrin (LD_{50} = 35.00～150.00 mg/kg weight) are moderate poisonous (LD_{50} = 51.0～500.0 mg/kg weight) to mammals; while resmethrin (LD_{50} = 750.0～2000.0 mg/kg weight) are low toxic (LD_{50} = 501.0～5000.0 mg/kg weight) to mammals, and tetramethrin (LD_{50} = 15.0～22.0 g/kg weight) is actually non-toxic (LD_{50} =

5000. 0 mg/kg weight) to mammals. The LD_{50} of the above compounds in rats/mice are calculated as the lowest value. Moreover, after administration, the longer the observation period, the smaller the LD_{50}. And the larger the age of the mouse, the larger the LD_{50}. For example, when the observation period is 1, 7, 14, 21, 28 d, the LD_{50} of cypermethrin for female Wistar rats of eight weeks old is 41. 70, 5. 96, 2. 98, 1. 99, 1. 50 mg/kg weight. When the age of the male Wistar rats is 1, 2, 3 weeks old, the LD_{50} of cypermethrin to rats is 14. 90, 27. 10, 49. 30 mg/kg weight, respectively.

2. Chronic Toxicity

In the chronic toxicity test, most of the drugs are administered by gavage. When the period is 18～60 d, the maximum dose does not exceed 1/10 LD_{50}, and when the dose is large, the period becomes shorter. Besides, the experimental design of administration by gavage mostly uses vegetable oil (like corn oil, olive oil, peanut oil and so on) as the solvent, which indicates that a number of pyrethroids are soluble in vegetable oil. Lastly and most importantly, the main symptoms of the chronic toxicity of pyrethroids (i) induce neurotoxicity, such as promote anxiety-like behavior, induce of oxidative stress, induce developmental neurotoxicity, lead to neurotoxic side effects, induce memory, emotional and tyrosine hydroxylase immunoreactivity alterations, cerebral necrosis in the hippocampus and the striatum, significantly increase the intensity of neurotoxicity and liver dysfunction, cause long-term decreases of neuronal sodium channel expression, markedly delayed growth development of neonatal offspring during lactation show clinical signs of neurotoxicity; (ii) induce lymphoid cells apoptosis, such as lymphoid cells, nerve cell, pubertal Leydig cell; (iii) lead to hepatotoxic; (iv) lead to nephrotoxic; (v) lead to reproductive toxicity, such as markedly delayed growth development of neonatal offspring during lactation, significantly decrease the fertility.

Toxicity to Aquatic Organisms

1. Acute Toxicity

One of the methods for testing toxicity to aquatic organisms is the fish embryo acute toxicity (FET) test adopted by the Organisation for Economic Co-operation and Development (OECD), 2013. Another alternative methods is to determine the acute toxicity of substances in zebrafish juveniles in accordance with the China national standard (GB/T 13267—1991 Water quality-Determination of the acute toxicity of substance to a freshwater fish) which is formulated and adopted with reference to the international standard (ISO 7346 1-3) and is the current standard in China.

The aquatic toxicity of pyrethroids is related to the type of pesticides and the type of aquatic organisms. The same pyrethroid has different toxicity to different aquatic organisms, and different types of pyrethroids have different toxicity to the same aquatic organism. The smaller the value of lethal concentration 50 (LC_{50}), the higher the toxicity of the compound. For example, the relative toxicity of permethrin to different aquatic organisms is: *Hyalella azteca* (96 h-LC_{50}= 0. 02 mg/L) > *zebrafish* (96 h-LC_{50}= 2. 50 mg/L) > *Procloeon* sp. (96 h-LC_{50}= 10. 45 mg/L). The relative toxicities of different pyrethroids to the same aquatic organism *H. Azteca* are: λ-cyhalothrin (96 h-LC_{50} = 0.64 mg/L) > permethrin (96 h-LC_{50} = 2.50 mg/L) > bifenthrin (96 h-LC_{50} = 9.3×10^{-3} mg/L) > tetramethrin (96 h-LC_{50} = 46.00 mg/L). Additionally, pyrethroids are hydrophobic (high K_{ow}) and tend to sorb to particulate matter (high K_{oc}), so they are commonly found in sediments of aquatic systems. This indicates that pyrethroids in water and sediment also have a certain toxic effect on aquatic organisms.

Besides, the pyrethroids of different spatial configurations have different toxicity to aquatic organisms, which is similar to the insecticidal activity of the compounds. *Trans*-permethrin (96 h-LC_{50} = 0.52, 0.74 mg/L) is slightly more toxic to aquatic invertebrates *Daphnia magna* and *Ceriodaphnia dubia* than *cis*-permethrin (96 h-LC_{50} = 0.54, 0.79 mg/L). This means that in screening for the insecticidal activity of pyrethroids, attention should also be paid to its toxicity to aquatic organisms in order to select novel pyrethroids with low aquatic toxicity. Also, the aquatic toxicity of pyrethroids is temperature dependent. Toxicity increases at low temperatures, and this has been incorporated into toxicity identification evaluation (TIE) procedures to determine the cause of ambient toxicity.

2. Chronic Toxicity

It is worth mentioning in this section that in addition to lethal/acute toxicity that sublethal effects such as the ones mentioned in mammals (e.g., neurotoxicity, reproductive toxicity) are also of concern in aquatic organisms. Moreover, pyrethroids not only deposit in sediments, but also bioretain in aquatic organisms. In addition, pyrethroids and their metabolites may also be endocrine disruptors and immune interference agents. Particularly, the metabolites of pyrethroids may have higher endocrine disruption capabilities than their parents. Besides, like other particulate organics, the presence of any particulate matter can also interfere with the effects of pyrethroid pesticides and other hydrophobic contaminants. And pyrethroids sorb to particulate material, reducing dissolved concentrations.

Selected from: Zhu Q, Yang Y, Zhong Y, *et al.* Synthesis, insecticidal activity, resistance, photodegradation and toxicity of pyrethroids (A review). Chemosphere, 2020, 254: 126779.

Words and Expressions

α-cyano-3-phenoxybenzyl α-氰基-3-苯氧苄基
tremor /'tremər/ *n.* [医] 震颤，颤动
convulsion /kən'vʌlʃn/ *n.* [医] 惊厥，动乱，震动
choreoathetosis /kɒɔːiːəuæ'θetəusɪs/ *n.* 舞蹈手足徐动症，舞蹈指痉病
cytotoxicity /ˌsaitəutɔk'sisiti/ *n.* 细胞毒性
neuroblastoma /ˌnjuərəublæs'təumə/ *n.* [肿瘤] 成神经细胞瘤
dopaminergic /ˌdəupəmi'nəːdʒik/ *adj.* 多巴胺能的
neuroblast /'njuərəblæst/ *n.* [胚] 成神经细胞

hepatoma /ˌhepəˈtəumə/ *n*. 肝癌，［肿瘤］肝细胞瘤（尤指恶性肿瘤）
intestinal /ɪnˈtestɪnl/ *adj*. 肠的
luciferase /ljuːˈsɪfəreɪs/ *n*.［生化］荧光素酶
estrogen /ˈestrədʒən/ *n*. 雌性激素
nausea /ˈnɔːziə/ *n*. 恶心，晕船，极端的憎恶
vomiting /ˈvɔmitiŋ/ *n*. 呕吐（vomit 的 ing 形式）
dyspnea /dispˈniːə/ *n*.［内科］呼吸困难
bronchospasm /ˈbrɔŋkəspæzəm/ *n*.［内科］支气管痉挛
paresthesia /ˌpærəsˈθiːʒə/ *n*.［内科］感觉异常
permethrin /ˈpəːmeθrin/ *n*. 苄氯菊酯，二氯苯醚菊酯
bifenthrin /ˈbaɪfenθrɪn/ 联苯菊酯
cypermethrin /saɪpəˈmeθrɪn/ 氯氰菊酯
neurotoxicity /ˌnjuərəutɔkˈsisəti/ *n*.［农药］神经中毒性，神经毒性
cumbersome /ˈkʌmbəsəm/ *adj*. 笨重的，累赘的，难处理的
β-cyfluthrin *β*-氟氯氰菊酯
deltamethrin /delˈtæmeθrɪn/ 溴氰菊酯
resmethrin /rezˈmiːθrin/ *n*. 苄呋菊酯
tetramethrin /ˌtetrəˈmeθrin/ *n*.［农药］胺菊酯
gavage /gæˈvɑːʒ/ *n*. 填喂法，强饲法
tyrosine hydroxylase 酪氨酸羟化酶
immunoreactivity /ˌɪmjunəurɪækˈtɪvətɪ/ *n*. 免疫反应性
cerebral /səˈriːbrəl/ *adj*. 大脑的，脑的
necrosis /neˈkrəusɪs/ *n*. 坏死，坏疽，骨疽
hippocampus /ˌhɪpəˈkæmpəs/ *n*. 海马体，海马
striatum /strˈɪeɪtəm/ *n*. 纹状体，终脑的皮层
vegetable oil 植物油
lactation /lækˈteɪʃn/ *n*. 哺乳，哺乳期，分泌乳汁
apoptosis /ˌæpəpˈtəusɪs/ *n*. 细胞凋亡，细胞死亡
hepatotoxic /ˌhepətəuˈtɔksik/ *adj*. 肝毒素的
nephrotoxic /ˌnefrəˈtɔksik/ *adj*. 足以危害肾脏的，对肾脏有害处的
Hyalella azteca 端足虫，片脚类动物
λ-cyhalothrin *λ*-三氟氯氰菊酯
hydrophobic /ˌhaɪdrəˈfəubɪk/ *adj*. 狂犬病的，疏水的
trans-permethrin 反式氯菊酯
Daphnia magna 大型蚤
Ceriodaphnia dubia 模糊网纹蚤，网纹溞
cis-permethrin 顺式氯菊酯

Notes

［1］MTT：四甲基偶氮唑盐。MTT 法又称 MTT 比色法，是一种检测细胞存活和生长的方法。其检测原理为活细胞线粒体中的琥珀酸脱氢酶能使外源性 MTT 还原为水不溶性的

蓝紫色结晶甲臜（formazan）并沉积在细胞中，而死细胞无此功能。

[2] LDH：乳酸脱氢酶。在胞质内含量非常丰富，细胞处于正常状态下其不能通过细胞膜，但当细胞受到损伤或死亡时便可释放到细胞外，此时细胞培养液中 LDH 的活性与细胞的死亡数目呈正比，通过比色法测定并与靶细胞对照孔的 LDH 活性进行比较，可计算出效应细胞对靶细胞的杀伤百分数。

[3] SH-SY5Y cells：人神经母细胞瘤细胞。该细胞的密度可高达 1×10^6 cells/cm^2，具有中等水平的多巴胺 β 羟化酶的活性。

[4] HepG2 细胞来源于一个 15 岁白人的肝癌组织。该细胞分泌多种血浆蛋白：清蛋白、α2-巨球蛋白、血纤维蛋白溶酶原、铁传递蛋白等。

[5] Caco-2 细胞模型是一种人克隆结肠腺癌细胞，结构和功能类似于分化的小肠上皮细胞，具有微绒毛等结构，并含有与小肠刷状缘上皮相关的酶系，可以用来进行模拟体内肠转运的实验。在细胞培养条件下，生长在多孔的可渗透聚碳酸酯膜上的细胞可融合并分化为肠上皮细胞，形成连续的单层，这与正常的成熟小肠上皮细胞在体外培育过程中出现反分化的情况不同。细胞亚显微结构研究表明，Caco-2 细胞与人小肠上皮细胞在形态学上相似，具有相同的细胞极性和紧密连接。胞饮功能的检测也表明，Caco-2 细胞与人小肠上皮细胞类似。

[6] up-down procedure：上下法，由限度试验和主试验组成。限度试验又分为 2000 mg/kg 和 5000 mg/kg 剂量水平，用于受试物毒性可能较小的情况。两种剂量的限度试验具体步骤有细微差别，但最多仅用 5 只动物。设计限度剂量 5000 mg/kg，将受试物按 5000 mg/kg 给予 1 只动物。观察 48 h，如果该动物死亡，则进行主试验；如果该动物存活，依次将受试物给予另外 2～4 只动物。有 3 只或 3 只以上动物存活时，$LD_{50}>5000$ mg/kg；有 3 只或 3 只以上动物死亡时，$LD_{50}<5000$ mg/kg。主试验是一个预先设计的给药程序，在此程序中，每次给药一只动物，若该动物存活，第 2 只动物给予高一级剂量，若第 1 只动物死亡或出现濒死状态，第 2 只动物给予低一级剂量。在对每只给药动物仔细观察 48 h 后，可以决定是否对下一只动物给药，以及给药剂量。当最高剂量下有连续 3 只动物存活，或在 6 只动物中有 5 只连续发生在高一级剂量死亡、在低一级剂量生存的生死转换等情况时，可以停止试验。根据终止时所有动物的状态，用最大可能性法计算 LD_{50} 值。

[7] the Organisation for Economic Co-operation and Development（OECD）：经济合作与发展组织（简称经合组织）。

Exercises

Ⅰ Answer the following questions according to the text.

1. What are the methods for assessing cytotoxicity? please give examples.
2. What symptoms can be caused by exposure to pyrethroids in humans?
3. What are the design methods of acute toxicity test?
4. What are the main symptoms of chronic toxicity of pyrethroids?

Ⅱ Translate the following English phrases into Chinese.

sodium channel　immunoreactivity　acute symptom　reproductive toxicity
gene expression　acute oral toxicity　lymphoid cell　aquatic toxicity

Ⅲ Translate the following Chinese phrases into English.

类固醇受体　水生生物　间质细胞　生殖激素　毒性症状　活性氧

内分泌干扰　优先目标　细胞毒性　神经毒性　膳食摄入　肝功能异常

Ⅳ Choose the best answer for each of the following questions according to the text.

1. What methods do use living cells to assess cytotoxicity?

A. MTT assay　B. LDH assay　C. bioassay　D. all of the above

2. according to toxicity stands, the toxicity of deltamethrin is ________.

A. very toxic　B. moderate poisonous

C. low toxic　D. non-toxic

3. which of the follow are belong to chronic toxicity?

A. hepatotoxic　B. nephrotoxic

C. reproductive toxicity　D. all of the above

4. Which pesticide is lower toxic as the temperature rises?

A. organophosphates　B. carbamates

C. pyrethroids　D. neonicotinoids

5. What are the targets for evaluating the toxicity of pesticides to aquatic organisms?

A. *Hyalella azteca*　B. zebrafish

C. *Procloeon* sp.　D. all of the above

Ⅴ Translate the following short paragraphs into Chinese.

1. Pyrethroids are divided into type Ⅰ and type Ⅱ depending on whether or not they contain α-cyano-3-phenoxybenzyl. Type Ⅰ pyrethroids induce tremors or convulsions, while type Ⅱ pyrethroids exposure primarily causes choreoathetosis.

2. In humans, exposure to pyrethroids can cause acute symptoms such as nausea and vomiting, dyspnea, cough, bronchospasm, dermal effects.

3. The longer the observation period, the smaller the LD_{50}. And the larger the age of the mouse, the larger the LD_{50}.

4. Also, the aquatic toxicity of pyrethroids is temperature dependent. Toxicity increases at low temperatures, and this has been incorporated into toxicity identification evaluation (TIE) procedures to determine the cause of ambient toxicity.

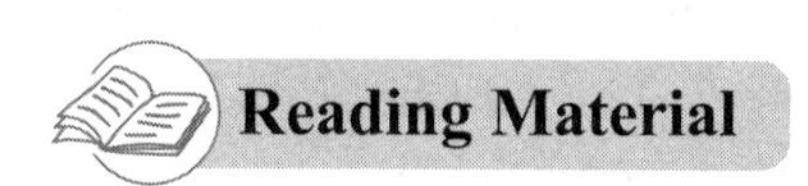

Developments in Toxicology

Toxicology and—later—ecotoxicology and endocrinology have become more and more sophisticated to assess the consequences for health and the environment with the use of plant protection products (PPPs). The awareness of the contamination potential of PPPs is now widespread among scientists, regulators and citizens alike, but this has not always been the case. When synthetic pesticides were introduced in the 1940s, toxicology was an emerging discipline, rapidly disentangling itself from pharmacology. "Safety" was defined in narrow terms as absence of acute toxicity and tests focused on a main outcome: mortality. Indeed, the first important standard set by toxicologists was the so-called Medial Lethal Dose

(LD_{50}), namely the lethal dose for half of the sample of laboratory animals. Such knowledge about "the dose that makes a thing a poison" —to recall the basic tenet of toxicology—was essential to provide recommendations on safe application and management. From this perspective, poisoning and contamination were thought to result from accidents or negligence leading to acute exposure or to spillage in the environment.

Over time the efforts of toxicologists allowed for the development of more sophisticated methods and standards to assess a variety of adverse effects of chemicals, creating the current "alphabet soup" that includes No Observed Adverse Effect Level (NOAEL), No Observed Effect Level (NOEL), Acceptable Operator Exposure Level (AOEL), Acceptable Daily Intake (ADI) among others. First and foremost, tests highlighted that long-term effects of correct usage may be harmful. Low exposure over a protracted period might lead to chronic health conditions in workers, bystanders and residents of agricultural areas. The negative cumulative effects of pesticides might include cancer, neurological diseases like Parkinson's disease, fertility and reproductive effects, chronic asthma, etc. Second, the appraisal of effects on the environment gained increasing relevance. The poisoning of non-target animals (butterflies, frogs) and beneficial insects (bees and other pollinators) has been recognised as a serious threat to the overall level of biodiversity in agricultural areas and ultimately on productivity. Further, PPPs can be very persistent in the environment and cause pollution of soil and of groundwater resources. In a recent report, Greenpeace summarised "ultimately, what is at stake are the diverse ecosystem services, such as pollination, natural pest control, cleaning of drinking water, nutrient cycling and soil fertility, which are provided by a fully functioning and fully functional ecosystem". Third, in the last 20 years PPPs have been linked with endocrine disruptions, namely the interference of chemicals with the hormone system, opening an entire new area of scientific inquiry.

An additional trend is worth mentioning here for its relevance to public policy, namely the efforts towards the standardisation of toxicological tests and laboratory practices for regulatory purposes. Back in the 1940s and indeed for many decades to follow, methods for testing varied a lot from context to context and even from laboratory to laboratory. For example, Hough explains that studies on DDT were numerous but somehow inconsistent, since "there was little continuity linking one study to another". The OECD "Program on Pesticides and Sustainable Pest Management" started in 1992 to develop protocols for laboratory testing. Today there are around 150 guidelines on testing methods, covering physical-chemical properties, degradation and accumulation in the environment and health effects. The expectation is that "OECD-wide accepted data requirements, test guidelines and documentation standards for country evaluation reports should lead to mutual trust and full acceptance of evaluations based on good science". Even more ambitiously, the OECD recommends to deal with dossiers on pesticides at a global level, adhering to the ideal of "one substance, one toxicological assessment".

In short, modern toxicology progressed from testing acute toxicity leading to poisoning and mortality towards the assessment of chronic toxicity linked to a large—and expanding—range of pathologies. It seems safe to argue that scientific assessments developed by adding complexity and becoming multidisciplinary-including insights from toxicology, environmental sciences and endocrinology.

This oversimplified account of toxicology should not suggest a linear progression of discoveries leading to an orderly accumulation of evidence. The proclamation of the idea of globally valid assessments does not imply that toxicologists agree on developments in their field. Rather, contradictions and contrasts among competing paradigms abound in the history of the scientific appraisal of chemicals, which shows "a back and forth of forgetting, remembering, contest, and disagreement". Some questions, like the choice of endpoints, the comparison of different species and the extrapolation from animal studies to humans are still fiercely debated today, as the section on endocrine disruption will show. Also, over time more and more previously "unexpected" effects and modes of action of chemicals have been discovered, although controversies arise since attribution of causality is difficult in complex ecosystems.

Developments in toxicology are extremely relevant since they are intertwined with regulatory processes. It would be misleading to suggest that developments in regulation mirror those in science, or to expect that discoveries of new risks translate directly into policy. However, in general terms, pesticide regulations present a trend from basic to extremely complex, based on increased awareness and recognition of an expanding range of potential adverse effects of chemicals.

Notably, the first laws in the USA and in European countries did not require any pre-market testing of the health and environmental effects of pesticides. In the USA, the most relevant provision introduced by regulators consisted of legal requirements on labelling and on of mismanagement. Today, a complex set of regulatory provisions for the authorisation, commercialisation and use of pesticide is in place. Health and environmental concerns gained increasing relevance, though national differences on how to balance food security and safety are huge.

Selected from: Bozzini E. Pesticide Policy and Politics in the European Union. Switzerland: Springer International Publishing AG, 2017: 11-14.

Words and Expressions

endocrinology /ˌendəʊkrɪˈnɒlədʒi/ *n*. 内分泌学，内分泌科

medial lethal dose 致死中量

No Observed Adverse Effect Level (NOAEL) 未观察到有害作用剂量水平

No Observed Effect Level (NOEL) 未观察到作用剂量水平

Acceptable Operator Exposure Level (AOEL) 可接受操作者接触水平

Acceptable Daily Intake (ADI) 人体每日允许摄入量

chronic asthma 慢性哮喘

PART XIII
RISK ASSESSMENTS

Unit 25 Toxicity Assessment

The enormous amount of chemical contaminants that are currently found in air, soil, water and sediments of our planet, mostly as a result of human activities, call for an evaluation of their risks to the web of life. The contaminants that potentially result in adverse biological effects—whether at individual, population, community or ecosystem level—the so-called pollutants, are of concern. For obvious reasons, pollutants that relate to our food resources are of particular interest and require priority in risk assessment. Among the latter chemicals are those used in agricultural production, also called agrochemicals, which include pesticides of various kinds, plant growth regulators, repellents, attractants and fertilisers. These chemicals are intentionally added to the environment for controlling pests and diseases of crops, requiring an accurate assessment of the maximum amounts that can be applied.

Because of their huge market in all continents, agrochemicals are among the most common pollutants in one fifth of the Earth's land. Also, given the toxicity of most insecticides, herbicides and fungicides to their target and nontarget organisms, their negative impacts on the environment cannot and must not be ignored. Indeed, pesticides constitute one of the main drivers of population decline in some wildlife species. Therefore, proper evaluations of the risk that pesticides and other agrochemicals pose to organisms and ecosystems should be conducted in a scientific manner.

The first condition for a risk assessment is to know whether a chemical is toxic or not. Multiple lines of evidence are used for that purpose, including bioassays with standard species that are representative of various environmental media, biomarkers and others.

For decades, acute and chronic bioassays have helped determine the toxicity levels of most agrochemicals to a range of organisms. For all chemicals, lethal toxicity to mammals is based on rats or mice bioassays, from which the oral LD_{50} and contact LC_{50} can be determined. In the past, no-observed-effect concentrations or levels (NOEC or NOEL) and the lowest-observable effect concentration (LOEC) were also determined, mainly for human risk assessment. However, ecotoxicologists have been arguing that the latter measures are statistically flawed and consequently should not be used any longer.

Instead, the estimation of LD_{10} or LD_{20} from standard bioassays is now preferred to define the low levels of effect, given that populations of organisms in nature always have a small proportion of casualties due to diseases, lack of fitness, starvation and other factors, which cannot be distinguished from chemical toxicity. Similar bioassays for acute toxicity have been used with a range of organisms, including birds, fish, amphibians, earthworms, springtails, bees, flies and a variety of insect pests, mosquito larvae, midge larvae, dragonfly and mayfly nymphs, several types of crustaceans and zooplankton, clams, oysters, mussels, algae and plant species. The need for standardisation of tests has produced a large body of literature, where some species are preferred based on ecological relevance, sensitivity, commonality, easiness to culture in the laboratory or other convenient traits. The commercial Microtox® assay system is also an acute bioassay designed to detect basal toxicity to the marine luminescent bacteria *Photobacterium phosphoreum*, mostly from some organic substances.

Many other standardised *in vitro* cytotoxicity tests have been developed to test a large number of chemicals. Acute lethality is typically determined for single doses at 24 h in terrestrial organisms and after a fixed time of exposure in aquatic organisms: from 24 or 48 h in small organisms to 96 h for larger crustaceans and fish. Chronic toxicity bioassays are carried out to detect possible biological effects after repeated or continuous exposure to sublethal doses or concentrations of a chemical. The endpoints in this case may vary from lethality to carcinogenicity and mutagenicity or from reproduction impairment to hormonal imbalance and developmental effects. Any negative or deleterious effect can be considered a chronic toxicity endpoint, not just the reproduction effects—the criterion for chronic toxicity is long, repeated exposure to chemical amounts that are not lethal within the short timeframe of the acute bioassays. Not all chronic endpoints are considered ecologically relevant, and hence, such endpoints as growth and reproduction would have a higher weighting than others such as biomarkers (unless, of course, these can be linked to community and population effects). Each one of the above tests is considered a line of evidence for the risk assessment of a chemical.

Current information on acute and chronic toxicity for all agrochemicals can be accessed online through the ECOTOX database compiled by the US Environmental Protection Agency (http://cfpub.epa.gov/ecotox/) and other databases with a narrower scope, e.g. AGRITOX (http://www.agritox.anses.fr/index.php), Footprint Database (http://sitem.herts.ac.uk/aeru/iupac/), the Pesticide Manual and others.

Toxic action may depend not only on dose or concentration but also on the duration of exposure. It has long been suggested that toxicity tests should include information not only on doses or concentrations but also on exposure time, with time-to-effect bioassays being a practical way to achieve this. Toxicokinetics are necessary to interpret the toxicity of a compound with time of exposure, because they can relate time-dependent toxicity to target organ concentrations. For some compounds, toxic effects may depend almost exclusively on the total dose, whereby the product of dose, or concentration, and exposure time produces the same biological effect (Haber's rule). For pesticides that block specific metabolic or physiological pathways, the toxicodynamics of the compounds determine the time dependency of the effect. In this case, toxicity is better described as a function of the time of exposure in addition to the dose. Understanding of the mode of action of such compounds is, therefore, essential for a correct prediction of their toxic effects. Thus, whenever the interactions of chemicals with critical receptors are either slowly reversible or irreversible and the resulting biological effects are also irreversible (i. e. death), such effects are time cumulative and long lasting. Examples of such chemicals are the neonicotinoid insecticides, which impair cognition by blocking nicotinic acetylcholine receptors in the central nervous system of insects and other arthropods.

One particular group of compounds that have attracted special attention in recent times is the endocrine disruptor chemicals (EDC), among which are some agrochemicals: amitrole, atrazine, DDT, lindane, mancozeb, maneb, metiram, metolachlor, pentachlorophenol, vinclozolin and ziram. EDCs alter the homeostatic balance in organisms by either mimicking the activity of hormones or influencing the metabolism of natural hormones as a side effect. Because hormones are substances that effectively induce or suppress physiological mechanisms, often complex and interlinked, exposure to EDCs can have unforeseen effects, certainly unrelated to the specific mode of action of the ED compound. For example, some organochlorine pesticides induce feminisation in alligators. EDCs should not be confused with hormonal pesticides, which by their very nature mimic natural hormones in plants (e. g. auxinlike herbicides such as 2, 4-D) or insects (e. g. ecdysone mimics such as tebufenozide). Identification of EDCs has been given high priority in some developed countries with the aim of restricting their use in view of the potential consequences for human health, even if the research needed to characterise and elucidate their impacts is lagging behind. This is a case where regulation appears to have run ahead of scientific evidence, which in some cases is controversial.

All the above is for direct toxicity, which is the only way to assess the toxic potency of a chemical compound. Assessment of the toxicity of mixtures of chemicals started in the 1970s; this can be determined using the same protocols for acute and chronic toxicity or using genomics. Apart from determining the total toxicity of the mixture, it is important to assess whether the compounds show additive, synergistic or antagonistic toxicity, and if so, try to estimate the synergistic/antagonistic ratios. Additive toxicity occurs with compounds that have the same or similar mode of action, e. g. organophosphorus and carbamate insecticides, pyrethroids, neonicotinoids, triazines, etc. In such cases, a toxic equivalent (TE) with reference to a standard compound in the same group of chemicals is estimated. TEs are then used for assessing the total toxicity of such mixtures in tissues, sediments or another matrix. Among synergistic mixtures, the best and well known is that of piperonyl butoxide with pyrethroid insecticides, which occurs because of inhibition of the mono-oxygenase detoxification system in insects. Recently, mixtures of ergosterol-inhibiting fungicides with cyanosubstituted neonicotinoids have also shown enhanced synergism of these insecticides in honeybees.

Indirect impacts of chemical toxicity occur in nature. Although it is often difficult to determine the causal relationship that exists between chemicals and the indirect effects they may have in ecosystems, well-designed microcosms and mesocosms, as well as field studies, have often proved such effects. Indeed, ecological theory predicts that the elimination of predators inevitably results in a bloom of the primary prey species they control. Equally, the elimination of key primary consumers leads to the uncontrolled growth of their feeding sources (e. g. weeds, algae), and this occurs when insecticides eliminate grazing planktonic species, but not necessarily to the demise of their predators, as the latter organisms can switch prey preferences. Fungicides can eliminate fungal communities that have the important role of recycling organic wastes in nature and thus reduce the efficiency of the litter decomposition and mineralisation in soil. Reduction of the insect food source after application of imidacloprid in agricultural environments over many years has led to the reduction of several bird species in the Netherlands. Even biological insecticides, such as Bt (*Bacillus thuringiensis*) applied to control mosquitoes, flies and other nuisance insects in wetland areas, inevitably reduce the bird populations that feed on those insects. One of the practical applications of this theory in agriculture is the integrated pest management (IPM). Indeed, IPM is mindful of the disastrous consequences of pesticide application for pest control without consideration of the impacts on the whole ecosystem. Extreme examples are the massive explosion of brown plant hoppers (*Nilaparvata lugens*) in Indonesia during the 1970s following excessive application of insecticides that eliminated spiders and other predators of this rice pest and the uncontrolled population growth of red-billed quelea (*Quelea quelea*) in areas of Botswana with high residue loads of fenthion, a persistent organophosphorus insecticide applied for pest control in agriculture.

Selected from: Sánchez-Bayo F, Tennekes H A. Environmental Risk Assessment of Agrochemicals — A Critical Appraisal of Current Approaches. Toxicity and Hazard of Agrochemicals, 2015: 7-9.

Words and Expressions

agrochemical /ˌæɡrəʊˈkemɪkl/ *n.* 农用化学品，农药
biomarker /ˈbaɪəʊmɑːkər/ *n.* 生物标志
casualty /ˈkæʒuəlti/ *n.* 意外事故，伤亡人员，急诊室
amphibian /æmˈfɪbiən/ *n.* [脊椎] 两栖动物；*adj.* 两栖类的，[车辆] 水陆两用的
springtail /ˈsprɪŋteɪl/ *n.* 跳虫，弹尾虫
midge larvae 孑孓
dragonfly /ˈdræɡənflaɪ/ *n.* [昆] 蜻蜓
mayfly nymphs 蜉蝣稚虫
crustacean /krʌˈsteɪʃn/ *n.* 甲壳纲动物
zooplankton /ˈzuːəˌplæŋktən/ *n.* 浮游动物
clam /klæm/ *v.* 挖蛤，保持沉默；*n.* 蛤蜊
oyster /ˈɔɪstər/ *n.* 牡蛎，[无脊椎] 蚝，沉默寡言的人
mussel /ˈmʌsl/ *n.* 蚌，贻贝，淡菜

microtox　氨苯磺胺
luminescent　/ˌluːmɪˈnesnt/ *adj*. 发冷光的，冷光的
Photobacterium phosphoreum　发光杆菌，明亮发光杆菌
carcinogenicity　/ˌkɑːsinəudʒəˈnisəti/ *n*. 致癌性，致癌力
mutagenicity　/ˌmjuːtədʒəˈnɪsətɪ/ *n*.［遗］诱变（性）
toxicokinetics　/tɒksɪkəukaɪˈnetɪks/ 毒物代谢动力学
toxicodynamics　/tɒksɪkɒdaɪˈnæmɪks/ 毒物作用动力学
neonicotinoid　新烟碱，烟碱类农药
cognition　/kɒgˈnɪʃn/ *n*. 认识，知识，认识能力
nicotinic acetylcholine receptors　烟酰胺乙酰胆碱受体
amitrole　/ˈæmitrɔl/ *n*. 杀草强
mancozeb　/mænˈkʌzb/ *n*.［农药］代森锰锌
maneb　/ˈmæneb/ *n*.［农药］代森锰
metiram　/ˈmetiræm/ *n*. 代森联
metolachlor　/mɪtɒ ˈlæklɔː/ 异丙甲草胺
pentachlorophenol　/ˌpentəˌklɔːrəuˈfiːnɒl/ *n*.［有化］［农药］五氯苯酚
vinclozolin　/vɪŋkˈləuəlɪn/ 乙烯菌核利
ziram　/ˈzaiəræm/ *n*.［农药］福美锌
feminization　/femɪnaɪˈzeɪʃn/ 女性化，雌性化
alligator　/ˈælɪgeɪtər/ *n*. 短吻鳄；*adj*. 鳄鱼般的，鳄鱼皮革的；*vi*. 皱裂，裂开
ecdysone　/ˈekdɪsəun/ *n*. 蜕化素，蜕皮激素
mimics　/ˈmɪmɪk/ *vt*. 模仿，模拟；*n*. 模仿者，仿制品；*adj*. 模仿的，假的
tebufenozide　/tɪbjuːfeˈnɒzaɪd/ 虫酰肼
piperonyl butoxide　/ˈpipərənilbjuˈtɔksaid/ 增效醚（一种杀虫剂），胡椒基丁醚
ergosterol-inhibiting　麦角甾醇抑制剂
planktonic　/ˌplæŋkˈtɔnik/ *adj*. 浮游生物的，浮游的
imidacloprid　/ɪmɪdækˈlɒprɪd/ *n*. 吡虫啉
Nilaparvata lugens　褐飞虱，褐稻虱
Quelea quelea　红嘴奎利亚雀
fenthion　/fenˈθaɪən/ *n*.［农药］倍硫磷

Notes

［1］no-observed-effect concentrations or levels（NOEC or NOEL）：未观察到效应浓度或水平。

［2］the lowest-observable effect concentration（LOEC）：最低可观测效应浓度。

［3］ECOTOX database：生态毒理学知识库，它是一个综合性的、公开的知识库，提供有关水生生物、陆生植物和野生动物的单一化学环境毒性数据。

［4］Haber's rule：20 世纪初德国科学家弗里茨·哈伯（Fritz Haber）提出，表述物质浓度（c）和给药时间（t）的乘积对给定终点产生固定水平的效应。

Exercises

Ⅰ Answer the following questions according to the text.

1. Why do we do risk assessment? How do we do risk assessment?

2. What are direct toxicity? what are indirect toxicity?

3. What is the content of chronic toxicity bioassay?

4. What is the relationship between toxic action and dose, concentration, and exposure time?

5. Which compounds are endocrine disruptors? please give examples.

Ⅱ Translate the following English phrases into Chinese.

synergistic toxicity　direct impacts　risk assessment　additive toxicity

population decline　exposure time　critical receptor　plant growth regulator

Ⅲ Translate the following Chinese phrases into English.

生物效应　毒性评价　毒性反应　中枢神经系统　拮抗毒性

负面影响　致死毒性　毒杀作用　体内平衡状态　毒性当量

Ⅳ Choose the best answer for each of the following questions according to the text.

1. The first requirement for risk assessment is to know whether a pesticide is ________.

A. toxic　B. chemical　C. natural　D. useful

2. What does the toxic action depend on?

A. dose　B. concentration　C. exposure time　D. all of the above

3. which do the toxicity of pesticide mixtures include?

A. additive toxicity　B. synergistic toxicity

C. antagonistic toxicity　D. all of the above

4. the elimination of key primary consumers leads to the uncontrolled growth of their feeding sources (e. g. weeds, algae), which belong to ________ impacts.

A. derect　B. indirect　C. slight　D. seriour

5. In addition to toxicity biometrics, ________ is becoming an increasingly important first choice for bioassays.

A. endocrine disruption　B. indirect

C. exposure time　D. all of the above

Ⅴ Translate the following sentences into Chinese.

1. The enormous amount of chemical contaminants that are currently found in air, soil, water and sediments of our planet, mostly as a result of human activities, call for an evaluation of their risks to the web of life.

2. For all chemicals, lethal toxicity to mammals is based on rats or mice bioassays, from which the oral LD_{50} and contact LC_{50} can be determined.

3. Chronic toxicity bioassays are carried out to detect possible biological effects after repeated or continuous exposure to sublethal doses or concentrations of a chemical.

4. Apart from determining the total toxicity of the mixture, it is important to assess whether the compounds show additive, synergistic or antagonistic toxicity, and if so, try to estimate the synergistic/antagonistic ratios.

5. Additive toxicity occurs with compounds that have the same or similar mode of action, e. g. organophosphorus and carbamate insecticides, pyrethroids, neonicotinoids, triazines, etc.

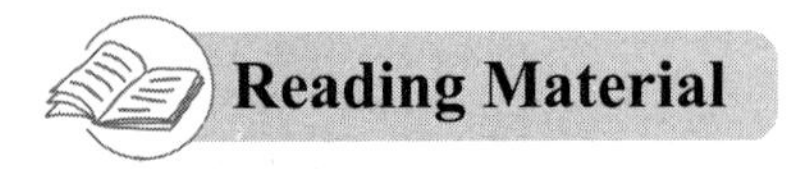

Reading Material

Framework for Ecological Risk Assessment

Ecological risks of agrochemicals are evaluated by standard procedures, as it is done with other chemicals. This approach is similar to the evaluations of risks for humans, except for one main difference: human risk assessments aim at protecting individuals, whereas ecological risk assessments aim at protecting ecological structures, whether populations, communities or entire ecosystems. Another difference is the preferential use of median effect concentrations (EC_{50}), in particular the median lethal concentrations (LC_{50}) and lethal doses (LD_{50}) in risk assessments of agrochemicals, rather than no-observed-effect concentrations or levels (NOEC or NOEL). This is because the data on the latter metrics are less readily available for the surrogate species used in toxicological testing of agricultural compounds; typically, only mammal toxicity is measured at the lowest levels. Besides this difficulty, NOEC and NOEL are statistically unreliable—it is almost impossible to prove that there are no effects in natural populations of organisms from exposure to low concentrations of chemicals.

In essence, all risk assessments rely on the framework shown in Figure 1, which is based on standard procedures of the US EPA and has been adopted, with some modifications, by most regulatory authorities in OECD countries. The two main components of the risk framework are the toxicity assessment and the exposure assessment. Without

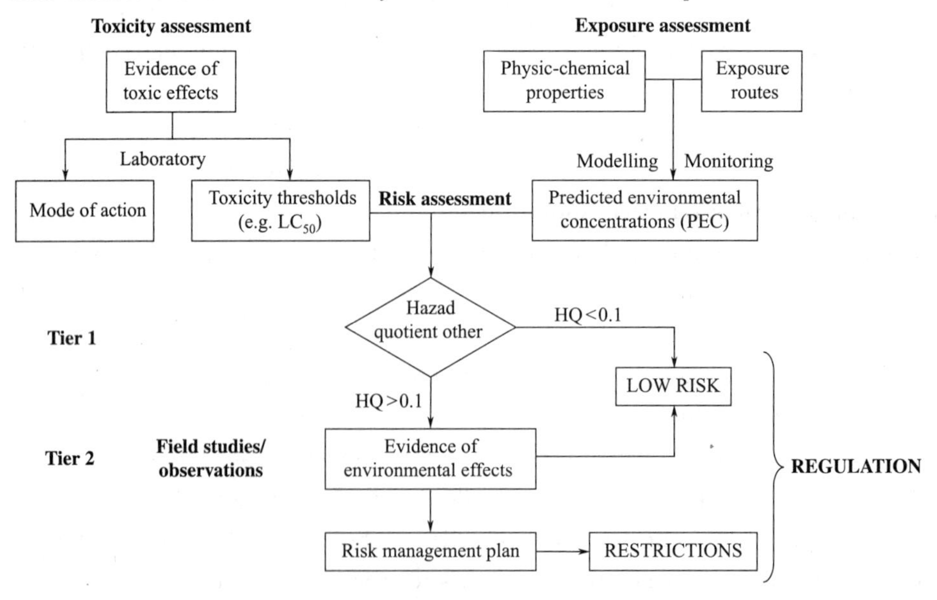

Figure 1 Framework for ecological risk assessment of agrochemicals

toxicity, there is no risk, but if organisms are not exposed to a toxic chemical, they are under no risk either. Therefore, it is only when organisms are exposed to potentially toxic chemicals that a risk assessment must be undertaken.

Toxicity levels of a chemical (either concentrations in aquatic media or doses in terrestrial animals) must be determined in the laboratory, as effects from exposure under field conditions may be confounded by other factors. Standard toxicity tests, carried out on several taxa of plants and animals, are primarily designed to establish the acute lethality: either the LC_{50} for aquatic organisms after a given time of exposure or the LD_{50} for doses applied to or ingested by terrestrial organisms. In addition, chronic toxicity after prolonged and repeated exposure (whether lethality or another effect) and reproductive endpoints after sublethal exposure over time are often determined as well. Knowledge about the mode of action of the chemical should help understand its toxic effects.

The exposure characterisation is complex as it usually involves some sort of modelling to determine the possible concentrations of chemical in the various media in which the organisms live. Inputs for the models include all relevant physicochemical properties of the chemical under consideration, e. g. water solubility, lipophilic behaviour, volatility and degradation constants in water, soil, light conditions, plants and animals. The latter constants are important to understand the persistence of chemical residues in the environment. In site-specific the ecological risk assessment (ERA), the exposure should also include monitoring data to validate the modelling used. If monitored and modelled data differ markedly, an explanation should be sought for the discrepancies.

In countries of the OECD, assessment of the two components of risk follows a tiered-structured process (Fig. 1). In the first step, rough estimates of risk are evaluated by a simple ratio, the hazard quotient (HQ) between the predicted environmental concentrations (PECs) in a particular medium and the acute lethality (LC_{50} or LD_{50}) to the standard test organisms found in that medium. HQ values above 1 are unacceptable, and risk thresholds are commonly set at 0.1, meaning that assessments that result in $HQ < 0.1$ pass the first tier and can be regarded as "safe" for a particular environment. HQ values > 0.1 require evaluation in a second tier. However, a given chemical may be considered safe to terrestrial organisms such as rats, but not safe for aquatic organisms such as midge larvae. Even within the same environmental medium, some species in the field are more sensitive than the surrogate species used in the laboratory tests. To overcome this problem, probabilistic risk assessments (PRAs) have been derived to identify the fraction of organisms that would be negatively affected by the PEC in a given environmental medium. Risk in a PRA is defined as a feasible detrimental outcome of an activity or chemical, and it is characterised by two quantities: (i) the magnitude (severity) of the possible adverse consequences and (ii) the likelihood of occurrence of each consequence. Species sensitivity distributions (SSDs) are typically used for that purpose.

HQ is the preferred method used in regulatory risk assessments of agrochemicals, whereas PRA is mostly used in site-specific ERA, which in most jurisdictions amounts to a second tier. Whether HQs or PRAs are used, if the chemical does not pass the first tier, a further evaluation must be done. The second tier involves gathering evidence of effects under realistic environmental conditions. Field trials and mesocosms are typically carried out in order to obtain that information.

Selected from: Sánchez-Bayo F, Tennekes H A. Environmental Risk Assessment of Agrochemicals — A Critical Appraisal of Current Approaches. Toxicity and Hazard of Agrochemicals, 2015: 4-6.

Words and Expressions

surrogate /'sʌrəgət/ *adj*. 替代的；*n*. 代理人；*v*. 取代，替代
US EPA 美国国家环境保护局
hazard quotient 危险系数，危害商
mesocosm /mesəʊ'kɒzəm/ 中宇宙，围隔实验

Unit 26 Exposure Assessment

The first step in the exposure assessment starts with characterisation of the chemical properties of a substance, as this will determine the main routes of exposure for organisms. Among the physical and chemical properties relevant for environmental risk are solubility in water and organic solvents; partitioning coefficients, in particular the octanol-water coefficient (K_{ow}); soil adsorption constants (e. g. K_{oc}); vapour pressure and volatilisation (Henry's law); dissociation constants (pK); and half-lives in several environmental matrices (air, water, soil, sediments and plant tissues) due to different processes: hydrolysis, photolysis or metabolism. In addition, mobility and leaching potential should be assessed, and this can be done by means of indices such as the ground ubiquity score, which combines soil adsorption properties and degradation in soil. The databases [i. e. AGRITOX (http://www.agritox.anses.fr/index.php), Footprint Database (http://sitem.herts.ac.uk/aeru/iupac/), and Pesticide Manual] are comprehensive sources of such data. Manufacturers of new products must provide this information—different from the material safety data sheets (MSDS)—to their regulatory authorities for risk assessment. Numerous laboratory measurements and tests must be carried out to obtain this kind of data, although in some cases, quantitative structure-property relationships (QSPRs) are used to derive approximate values when compounds have similar chemical structures.

Degradation pathways in animals are also important, as some chemicals are amenable to metabolic breakdown and produce metabolites, usually of less toxicity, that can be eliminated in urine and faeces. However, metabolites can also be as toxic as the parent compounds, in which case, they need to be assessed as independent toxicants. Often, a chemical may be degraded easily in water or plants but may be persistent in soil or vice versa. Recalcitrant chemicals are the ones that do not degrade readily in any of the environmental media, e. g. chlorinated compounds, including some neonicotinoid insecticides and herbicides. Persistent agrochemicals require special attention because their effects are not restricted to the time of their application to crops but also during a long period of time afterwards, which could be years. For example, in many developed countries, residues of DDE in soil are still causing eggshell thinning in birds, even if the parent compound (DDT) has not been applied for more than 30 years.

Once the basic data are available, concentrations of the chemical in various matrices can be predicted after its application to a crop, using mathematical models developed for specific purposes. Transport models can refer to a single medium (e. g. air) or to the fate and movement between media. Examples of the first type are the dispersion of particles and movement through air, which can be modelled by AgDRIFT and Gaussian models, soil erosion and sedimentation of runoff and leaching of water-soluble fractions into the soil profile, e. g. Pesticide Root-Zone Model (PRZM) and others. By contrast, multimedia models such as fugacity aim at predicting the concentrations of chemicals in all environmental media. In any case, model predictions should be validated by comparing the outputs of the model to the actual measurements of residues in water, air, soil and organisms.

However, for new chemicals this is not feasible, as residue amounts could be negligible or nonexistent, so current regulatory risk assessment relies entirely on the predicted environmental concentrations (PECs) that result from such models. For old agrochemicals, on the other hand, measured residue levels would be preferable for two reasons: (i) a large body of actual residue data exists for different regions and times, as comprehensive monitoring surveys are carried out over many years and are available from the open literature, and (ii) relevant physicochemical and degradation data are often incomplete for many old compounds, thus hampering their environmental modelling. Each model output and survey data is considered a line of evidence for the risk assessment.

Essential information in regard to prediction of modelled data is the usage pattern of the chemical under consideration; this information is often not readily available, so it is often necessary to infer usage from sales figures. The total amounts of agrochemical applied determine ultimately the total load of residues to which organisms will be exposed. For persistent chemicals, residue loads will increase with time, so it is important to know the period of time they have been used. When all these factors are put together, the margins of error necessarily increase, and so variations of one order of magnitude in PECs are acceptable. The situation is no better for monitoring data, as the variability in residue loads between sites, regions, times of the year and among years can be even larger. Therefore, sensible exposure assessments should be based not so much on accurate concentration levels but on a range of concentrations for specific scenarios and situations. Statistical distributions of such residues help determine their probability of exposure in site-specific ERA.

PECs by themselves are not sufficient to estimate exposure within reasonable limits. The availability of chemicals is paramount, since only chemicals that are taken up by organisms can cause harm. Routes of exposure differ for aquatic and terrestrial organisms, and the estimated PECs should refer not only to the levels of chemicals in the environmental matrices but also the uptake by the organisms living there. In the aquatic environment, organisms are embedded in a matrix that contains pollutants, and their uptake through the gills and epidermis is almost constant. In this case, time of exposure really influences the chemical uptake, while toxicokinetics are influenced by surface area, porosity of the surface and volume/surface ratio. Also, the internal doses and effects with time can be estimated in accordance with Haber's rule. Ingestion of residues in food appears to be a lesser route of exposure for zooplankton and possibly other aquatic organisms. For terrestrial organisms, however, a more complex scenario is envisaged. Uptake of air pollutants may be constant through inhalation, but their residues in water and food sources (e. g. pollen, plant material, insects) are taken intermittently in small, discrete amounts and at different times. Dermal exposure, by contrast, is most relevant to animals living in the soil and sediment (e. g. earthworms, invertebrate larvae, etc.) or having unprotected skins (i. e. amphibians), whereas reptiles, birds and mammals would be somehow protected by the scales, feathers and fur that cover their bodies. Nevertheless, the lethal effects that follow deposition of pesticide spray droplets on birds suggest that concentrated hydrophobic compounds can enter easily their bodies through this route of exposure.

One way of finding out the availability of pollutants to organisms is by using biomarkers of exposure. Many biomarkers have been developed in the last two decades with the aim of measuring the extent of chemical exposure in organisms. Usually they measure the physiological response to chemicals with the same mode of action, i. e. the cholinesterase assay determines the proportion of acetyl-cholinesterase enzyme bound to cholinesterase inhibitor insecticides. Biomarkers based on inhibition of the detoxification mechanisms (P450 monooxygenases, glutamate-S-transferases, etc.) are less specific, as many different compounds can trigger the same response. Unless the chemical source is known, biomarkers can only provide evidence of the existence of a group of contaminants that are taken up by the organisms. For this reason, they are useful to prove the availability of such chemicals, not for the identification of the individual compounds.

One area that will always be crucial to any chemical risk assessment is the accuracy and reliability of chemical and toxicity data. Without proper data, the evaluations would fail to determine realistic risks under possible scenarios. In this context, modelled predictions of exposure must be validated with factual data, whatever the conditions.

Selected from: Sánchez-Bayo F, Tennekes H A. Environmental Risk Assessment of Agrochemicals — A Critical Appraisal of Current Approaches. Toxicity and Hazard of Agrochemicals, 2015: 9-12.

Words and Expressions

octanol-water coefficient (K_{ow}) 正辛醇-水分配系数
volatilisation /vɒlətɪlaɪ'zeɪʃən/ *n*. 挥发
quantitative structure-property relationship 定量结构性质关系
vice versa /ˌvaisi'vəːsə/ 反之亦然
recalcitrant /rɪ'kælsɪtrənt/ *n*. 顽抗者，不服从的人；*adj*. 反抗的，反对的，顽强的
fugacity /fjʊ'gæsɪtɪ/ *n*. 无常，易逃逸，不安定
paramount /'pærəmaʊnt/ *n*. 最高统治者；*adj*. 最重要的，主要的，至高无上的
gill /gɪl/ *n*. 鳃，沟壑；*vi*. 被刺网捕住；*vt*. 用刺网捕鱼，去除内脏
epidermis /ˌepɪ'dɜːmɪs/ *n*. 上皮，表皮
zooplankton /'zuːəˌplæŋktən/ *n*. 浮游动物
glutamate-S-transferases 谷胱甘肽-S-转移酶

Notes

[1] Henry's law：亨利定律，物理化学的基本定律之一，是英国的 Henry（亨利）在 1803 年研究气体在液体中的溶解度规律时发现的。表述为：在等温等压下，某种挥发性溶质（一般为气体）在溶液中的溶解度与液面上该溶质的平衡压力成正比。

［2］ AgDRIFT（R）模型是通过将模型模拟与喷雾漂移工作组收集的现场试验数据进行比较得出的。

［3］ Pesticide Root-Zone Model（PRZM）：农药根区模型，模拟农药挥发、土壤温度、土壤气相农药的传输，模拟微生物转化、灌溉并进行特定算法以消除数值弥散。

Exercises

Ⅰ Answer the following questions according to the text.

1. What data are required for exposure assessment?

2. What are the advantages of the exposure assessment of the old compounds over the new chemicals?

3. How to predict environmental concentrations of pesticides?

4. What are biomarkers? What are its functions?

Ⅱ Translate the following English phrases into Chinese.

exposure assessment　vapour pressure　statistical distribution　less toxicity

dissociation constant　dermal exposure　partitioning coefficient　leaching potential

Ⅲ Translate the following Chinese phrases into English.

土壤吸附　管理机构　降解途径　疏水化合物

土壤侵蚀　水生环境　半衰期　定量结构关系

Ⅳ According to the text，whether are the following statements true or false?

1. Exposure assessment requires a lot of data.

2. Pesticides enter animals and are metabolized into low or nontoxic substances.

3. DDT has a long half-life.

4. Mathematical models are important in exposure assessment.

5. It is difficult to obtain a good model for new compounds due to inadequate experimental data.

6. Biomarkers laterally reflect evidence of organic residues.

7. Pesticides in the environment do not enter living organisms and generally cause little harm.

Ⅴ Translate the following sentences into Chinese.

1. Degradation pathways in animals are also important，as some chemicals are amenable to metabolic breakdown and produce metabolites，usually of less toxicity，that can be eliminated in urine and faeces.

2. A large body of actual residue data exists for different regions and times，as comprehensive monitoring surveys are carried out over many years and are available from the open literature.

3. Sensible exposure assessments should be based not so much on accurate concentration levels but on a range of concentrations for specific scenarios and situations.

4. One area that will always be crucial to any chemical risk assessment is the accuracy and reliability of chemical and toxicity data. Without proper data，the evaluations would fail to determine realistic risks under possible scenarios.

Risk Assessment: First and Second Tier

First Tier

One way is to evaluate representative species of each environmental media, and this has traditionally been done with the hazard quotient (HQ). OECD hazard assessments require at least one species from three different taxonomic groups (e. g. alga, crustacean, fish) for this first tier. Thus, for each environmental medium, one to three HQ values may help define the range of negative impacts. These HQ values should be derived for different exposure routes in the case of terrestrial organisms (e. g. dietary, inhalation or contact by dermal exposure), whereas a single value may be sufficient for aquatic organisms. Two shortcomings are found with this approach. Firstly, it is simplistic and often unrealistic, given that PECs for a range of scenarios typically vary up to one order of magnitude, and this variability can introduce a swing from being "safe" to "unsafe" if the resulting HQ values change from <0.1 to $0.1\sim1.0$. Thus, there is an inherent large uncertainty in this approach even if the toxicity endpoints used are consistent and accurate. Secondly, a given chemical may be considered safe for the representative organism on which the toxicity data are based, but not safe for other organisms living in the same environment. This highlights the problem of using data for standard test species in the first tier to make decisions regarding the registration of a chemical; and yet most regulatory assessments still use this method.

A second and more plausible approach is to use all the toxicity data available for a chemical at once, by constructing an species sensitivity distributions (SSD)—risk assessments using this approach are termed probabilistic risk assessments (PRA). For agrochemicals, SSDs are usually built using acute LC_{50} or LD_{50} data. Often an assessment factor is applied to give a chronic equivalent when the NOEL values are not available. In such cases, a factor of 10 or a derived acute-to-chronic ratio is commonly used. The SSD shows the overall range of toxicities for all species tested in the same environmental medium (e. g. aquatic), so the assessor can determine the levels of chemical that would not have a serious impact on the majority of species. An advantage of the SSD is that it allows estimation of the number of species within a particular medium that would be seriously affected for any residue concentration of a given chemical. Since the latter residue levels can vary spatially, seasonally and with the passing of time, this relationship between residues in the environment and species affected (which includes their identity as well) can be very useful for risk management in site-specific the ecological risk assessment (ERA). In regulatory assessments, the threshold concentration hazardous to 5% of species (HC_5) is defined and compared to the PECs in that particular medium. The term protective concentration for 95% of species (PC_{95}), which equates to HC_5, is also used.

Nevertheless, the probabilistic approach is rarely used for registration of new agrochemicals, since the range of toxicity data available for such compounds is usually insufficient to generate a valid SSD. They are mostly used for site-specific risk assessments, in which case, actual residue data from monitoring at the site or region are also used to complement the predicted exposures.

A third approach is to use risk indices that combine toxicity values of a chemical to several organisms, its persistence in the environment, the frequency of detection of its residues, the likelihood of drift and other characteristics that are deemed important. Indices of risk were first introduced by Metcalf to rank the risk of pesticides to fish, birds, mammals and bees. They are useful for comparing the potential risks of several compounds applied to the same environment, so that farmers, pest control operators and other users of pesticides may choose chemicals that have the least risk to the environment. Examples of indices are the Pesticide Impact Rating Index (PIRI), used by managers and pesticide users as a simple guide for selecting agricultural chemicals that are less damaging to the environment; the ecological relative risk (EcoRR), designed to discriminate between chemicals that may have ecological impacts at a site-specific environment.

Second Tier

Unlike the straightforward mathematical methods used to analyse information for the first tier of the risk assessment, there aren't any mathematical procedures to deal with the risks derived from sublethal effects, indirect effects on communities, endocrine disruption, etc. Consequently, all of the evidence gathered for this second tier must be evaluated using the logical methods commonly used in the weight of evidence (WOE) approach. The exception, perhaps, is the PERPEST model, which was developed to predict risk of pesticides in freshwater ecosystems. It simultaneously predicts the effects of a particular concentration of a pesticide on various community endpoints. The model relies on case-based reasoning, a technique that solves new problems by using past experience (e. g. published microcosm or mesocosm experiments). In order for the model to do that, empirical data extracted from the literature are stored in a database of freshwater ecotoxicity studies, which are updated regularly.

Apart from PERPEST, whenever there is some evidence of negative effects of a chemical on a species or an ecosystem, risk assessors must weigh that evidence against the benefits that the chemical may have for human life, the general environment and agriculture—there may be some negative effects to certain organisms or certain areas, but not necessarily to entire ecosystems. How serious are the effects and how widespread their threats are questions for the assessors to ask and evaluate using their best professional judgement. Here, the purpose of the risk assessment influences the assessor's evaluation. For example, in regulatory assessments, the aim is to minimise the impact of a particular chemical on ecosystems to levels accepted by the community, which is not necessarily the same as protecting the integrity of the ecosystems. Thus, some chemicals can be tolerated if the benefits they have for our lifestyle offset their impacts to specific environments, i. e. agricultural areas that are sacrificed for the "common good". The task of assessing a chemical's risk in this context is not easy, as the value of the environment cannot be measured directly in terms of money, whereas the benefits of agricultural and industrial production are more tangible. Not surprisingly, there are many discrepancies between the regulations of various countries, even for the same agricultural chemical because they are based on different human judgements. In practice, many hazardous agrochemicals are registered as long as certain precautions and management options are put in place. This similarly occurs with pharmaceuticals and industrial chemicals.

Other important questions to be asked during the evaluation process are how much of the chemical will be used and how often, since the concentrations or doses in the environment will depend entirely on the application rate and frequency of usage. Residue loads of persistent agrochemicals may accumulate in soil over the years until their concentrations or doses may reach sublethal or lethal levels to certain organisms and, in the case of herbicides, some agricultural crops. In such cases, regulators may suggest to restrict the application of such products to avoid their accumulation in the environment.

A final question, often raised by risk assessors of agrochemicals, refers to the different behaviours that a given compound may have in regions of the world that vary in climatic and other environmental conditions. Should the special conditions of those environments be taken into consideration at the time of allowing a pesticide to be used in a country? Currently, the bulk of insecticides are being used in developing countries, most of which are located in the tropical or subtropical regions of the world, where warm and humid conditions may affect their dissipation and exposure to organisms. Would the risk of such chemicals in tropical countries be lesser than their risk in temperate and cold countries? It appears that most chemicals pose a similar risk in either region because the increasing losses by microbial degradation or volatilisation are usually counterbalanced by greater desorption and movement of residues into the aquatic environment. Using tropical taxa for toxicity testing in those countries has also been suggested, arguing that tropical species may differ in susceptibility to pesticides. However, most tropical species do not differ in sensitivity from their temperate counterparts in regard to agrochemicals, unlike what happens with metal contaminants. Overall, there is no convincing evidence that, in regard to environmental impacts of pesticides, agrochemical products used in tropical countries should be evaluated differently from any other country.

Selected from: Sánchez-Bayo F, Tennekes H A. Environmental Risk Assessment of Agrochemicals — A Critical Appraisal of Current Approaches. Toxicity and Hazard of Agrochemicals, 2015: 12-22.

Words and Expressions

species sensitivity distributions (SSD) 物种敏感度分布
probabilistic risk assessments (PRA) 概率风险评估
the ecological risk assessment (ERA) 生态风险评估
the Pesticide Impact Rating Index (PIRI) 农药影响评价指数
the ecological relative risk (EcoRR) 生态相对风险
weight of evidence (WOE) 证据权值

PART XIV

PESTICIDE MANAGEMENT

Unit 27 Pesticide Management in China

As the world's largest country for pesticide production and use, China plays an important role in reducing pesticide use and pollution. Actually, China has banned some highly toxic and residual pesticides since 2007. For example, the five highly toxic organophosphorus insecticides (methamidofos, parathion, methyl parathion, monocrotophos, and phosphamidon) are prohibited from use on fruits and vegetables. Since 2015, China has launched zero growth action in the use of pesticides and has promoted reduction and efficiency enhancement of pesticide use. According to an investigation in 2017, the utilization rate of pesticides for the three major grain crops rice, corn and wheat was 38.8%, an increase of 2.2% over 2015, equivalent to a reduction of 30000 tonnes of pesticide use. Consequently, the use strength of pesticide declined from 2015.

Pesticide Registrations

By the end of 2016, there were 35604 registered pesticide products and 665 registered active ingredients in China. Among them, insecticides, fungicides & bactericides, and herbicides accounted for 89.7% of the total registrations. Pesticide registrations rank first, accounting for 40% of the total registrations, with a total of 14233 registrations, followed by fungicides & bactericides (9121 registrations), and herbicides (8584 registrations).

The registered products were mainly low-toxic pesticides (26520 registrations), accounting for 74.5% of the total; the moderate ones accounted for 16.4% (5840 registrations); the slightly-toxic pesticides accounted for 7.8% (2767 registrations). Highly toxic pesticides accounted for 1.4% only. In terms of formulations, emulsifiable concentrates accounted for 30% of the total (9414 registrations); wettable powder accounted for 21.4% (6731 registrations); suspension concentrates, water concentrates, and water dispersible granules held 3527, 2143 and 1643 registrations, respectively.

The top 20 registered pesticides of China before 2017 are counted. With a total of 9200 registrations, rice is the hottest crop for pesticide registrations, accounting for 29.3% of the total preparations registered, followed by cotton, wheat, citrus, apple, corn, cabbage, cucumber, health insecticides, and cruciferous vegetables.

Among the newly registered products in 2016, the top ten varieties were pyraclostrobin, atrazine, mesotrione, thiamethoxam, difenoconazole, azoxystrobin, avermectin, tebuconazole, nicosulfuron and glufosinate. In 2016, the newly registered active ingredients were guadipyr, triflumezopyrim, flupyradifurone, *Sophora alopecuroides* alkaloids, *Metarhizium anisopliae* CQMa421, *Bacillus methylotrophicus*, terpenes, *Bacillus amyloliquefaciens*, 2,4-dichlorophenoxybutyric acid sodium butyrate, halauxifen-methyl, benzobicylon, and pyraclonil, etc.

Pesticide Export

Pesticide export of China were 1.4088, 1.5994, 1.6219, 1.6417, 1.5095, 1.3725 million tonnes and pesticide import were 43.9, 53.5, 62.2, 67.2, 57.6, 39.1 thousands of tonnes for the years 2011 to 2016. In 2016, China exported over 400 varieties of pesticides, of which the top 10 were glyphosate paraquat, atrazine, chlorothalonil, chlorpyrifos, imidacloprid, acephate, carbendazim, clethodim, and sulfentrazone.

In 2016, the pesticide originals of China were mainly exported to USA, Brazil, India, Australia and Argentina. The preparations were mainly exported to Brazil, Australia, Thailand, Indonesia and Vietnam. In terms of quantities, China's pesticides were mainly exported to Asia (33%) and South America (28%), followed by Africa (14%), Europe (9%), North America (8%) and Oceania (8%).

Amendment of the Pesticide Management Regulations

The legislative management of Chinese pesticides began from the promulgation of the Regulations on Pesticide Administration of the People's Republic of China in 1997. The implementation of this regulation is a milestone in China's history of pesticide management. Since then, the pesticide registration system has been implemented. In 1999, the Ministry of Agriculture of the People's Republic of China (MOA) issued Measures for the Implementation of Regulation on Pesticide Administration of the People's Republic of China as its legal basis and guidance. In 2001, The decision to amend the Regulations on Pesticide Administration of the People's Republic of China was passed by the State Council of China. In 2006, China promulgated the Agricultural Product Quality and Safety Law of the People's Republic of China to elevate the quality and safety of agricultural products from a legal dimension. Three years later, the Food Safety Law of China was introduced.

In order to adapt to changes in the national industrial management and operation pattern of agricultural production, in 2007 the Legislative Affairs Office of the State Council and the Ministry of Agriculture started to revise the 1997 version of the pesticide management regulations. The revised draft was intended to cancel the provisional registration of pesticides and to change to a pesticide business license system in terms of the pesticide marketing regulations. It also inserted the requirement for a pesticide traceability system, encouraging the use of low toxicity biological pesticides, and strengthening the obligation of a business to deliver guidance on pesticide use. It also sought to cancel the provisional registration of pesticide, which allows marketing of a pesticide product without submission of the residue data for dietary risk assessment and MRL elaboration prior to the application for registration. The residue chemistry data for a pesticide product would now need to be committed during initial registration. This would reduce the risk of residual contamination, even with excessive residue from the agro-products.

In 2017, the new Regulations on Pesticide Administration of the People's Republic of China was passed by the State Council of China. Strengthening the pesticide marketing management would regulate pesticide sales and technical service, improve the level of pesticide application technology, and supervision management, especially, and adapt to the intensive, professional, and improved change in the current cultivation system. It is hoped to reduce pesticide residue pollution, and promote the level of quality and safety of agricultural products.

The Pesticide Reduction Plan

Under pressure from the increasing demand of food, Chinese agriculture is on the way to a higher level of intensification. What is remarkable is the increasing trends for the input of fertilizer and pesticide. However, due to the agricultural market environment, family decentralized management, and poor training of the agricultural labor force, space to improve the level of agricultural technology is limited. All these build a constraint on the quality and safety of agricultural products. Over the years, on account of the expanding crop acreage and increasing difficulty of pest control and prevention, the use of pesticides overall is on an upward trend. According to statistics, from 2012 ~ 2014, the average annual pesticide use on crop pest control and prevention reached 311000 tons. Compared to the 2009~2011 period, there was an increase of 9.2%.

The extensive use of pesticides, the low level of pesticide application technology, and the poor quality of pesticide spreading machinery, has resulted in increasing production costs, high residue levels in agricultural products, crop phytotoxicity, and environmental pollution. In 2015, the Ministry of Agriculture prepared and issued the Zero-Growth Action Plan for pesticide use for the following five years until 2020. In the same year, the Ministry of Agriculture, with the National Development and Reform Commission, the Ministry of Science, the Ministry of Finance, the Ministry of Land Resources, the Ministry of Environmental Protection, the Ministry of Water Resources, and the State Forestry Administration, jointly issued the "National Sustainable Agricultural Development Plan (2015—2030)". It aims to overcome the current contradiction in the crop industry between agricultural production and environment protection, improve food quality, and achieve a sustainable effective supply of high quality agricultural products. In the next few years, in accordance with the guidance for changing the agricultural development pattern, control of agricultural non-point source pollution, cost savings and an increase in effective agriculture, the phasing out of highly toxic pesticides, reduction of the overall risk of relying on new agricultural business entities, vigorous promotion of large-scale pest control and prevention, and professional services, guidance of farmers to use pesticides scientifically in a rational way, and improvement of pesticide utilization will all be achieved. The measures are expected to gradually realize the goal of control of total pesticide use at nil growth levels compared to the average usage levels between 2012 and 2014.

Pesticide Regulation and Management Systems

The scope of pesticide management in China is to regulate a substance or a mixture of substances chemically synthesized or originating from biological and other natural substances and the formulations made from these substances used for (i) preventing, destroying or controlling diseases, pests, weeds, and other harmful organisms detrimental to agriculture, forestry, and public health; (ii) regulating the growth of plants and insects, such as insecticides, antiseptics, herbicides, plant growth regulators, rodenticides, hygiene pesticide; (iii) GMO products and natural enemies.

There are four main goals of pesticide management of China: (i) controlling crop pests to ensure agricultural production; controlling disease-bearing insects to prevent breakout of epidemics; (ii) minimizing the adverse effects to protect human health and environmental safety; (iii) building a fair and competitive environment for the pesticides market; (iv) promoting import and export of pesticide.

The development of Chinese pesticide management can be divided into five stages: (i) in the 1950s~1960s, focus on preventing actual poisoning and product quality control; (ii) in the 1970s, emphasis on the safe use of pesticides; (iii) in the 1980s, establishment of the registration system (1982); (iv) in 1990s, production premising system set up; distribution premising system regulatory residue monitoring system, import and export management system, Regulation on Pesticide Administration issued in 1997; (v) 21st century: priority in safety management.

Future Direction of Pesticide Regulation in China

Philosophy and emphasis of pesticide management has changed in China. The priority of pesticide management in China is changing from quality control to safety management, also from supervision only to service and guidance also. The main tasks are (i) implementation of legal framework to improve regulations on safety management such as strengthening marketing permits; (ii) strengthening legal infrastructure, rigorously enforcing registration rules and regulations, enhancing safety requirements thus strengthening legislative and safety management; (iii) improving risk assessment evaluation procedures for more scientific, fair transparence; (iv) participating in international activities, especially on the international harmonization of pesticide management system and requirements. After 35 years of efforts, China has made great progress on pesticide management. As a large and diverse country of pesticide production and usage, China still faces many challenges to establish pesticide regulatory system that promotes agricultural development, and adequate protection to the environment, farmers, and consumers.

The principle of China's pesticide management in the future is to fulfill the following five requirements: (i) requirement of reforming agricultural economical structure; (ii) requirement of enhancing pesticide quality/safety and human healthy; (iii) requirement of promoting pesticide industry progress; (iv) requirement of developing sustainable agriculture; (v) requirement to increase China's competitiveness.

Selected from: Ohkawa H, Miyagawa H, Lee P W. Pesticide Chemistry. Weinheim: WILEY-VCH Verlag GmbH, 2007: 23-28.

Qiao, X. Pesticide Residues. Food Safety in China, 2017: 201-218.

Zhang W J. Global pesticide use: Profile, trend, cost/benefit and more. Proceedings of the International Academy of Ecology and Environmental Sciences, 2018, 8 (1): 1-27.

Words and Expressions

methamidofos 甲胺磷
parathion /ˌpærəˈθaɪən/ *n*.[农药] 对硫磷
methyl parathion 甲基对硫磷
monocrotophos /mɒnəukrəutəˈfəuz/ 久效磷
phosphamidon /fɒsˈfæmɪdən/ *n*.[化] 磷胺
pyraclostrobin /paɪrəkˈlɒstrəubɪn/ 吡唑醚菌酯
mesotrione /meˈsətriən/ 硝磺草酮
thiamethoxam /θaɪəmɪðɒkˈsæm/ 噻虫嗪
difenoconazole /daɪfenəuˈkənəzəuliː/ 苯醚甲环唑
azoxystrobin /əzɒksɪstˈrəubɪn/ 嘧菌酯
avermectin /əvəˈmektɪn/ 阿维菌素
tebuconazole /tiːbuːˈkənəzəul/ 戊唑醇
nicosulfuron /nɪkəuˈsʌlfuərɒn/ 烟嘧磺隆
guadipyr 戊吡虫胍
triflumezopyrim 三氟苯嘧啶
flupyradifurone 氟吡呋喃酮
Sophora alopecuroides 苦豆子
Metarhizium anisopliae 绿僵菌，金龟子绿僵菌
Bacillus methylotrophicus 甲基营养型芽孢杆菌
Bacillus amyloliquefaciens 解淀粉芽孢杆菌
2,4-dichlorophenoxybutyric acid sodium butyrate 2,4-二氯苯氧丁酸钠
halauxifen-methyl 氟氯吡啶酯
benzobicylon 双环磺草酮
pyraclonil 双唑草腈
chlorothalonil /ˌkləurəˈθælənil/ *n*.[农药] 百菌清
chlorpyrifos /kˈlɔːpɪrɪfəuz/ 毒死蜱
imidacloprid /ɪmɪdækˈlɒprɪd/ 吡虫啉
acephate /əˈsefət/ 乙酰甲胺磷
carbendazim /kɑːˈbendeɪzɪm/ *n*. 多菌灵

clethodim /kliːˈθɒdɪm/ 烯草酮
sulfentrazone 甲磺草胺
promulgation /ˌprɒmlˈgeɪʃn/ *n*. 颁布，公布，宣传，普及
antiseptics /ˌænti'septiks/ *n*.［助剂］防腐剂（antiseptic 的复数）

Notes

［1］the State Development and Reform Committee（SDRC）：国家发展与改革委员会。

［2］GMO（Genetically Modified Organism）：转基因生物，基因改造产品或基因食品。

［3］the Regulations on Pesticide Administration of the People's Republic of China：中华人民共和国农药管理条例，1997 年 5 月 8 日中华人民共和国国务院令第 216 号发布，2001 年 11 月 29 日决议修改《农药管理条例》，2017 年 2 月 8 日国务院第 164 次常务会议通过修订。

［4］the Agricultural Product Quality and Safety Law：《中华人民共和国农产品质量安全法》，2006 年颁布实施，2019 年农业农村部公布了《中华人民共和国农产品质量安全法修订草案（征求意见稿）》。

［5］Zero-Growth Action Plan："零增长的行动计划"。农业部 2015 年发布了《到 2020 年农药使用量零增长行动方案》。力争到 2020 年，主要农作物农药使用量实现零增长。

Exercises

Ⅰ Answer the following questions according to the text.

1. What are the five highly toxic organophosphorus pesticides? please gives examples.
2. Please briefly introduce the pesticide registration in China.
3. Please briefly analyze the import and export of pesticides in China.
4. Please briefly introduce the process of pesticide management in China.
5. What is Zero-Growth Action Plan of pesticide?
6. Please briefly introduce China's pesticide regulation and management system.

Ⅱ Translate the following English phrases into Chinese.

efficiency enhancement traceability system technical service pest control
provisional registration cultivation system hygiene pesticide poor quality
pesticide management utilization rate cost savings ratecruciferous vegetable

Ⅲ Translate the following Chinese phrases into English.

零增长 管理条例 残余污染 上升趋势 公共卫生
农产品 营业执照 应用技术 专业服务 分散经营

Ⅳ Choose the best answer for each of the following questions according to the text.

1. Where does China rank in the world in the production and use of pesticides?
   A. the first B. the secend C. the third D. the fourth
2. Which pesticides have the most registrations?
   A. insecticides B. fungicides C. herbicides D. bactericides
3. In 2016，which formulations was registered the most in China?

A. emulsifiable concentrates　　B. wettable powder

C. suspension concentrates　　D. water dispersible granules

4. What are the substances used for, in the scope of pesticide management in China to regulate?

A. preventing, destroying or controlling diseases, pests, weeds, and other harmful organisms detrimental to agriculture, forestry, and public health

B. regulating the growth of plants and insects, such as insecticides, antiseptics, herbicides, plant growth regulators, rodenticides, hygiene pesticide

C. GMO products and natural enemies

D. all of the above

5. "Implementation of Regulations on Pesticide Administration" was issued by MOA in the year of ________.

A. 1997　　B. 1998　　C. 1999　　D. 2001

6. How many times did the Regulations on Pesticide Administration amended?

A. one　　B. two　　C. three　　D. four

Ⅴ Translate the following short paragraphs into Chinese.

1. The revised draft was intended to cancel the provisional registration of pesticides and to change to a pesticide business license system in terms of the pesticide marketing regulations. It also inserted the requirement for a pesticide traceability system, encouraging the use of low toxicity biological pesticides, and strengthening the obligation of a business to deliver guidance on pesticide use.

2. In 2015, the Ministry of Agriculture prepared and issued the Zero-Growth Action Plan for pesticide use for the following five years until 2020. In the same year, the Ministry of Agriculture, with the National Development and Reform Commission, the Ministry of Science, the Ministry of Finance, the Ministry of Land Resources, the Ministry of Environmental Protection, the Ministry of Water Resources, and the State Forestry Administration, jointly issued the "National Sustainable Agricultural Development Plan (2015—2030)".

3. The principle of China's pesticide management in the future is to fulfill the following five requirements: (i) requirement of reforming agricultural economical structure; (ii) requirement of enhancing pesticide quality/safety and human healthy; (iii) requirement of promoting pesticide industry progress; (iv) requirement of developing sustainable agriculture; (v) requirement to increase China's competitiveness.

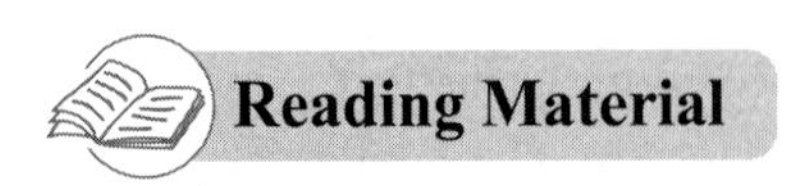

International Conventions

The Basel Convention on the Control of Transboundary Movements of Hazardous Wastes and Their Disposal was adopted in 1989 in response to concerns about toxic waste from industrialized countries being dumped in developing countries and countries with economies in transition. The Convention's principal focus was to elaborate the controls on the "transboundary" movement of hazardous wastes and the development of criteria for environmentally sound management of wastes. More recently, the work of the Convention has emphasized full implementation of treaty commitments, promotion of the environmentally sound management of hazardous wastes and minimization of hazardous waste generation. The Convention became effective on 5 May 1992.

Rotterdam Convention

One of the concerns about movement of highly hazardous pesticides was that some countries were importing pesticides that were likely to cause health problems unless the spray operators were adequately protected while spraying. The FAO set up a voluntary Prior Informed Consent (PIC) procedure in 1989 to ensure that the importing country had full details of the product and how it should be applied before consent was given for importing it. This led to the Rotterdam Convention in 1998, which established legally binding standards of conduct involving information exchange to enable an importing country to follow the PIC procedure if it needed to use a highly hazardous pesticide. The Convention became effective on 24 February 2004.

Stockholm Convention

Following Rachel Carson's book Silent Spring, use of organochlorine insecticides, especially DDT, was prohibited, but considering the wider problem of persistent chemicals in the environment, the Stockholm Convention on Persistent Organic Pollutants was adopted on 22 May 2001 and came into force as a global treaty on 17 May 2004. The aim of the treaty is "to protect human health and the environment from chemicals that remain intact in the environment for long periods, become widely distributed geographically, accumulate in the fatty tissue of humans and wildlife, and have harmful impact on human health or the environment". This requires a global effort to eliminate or reduce the release of POPs into the environment. In relation to pesticides, a key development has been efforts to ensure that stockpiles and waste contaminated with POPs are managed safely and in an environmentally sound manner, which has involved identifying obsolete stocks, repackaging them where necessary and transporting them to suitable incinerators, following relevant international rules, standards and guidelines. Initially, much attention was given to disposal of dieldrin, which had been strategically stored in areas where locust outbreaks were likely to occur. The ban on organochlorines created the problem of disposal of the existing stocks of dieldrin, where, as previously, any withdrawal of approval required using up existing stocks within two years. One benefit of the ban was the funding of research on a biological mycoinsecticide based on *Metarhizium acridum* that was effective against locusts.

Montreal Protocol

The Montreal Protocol, signed in 1987 and since amended many times, relates to the use of ozone-depleting chemicals. Banning the use of chlorofluorocarbons (CFCs) used in aerosol spraycans was possible by substituting alternatives. The main problem has been in relation to protection of stored grain and fumigation of soils where methyl bromide was extremely effective. Exemptions have allowed its use in quarantine and pre-shipment of goods, and research has continued to find alternative methods of control.

Minamata Convention

Mercuric chloride had been recommended in the 19th century in a wash with soap applied to the base of apple trees to prevent borers attacking the trees. It was also used as a fungicide to control scab on potatoes. In 1929, it was replaced with mercurous chloride, as it had a lower mammalian toxicity. Mercuric oxide was also commercially available to treat wounds on trees after pruning and control of canker on fruit, rubber and other trees and shrubs. Their use was discontinued, later reinforced by the Minamata Convention to protect human health and the environment from anthropogenic emissions and releases of mercury and mercury compounds. The Convention is named after the Japanese city Minamata, which had suffered a devastating incident of mercury poisoning.

Selected from: Matthews G A. A History of Pesticides. Boston: CABI Press, 2018: 229-232.

Words and Expressions

International Conventions 国际惯例，国际公约，国际条约
transboundary 跨境，跨界
Rotterdam Convention 鹿特丹公约
Prior Informed Consent (PIC) 事先知情同意
Stockholm Convention 斯德哥尔摩公约
Persistent Organic Pollutants 难降解有机污染物，持久性有机污染物
stockpile /'stɒkpaɪl/ *n.* 库存，积蓄；*vt.* 贮存，储蓄；*vi.* 积累，储备物资
dieldrin /'diːldrɪn/ *n.* [农药] 狄氏剂
Metarhizium acridum 绿僵菌
mercuric chloride 氯化汞，升汞
mercurous chloride 氯化亚汞，甘汞
Minamata Convention 水俣公约

Unit 28 Pesticide Regulation in USA

Pesticides used in agriculture to control pests, such as insects, weeds, and plant diseases, have been subject to considerable legislative, regulatory, and consumer scrutiny over the past few decades. Pesticides are, by their nature, toxic chemicals; since many pesticides may potentially leave residues on foods available for human consumption, there is much concern regarding the potential health risks of pesticides in the human diet. Concerns stem from possible risks of acute poisoning from exposure to large amounts of pesticides consumed in a short duration as well as from chronic risks from exposure to low levels of pesticide residues over extended periods of time. In the United States, particular concern has been raised as to the potential carcinogenic effects of pesticides consumed in the diet. More recent concerns have emerged regarding the potential of some pesticides to affect endocrine functions. Federal regulatory activities have also recently focused on ensuring that the risks faced by infants and children consuming pesticides are not excessive.

In addition to consumer concerns, pesticide use also presents risks to agricultural workers involved in the mixing, loading, or application of pesticides and to those working in fields treated with pesticides. Occupational illnesses and injuries resulting from pesticide use have been reported frequently and some epidemiological evidence has linked specific pesticides with increases in specific types of cancers among those exposed occupationally. Pesticides also present environmental concerns including water and soil contamination, air pollution, destruction of natural vegetation, reductions in natural pest populations, effects upon non-target organisms including fish, wildlife, and livestock, creation of secondary pest problems, and the evolution of pesticide resistance.

Advances in crop protection technology have actually resulted in a small decrease in overall pesticide use in the United States since 1979. Much of this reduction may have resulted from widespread adoption of Integrated Pest Management (IPM) techniques that stress the judicious use of pesticides in combination with other cultural, physical, and biological practices. Nevertheless, public concerns about pesticide residues in foods are still strong, with some surveys indicating that the vast majority of consumers consider pesticide residues in foods to represent a serious health threat. Such consumer concern has led some retail outlets to work with private laboratories to certify that their produce is free of detectable pesticide residues, and has likely contributed to the dramatic growth of the organic foods industry. While much attention has focused on the potential health and environmental risks posed by pesticides, the benefits of pesticides also require consideration. It is clear that pesticide use is frequently associated with increases in crop yield and reductions in crop loss. One estimate indicates that the economic benefits from the use of pesticides in developed countries range from \$3.50 to \$5.00 for every dollar spent on pesticides, and that 40% of the world's food supply would be at risk if pesticides were not available. In a developed nation like the United States, the use of pesticides undoubtedly has resulted in a greater availability of fruits, vegetables, and grains at lower consumer costs. Surveillance programs comprise one type of several different programs put in place by the US government and by US food producers in a comprehensive effort to regulate and manage pesticides; the results of such programs provide one indication of the effectiveness of such pesticide regulatory and management efforts.

Regulation of pesticides at the US federal level involves three agencies: the Environmental Protection Agency (EPA), the Food and Drug Administration (FDA), and the Department of Agriculture (USDA). Individual states also have the authority to regulate pesticides and are allowed to apply pesticide use restrictions that are more stringent than those established federally. The most comprehensive state pesticide regulatory program is in California; other large state pesticide regulatory programs exist in Florida, Texas, Michigan, and New York.

Agency Responsibilities

The primary law governing pesticide regulation in the United States is the Federal Insecticide, Fungicide, and Rodenticide Act (FIFRA) of 1947. FIFRA is primarily a risk/benefit balancing statute; if the benefits of specific uses of a pesticide are deemed to outweigh their risks, the pesticide is allowed for such specific uses. Some of the potential risks, as discussed above, include human health effects and environmental factors; benefits may include greater production efficiency, lower food costs, and public health protection.

The USDA was initially assigned the responsibility of implementing FIFRA. When the EPA was created in the early 1970s, it assumed FIFRA responsibility for approving and revoking pesticide registrations.

Currently, the EPA spends much effort reviewing toxicological and environmental fate data submitted by a pesticide's manufacturer in a request to have the pesticide registered for use. A full battery of toxicological tests must be performed before a pesticide can be registered for use on a food crop; such studies investigate the pesticide's acute, subchronic, and chronic exposure, carcinogenicity, teratogenicity, mutagenicity, and metabolic fate, among other things. The manufacturer of the pesticide must also provide studies on the pesticide's potential effects on non-target organisms, food residue studies, and studies of the environmental behavior of the pesticide. It may take ten years or more for the studies to be conducted and submitted by the manufacturer and reviewed by the EPA; the cost of completing the studies may often exceed $30 million.

If the use of the pesticide may have the potential to leave residues on foods, the EPA is also required to determine the maximum allowable residues on the crops. Tolerances are pesticide and commodity specific; the same pesticide may have different tolerances established on different commodities while the same commodity may have several different tolerances established for various pesticides that are allowed for use on the commodity.

The FDA is the primary US federal agency involved in pesticide residue monitoring and in enforcing pesticide tolerances for domestic and imported foods shipped in interstate commerce. The FDA performs its regulatory monitoring program to enforce pesticide tolerances. In this program, sampling is not random as the FDA targets the types and origins of commodities considered most likely to present violative residues. Violative residues may occur when pesticide residues are found that exceed the established tolerances, but are more commonly associated with the finding of a residue of a pesticide on a commodity for which no tolerance has been established.

The FDA also performs its annual Total Diet Study, in which food samples are collected as "market baskets" from four geographical regions in the United States and from three cities in each region. Each market basket comprises 257 or 258 different food items obtained from retail outlets and the food items are all prepared for consumption prior to analysis. Unlike the FDA's regulatory monitoring program, the Total Diet Study is not designed to enforce pesticide tolerances but rather to provide an estimate of dietary pesticide residue exposure to the general US population and to specific US population subgroups.

The USDA conducts two pesticide residue monitoring programs as well. USDA's National Residue Program obtains samples of meat, poultry, and raw eggs that are analyzed for pesticide residues as well as for animal drugs and environmental contaminants. The USDA initiated its Pesticide Data Program (PDP) in 1991. In contrast with FDA's regulatory monitoring programs that are designed primarily to enforce pesticide tolerances, PDP's sampling protocols are designed to more accurately reflect residue levels that reach consumers and, as such, provide a more reliable tool for assessing human dietary risk from exposure to pesticides. In 2001, ten states participated in collecting and analyzing samples in the PDP program.

Establishing Pesticide Tolerances

Since pesticide residue surveillance programs frequently are designed to enforce pesticide tolerances, it is critical to understand how pesticide tolerances are established and their significance with respect to human health or agricultural practices. Tolerances are frequently, and incorrectly, considered to represent safety standards while violative residues are frequently implied to represent unsafe residues. In actuality, tolerances have a valuable role as indicators of proper pesticide use but should not be considered as barometers of safety. The seemingly counterintuitive practices used to establish pesticide tolerances are described in detail by Winter and are summarized below. In general, tolerances are established at levels that represent the maximum residues that might be expected on commodities resulting from "worst-case" application conditions such as maximum allowed application rate, maximum number of applications per growing season, and harvest at the minimum legal interval following the final application. Pesticide manufacturers interested in obtaining tolerances for their pesticides on specific commodities perform a series of field studies in a variety of geographical locations. Once samples have been taken and the residues analyzed, the manufacturer petitions the EPA to establish a tolerance at or above the maximum residue detected from the "worst-case" field studies. As a result, tolerances should be considered as enforcement tools to indicate whether pesticide application practices have been performed in accordance with directions; residues detected in excess of tolerances most likely would occur only in cases in which pesticide applications were not made properly.

While the tolerance values themselves are not health-based, it should not be implied that the potential health effects from exposure to pesticide residues are not considered. The EPA, before approving a pesticide tolerance, will perform its own risk assessment to ensure that exposure to the pesticide from its proposed uses as well as from its existing uses, is at acceptable levels to ensure a reasonable certainty of no harm. If exposure is deemed to be acceptable, the EPA will accept the manufacturer's petition to establish the tolerance at 280 Pesticide, veterinary and other residues in food above the maximum level encountered from the field trials. If the expected exposure to the pesticide is determined to be excessive, the EPA will not approve the tolerance or will approve the tolerance only if other uses of the pesticide are eliminated to ensure that exposure is acceptable.

Prior to 1996, the risk assessments used to determine whether pesticide tolerances should be established considered individual pesticides on a case-by-case basis and considered only dietary exposure to pesticides. Passage of the 1996 Food Quality Protection Act (FQPA) expanded EPA's responsibility to consider the "aggregate" exposure to the pesticide from food, water, and from residential sources. In addition, in cases where several different individual pesticides share a common mechanism of toxicological action, the risk assessments must include consideration of the "cumulative" exposure to all of the pesticides sharing the common mechanism rather than just individual pesticides. Consideration of the potential increased susceptibility of infants and children to pesticides is also an important provision of FQPA. The EPA is required to consider all of these factors in its efforts to reassess all pesticide tolerances by August 2006; several pesticide tolerances have already been revoked resulting from the increased scrutiny required by FQPA. It should be emphasized, however, that the levels established for pesticide tolerances are still set to equal or slightly exceed the maximum residues found in the manufacturers' field trials. As such, the tolerances still represent enforcement tools and should not be confused as safety standards even though the EPA does consider possible health risks prior to establishing tolerances.

Selected from: Watson D H. Pesticide, Veterinary and other Residues in Food. Florida: CRC Press LLC, 2004: 278-281.

Words and Expressions

scrutiny /'skruːtəni/ *n*. 详细审查，监视，细看，选票复查

duration /dju'reɪʃn/ *n*. 持续时间，为期

chronic /'krɒnɪk/ *adj*. 慢性的，延续很长的

carcinogenic /ˌkɑːsɪnə'dʒenɪk/ *adj*. 致癌物（质）的

endocrine /'endəʊkrɪn; 'endəʊkraɪn/ *n*. 内分泌

occupational /ˌɒkjuˈpeɪʃənl/ *adj*. 职业的，占领的
epidemiological /ˌepɪˌdiːmiəˈlɒdʒɪkl/ *adj*. 流行病学的
vegetation /ˌvedʒəˈteɪʃn/ *n*. [植]（总称）植物，呆板单调的生活
judicious /dʒuˈdɪʃəs/ *adj*. 明智的
retail /ˈriːteɪl/ *n*. 零售；*adj*. 零售的；*vt*. 零售，转述；*vi*. 零售
stringent /ˈstrɪndʒənt/ *adj*. 严厉的，迫切的
deem /diːm/ *v*. 认为，相信
revoke /rɪˈvəʊk/ *vt*. 撤回，废除，宣告无效
teratogenicity /ˌterətəʊdʒɪˈnɪsətɪ/ *n*. [生] 畸形形成性，致畸性
mutagenicity /ˌmjuːtədʒəˈnɪsətɪ/ *n*. [生] 诱变（性）
interstate /ˈɪntəsteɪt/ *adj*. 州际的
barometer /bəˈrɒmɪtər/ *n*. 气压计
counterintuitive /ˌkaʊntərɪnˈtjuːɪtɪv/ *adj*. 违反直觉的
petition /pəˈtɪʃn/ *n*. 情愿书，诉状；*v*. 请求，恳求，请愿

Notes

[1] the Total Diet Study：全膳食分析。美国 30 余年来，一直由食品和药物管理局开展年度市场菜篮子监测，即全膳食分析，以评估食物摄入导致的农药残留。

[2] The Food Quality Protection Act：美国食品质量保护法，该法旨在保障农产品安全、保护儿童权益和解决法律体系不一致性的问题。

Exercises

Ⅰ Answer the following questions according to the text.

1. Give some examples to present the potential risks of pesticides to human's health and natural environment safety.

2. What are the benefits of the use of pesticides in agriculture? Please give some examples to illustrate some of these benefits.

3. Should tolerances be considered to represent safety standards? Why or why not?

4. How is a pesticide tolerance established in US? Please describe briefly the process that a tolerance is approved by the EPA.

Ⅱ Translate the following English phrases into Chinese.

carcinogenic effect　total diet study　residue surveillance　acceptable level
occupational illness　short duration　cumulative exposure　residue analyze
non-target organisms　health and environmental risk

Ⅲ Translate the following Chinese phrases into English.

急性中毒　慢性中毒　次生害虫　农药忍受度　农药使用限制
空气污染　违规残留　土壤污染　自然植被　自然有害物数量

Ⅳ Choose the best answer for each of the following questions according to the text.

1. According to the text, which of the statements is not true about the effects of

pesticide to human's health?

A. There exist possible risks of acute poisoning from exposure to large amounts of pesticides consumed in a short duration.

B. Risks don't exit from exposure to low levels of pesticide residues over extended periods of time.

C. Some pesticides consumed in the diet may have carcinogenic effects.

D. Some pesticides may affect endocrine functions.

2. Which is not the benefit of use of pesticides in agriculture?

A. Pesticide use increasingly leads to increases in crop yield and reductions in crop loss.

B. 40% of the world's food supply would be at risk without pesticides.

C. The use of pesticides undoubtedly has resulted in a greater availability of fruits, vegetables, and grains at lower consumer costs.

D. The use of pesticides has, to a great extend, been protecting the safety of environment.

3. As to the regulation of pesticide at the US federal and state level, which statement is true?

A. At present, the USDA has the responsibility of implementing FIFRA.

B. Approving and revoking pesticide registrations is EPA's responsibility.

C. Individual states are permitted to apply pesticide use restrictions but less stringent than those established federally.

D. Individual states also have the authority to regulate pesticides.

4. Before a pesticide can be registered and for use on a food crop, what kinds of work must be done by the manufacturer?

A. performing a full battery of toxicological tests

B. submitting the data of the toxicological and environmental fate to EPA

C. providing studies on the pesticide's potential effects on non-target organisms, food residue studies, and studies of the environmental behavior of the pesticide

D. All of the above

5. Which statement is true about the Total Diet Study?

A. In annual Total Diet Study, food samples are collected as "market baskets" from four geographical regions in the United States and from three cities in each region.

B. The Total Diet Study is designed to enforce pesticide tolerances.

C. The Total Diet Study is designed to provide an accurate dietary pesticide residue exposure to the general US population and to specific US population subgroups.

D. None of the above.

6. All of the following statements are true regarding the regulatory responsibilities of the federal level agencies except ________.

A. Currently, EPA spends much effort reviewing toxicological and environmental fate data submitted by a pesticide's manufacturer.

B. The FDA is the only US federal agency involved in pesticide residue monitoring.

C. The USDA also conducts two pesticide residue monitoring programs.

D. The purpose of FDA's regulatory monitoring programs is different from that of the USDA's.

7. Which of the following statements about FQPA is not correct?

A. FQPA expanded EPA's responsibility.

B. FQPA considers the potential increased susceptibility of infants and children to pesticides.

C. Food Quality Protection Act was established before 1996.

D. Because of increased scrutiny required by FQPA, several pesticide tolerances have already been revoked resulting.

Ⅴ Translate the following short paragraphs into Chinese.

1. Pesticides used in agriculture to control pests, such as insects, weeds, and plant diseases, have been subject to considerable legislative, regulatory, and consumer scrutiny over the past few decades. Pesticides are, by their nature, toxic chemicals; since many pesticides may potentially leave residues on foods available for human consumption, there is much concern regarding the potential health risks of pesticides in the human diet.

2. It should be emphasized, however, that the levels established for pesticide tolerances are still set to equal or slightly exceed the maximum residues found in the manufacturers' field trials. As such, the tolerances still represent enforcement tools and should not be confused as safety standards even though the EPA does consider possible health risks prior to establishing tolerances.

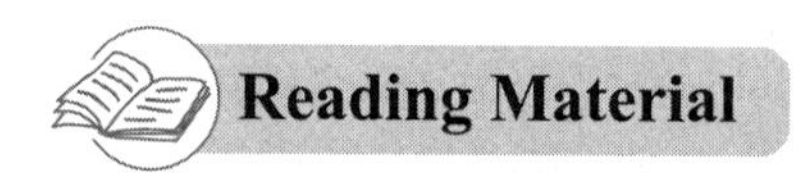

The Federal Insecticide, Fungicide, and Rodenticide Act

The regulation of pesticides by the federal government began in 1910 with the passage of the Federal Insecticide Act by Congress. This act was passed in response to concerns from the United States Department of Agriculture (USDA) and farm groups about the sale of fraudulent or substandard pesticide products, which was common at the time. Thus, the first federal pesticide legislation was to ensure the quality of pesticide chemicals purchased by consumers. Specifically, the act set standards for the manufacture of Paris green, lead arsenate, insecticides, and fungicides, and also provided for inspections, seizure of adulterated or misbranded products, and prosecution of violators.

Little change in the regulation of pesticides occurred until after World War Ⅱ, when the use of synthetic organic pesticides in agriculture became widespread. Because of the potential for damage to humans and wildlife by these more widely used chemicals, additional regulation of pesticides was deemed necessary by Congress. In 1947, the Federal Insecticide, Fungicide, and Rodenticide Act (FIFRA) was passed. This act extended the coverage of the

Federal Insecticide Act to include herbicides and rodenticides and required that all pesticide products be registered with the USDA before their sale in interstate or foreign commerce. It further required that products have certain information on their labels, such as the manufacturer's name and address, a list of ingredients, directions for proper use, and warning statements to protect users, the public, and nontarget species of plants and animals. The FIFRA of 1947 served to establish labeling standards for pesticides. However, it did not provide the USDA with an effective mechanism for removing misbranded or hazardous products from the market.

The FIFRA was amended in 1959 and 1964. The 1959 amendment added nematicides, plant regulators, defoliants, and desiccants to the definition of a pesticide or "economic poison". The amendment in 1964 mandated that pesticide products have a federal registration number and that labels contain signal words relating to the toxicity of the products (i. e., Warning, Caution, Danger). Also, the Secretary of Agriculture was authorized to suspend immediately registrations of pesticides believed to pose an imminent hazard to the public.

Publication of *Silent Spring* by Rachel Carson in 1962 focused public attention on the risks of pesticides such as DDT to human health and the environment, bringing about major changes in the regulation of pesticides. In 1970, the Environmental Protection Agency (EPA) was created as part of the Executive Branch of the federal government and charged with the administration of the FIFRA. Then, in 1972, the FIFRA was significantly changed under a new law titled the Federal Environmental Pesticide Control Act (FEPCA). The new law required that all pesticides be registered with the EPA and classified by the agency for general or restricted use. Certification was required for persons applying restricted use pesticides. Most importantly, FEPCA set a new registration standard for pesticides, allowing the registration of a pesticide only if it could be determined that the pesticide would not cause unreasonable adverse effects on the environment. With this law, the emphasis on pesticide regulation in the United States moved from quality assurance and adequate labeling of pesticide products to the protection of public health and the environment from their potential hazards.

Further amendments to the FIFRA occurred in 1975, 1978, 1980, and 1981. These amendments were designed to improve the pesticide registration process and provide for consideration of agricultural benefits of pesticides in regulatory decisions made by the EPA. The 1975 amendment required the EPA to notify the Secretary of Agriculture in advance of regulatory decisions affecting pesticides, consider the impact of pesticide cancellations on the production and costs of agricultural commodities, and establish a Scientific Advisory Panel to review proposed regulations. The amendment to the FIFRA in 1978 allowed the EPA, under certain circumstances, to grant conditional registrations of pesticides before all supporting data had been submitted by the registrant. This provision was intended to reduce the backlog of registration applications.

In 1975, the EPA began to review the registrations of pesticides issued before August 1975 to bring older chemicals up to current registration standards. A process called the Rebuttable Presumption Against Registration or RPAR was adopted to further review those pesticides found to pose a risk or concern. In this process, the agency weighs the risks versus the benefits of a chemical in making a decision on its continued registration. During the period from 1976 to 1988, the EPA's review or reregistration of older pesticides was proceeding at a much slower pace than originally anticipated by the EPA or Congress. Thus, the FIFRA was amended again in 1988 to establish a nine-year schedule for completion of the reregistration of pesticide active ingredients registered before November 1984 and to impose substantial fees on registrants to cover much of the costs of the reregistration. Only a few pesticide registrations have been canceled or suspended by the EPA as a result of reregistration; however, thousands have been canceled voluntarily by registrants because of the fees and data requirements. Pesticide reregistration likely will continue into the beginning of the next century.

As a result of existing regulations under the FIFRA, six to nine years and $50 million or higher are required for a new pesticide active ingredient to progress from the laboratory to the marketplace. Two to three years alone are needed to obtain a complete EPA pesticide registration. Data requirements for this registration can include 70 specific tests and cost the company registering the pesticide up to $10 million.

In 1992, the EPA revised the Worker Protection Standard for Agricultural Pesticides. The revisions to the worker protection standard required pesticide manufacturers to modify product labels to restrict the entry of workers into pesticide-treated areas, specify the use of personal protective equipment, and require notification of workers about areas treated with pesticides. Employers of agricultural workers and pesticide handlers were required to protect their employees by providing safety training, safety posters, pesticide health and safety information, decontamination sites, and emergency assistance when needed.

Selected from: Toth S J. Pesticide Impact Assessment Specialist. North Carolina Cooperative Extension Service, 1996.

Words and Expressions

fraudulent /'frɔːdjələnt/ *adj*. 欺诈的，欺骗性的，骗得的
paris green *n*. [化] 巴黎绿，乙酰亚砷酸铜
lead arsenate 砷酸铅
misbrand /mɪs'brænd/ *vt*. 贴错标签，贴假商标于
backlog /'bæklɒg/ *n*. 大木材，订货
rebuttable /rɪ'bʌtəbl/ *adj*. 可辩解的，可反驳的

PART XV
PESTICIDES DESIGN

Unit 29 Analogue Design of Pesticide

What are Analogues?

The term analogue, derived from the Latin and Greek analogia, has been used in natural science since 1791 to describe structural and functional similarity. Extended to pesticides, this definition implies that the analogue of an existing molecule shares chemical and active similarities with the original compound. This definition allows the establishment of three categories of analogues:

(i) analogues possessing chemical and active similarities

(ii) analogues possessing only chemical similarities

(iii) chemically different compounds displaying similar activities

The first class of analogues, those that have both chemical and active similarities, can be considered as direct analogues. These analogues correspond to the category of pesticides often referred to as "me-too" drugs. Usually they are improved versions of a "pioneer" pesticide with activites over the original. Direct analogue design involves straightforward molecular modifications, such as the synthesis of homologues, vinylogues, isosteres, positional isomers, optical isomers, modified ring systems and homodimers. As a rule, the basic scaffold is conserved or only slightly modified. Fine-tuning can be achieved by means of substituent effects.

The second class, comprising structural analogues, contains compounds originally prepared as close and patentable analogues of a novel lead, but for which biological assays revealed totally unexpected activites. for example, fluacrypyrin was originally designed as an analogue of the strobilurin fungicides, but it revealed acaricidal activity.

For the third class of analogues, chemical similarity is not observed; however, they share common biological properties. We propose the term functional analogues for such compounds. Functional analogues were, for a long time, the unpredictable result of fortuitous observations. for example, The insecticidal activity of ryania extract was the combination of ryanodine and the equipotent and more abundant 9,21-dehydroryanodine. there are now modern synthetic ryanoids, which include chlorantraniliprole, cyantraniliprole and flubendiamide.

Systematic Screening

This method consists of screening new molecules, whether they are synthetic or of natural origin, on an animal model or on any biological test without having in mind hypotheses about its activites potential. Systematic screening can be achieved in two different manners. The first one applies a very exhaustive activites investigation to a small number of chemically sophisticated and original molecules. This is known as "extensive screening". In contrast, the second one strives to find, among a great number of molecules, one that could be active in a given indication. This is "random screening". Now, High-throughput screening (HTS) has become an important technology to rapidly test thousands of biochemical molecules.

1. Extensive Screening

Extensive screening is generally applied to new chemical entities coming from an chemical research or from a natural source. For such molecules, the high investment in synthetic or extractive chemistry justifies an extensive activity to detect if there exists an interesting potential linked to these new structures. The extensive screening approach has often led to original molecules. It is, however, highly dependent on the skill and the intuition of the researchers. Besides the target-based pesticide discovery strategy followed by the researchers during the last decades, phenotypic screening has been revisited with the emergence of biochemical and molecular biology techniques allowing medium throughput screening on systems-based assays.

2. Random Screening

Random screening is one of the common approaches for pesticide discovery. This method screens many compounds to identify the compounds with relatively high biological activity. Thousands of compounds are often screened to obtain the lead compound. Although this method has a low hit rate, new lead compounds, new targets, and new functional mechanisms can be discovered. However, with the development of pesticide research, obtaining new varieties of pesticides by using this method is becoming more difficult.

3. High-Throughput Screening

The arrival of robotics in the 1980s along with the miniaturization of the in vitro testing methods meant it became possible to combine the two preceding approaches. In other words, screen millions of compounds on a large number of biological targets. HTS is usually applied to the displacement of radioligands and to the inhibition of enzymes. The present trend is to replace radioligand-based assays with fluorescence-based measurements. Primary sources can also be crudely purified vegetal extracts or fermentation fluids. In this latter case, one proceeds to the isolation and to the identification of the responsible active principle only when an interesting activity is observed.

Because it is now possible for a company to screen several million molecules simultaneously on 3050 different biochemical tests, the problem becomes feeding the robots interesting molecules and consequently designing "smart" compound collections. To this end, the role of the researcher is crucial in the preparation of the compound collection to the hit validation triage step and beyond; these steps involve multiple tasks where the interaction with chemoinformaticians plays a major role.

Virtual Screening

Systematic screening, despite being a fruitful method for identifying functional analogues, must be applied to real libraries made of hundreds or thousands of compounds. This is not the case for virtual screening, an approach based on computer-aided detection in a database of molecules that present similar pharmacophores.

Synthesis Methods

1. Isosterism

The search for analogues is, very often, aided by the concept of isosteric replacement and the derived concept of bioisosterism. Isosteric replacement requires substitution of one atom or group of atoms in the parent compound for another with similar electronic and steric configuration. For example, the oxygen atom of an analogue is successively replaced by the NH and CH_2 isosteric groups. This method has a strong focus and a high hit rate and is a popular method for pesticide discovery. However, due to a pesticide's structural similarities to the mother compound, its biological activity and target may also be similar; therefore, the pesticide may easily cause resistance, and sometimes property rights issues might be involved.

2. Bioisosterism

The term bioisosterism is used to describe replacements of fragments of the original molecule that are more significant in terms of overall affinity and function. Example is the replacement of the cyclopentene function of pyrethrins Ⅰ with open cycle group to give empenthrin with the high activities (Figure 2).

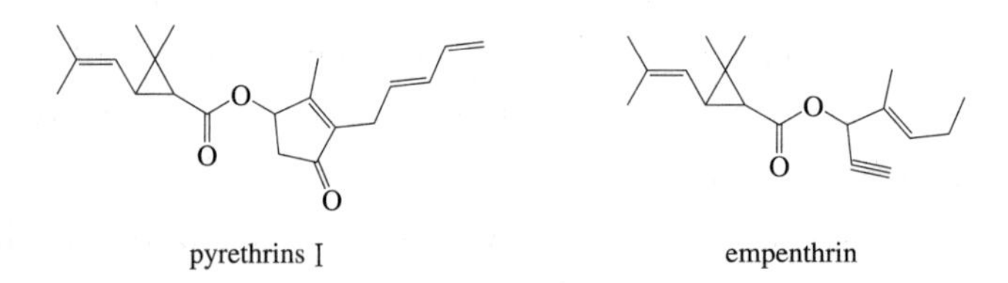

Figure 2 the structures of bioisosterism synthesis

3. Scaffold Hopping

Taken in its broadest meaning, bioisosterism can consist of a complete replacement of the initial molecular structure by another one, resulting in a structure that is isofunctional but based on a different scaffold. This approach, called scaffold hopping, can be considered an extreme application of the bioisosterism concept. It is illustrated by the conversion of fenvalerate, a broad-spectrum effective insecticide, to its functional analogues, thiofluoximate (Figure 3).

4. Biomimetic Synthesis

Biomimetic synthesis is the synthesis and modification of the unique substances of specific

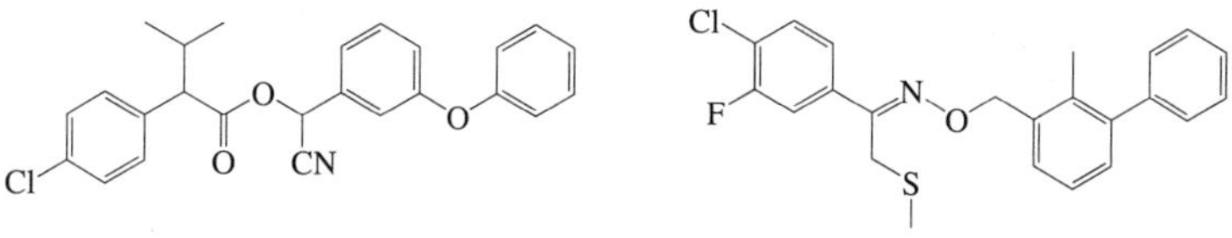

fenvalerate

thiofluoximate

Figure 3 the structures of scaffold hopping

plants, animals, and microorganisms, such as the unique substances generated in plants to avoid predation and competition, and the sex pheromones and aggregation pheromones generated in insects for attraction and aggregation. Using these natural substances and their metabolites, lead compounds were selected by activity screening, and then, the new pesticides were developed by structure identification, synthesis, and modification. Using this method, some compounds with novel structure, unique mechanism, and good environmental compatibility can be obtained.

5. Targeted Synthesis

Targeted synthesis, also known as biological rational design, is a method that uses reverse thinking to design compounds that fit a specific mechanism, mainly based on the known three-dimensional structure of the receptor, to synthesize the targeted compounds using computersimulation design. This method, which follows a "lock and key" process, can greatly improve development efficiency. Other methods are to try a large number of keys to unlock a lock. Targeted synthesis configures an appropriate key only after understanding the specific three-dimensional structure of the "lock", thereby greatly improving the efficiency of the "unlocking". This method of discovery is currently receiving much attention.

6. Asymmetric Synthesis

The chiral pesticides with a single enantiomer have the advantages of low dosage, high efficacy, and good environmental compatibility, which are in line with today's development trend for pesticides. Chiral asymmetric synthesis has been widely used in medicine, but because of the high cost of synthesis and technical difficulty, the production and application of chiral pesticides is not common but has a great development potential.

Selected from: Wermuth C G. Similarity in drugs: reflections on analogue design. Drug Discovery Today, 2006, 11 (7-8): 348-354.

Pan X L, Dong F S, Wu X H, *et al*. Progress of the discovery, application, and control technologies of chemical pesticides in china. Journal of Integrative Agriculture, 2019, 18 (4): 840-853.

Words and Expressions

analogue /'ænəlɒg/ *n*. 类似物，类似情况，对等的人；*adj*. 类似的，相似物的

me-too /ˌmiː'tuː/ *adj*. 仿效他人的，应声附和的，仿造的；*n*. 跟屁虫

straightforward /ˌstreɪt'fɔːwəd/ *adj*. 简单的，坦率的；*adv*. 直截了当地，坦率地

homologue /'hɒməlɒg/ *n*. 同族体，同系物，相同器官

vinylogue /'vɪnɪlɒg/ *n*. 插烯物

isostere /'aisəustiə/ *n*.［物］同电子排列体，（电子）等排物，等排性
homodimer /həumәu'daɪmə/ *n*.同（源）二聚体，同源双体
fine-tuning /ˌfaɪn 'tuːnɪŋ/ *n*.微调，细调；*v*.调整（fine-tune 的 ing 形式）
fluacrypyrin 嘧螨酯
strobilurin fungicide 甲氧基丙烯酸酯类杀菌剂
acaricidal /ækærɪ'saɪdl/ 杀螨的
fortuitous /fɔː'tjuːɪtəs/ *adj*.偶然的，意外的，幸运的
ryania /ra'ɪeɪniə/ 鱼尼丁属
ryanodine /raɪənəu'daɪn/ 鱼尼丁，兰尼碱
equipotent /ˌiːkwɪ'pəutənt/ *adj*.（能力、力量等）均等的
dehydroryanodine 脱氢鱼尼丁
ryanoid /'raɪənɔɪd/ 莱恩碱
chlorantraniliprole /klɔːrænt'rænɪlɪprəul/ 氯虫苯甲酰胺
cyantraniliprole 溴氰虫酰胺
flubendiamide /fluːbenda'ɪæmaɪd/ 氟虫双酰胺
systematic screening 系统筛选
exhaustive /ɪg'zɔːstɪv/ *adj*.详尽的，彻底的，消耗的
sophisticate /sə'fɪstɪkeɪt/ *v*.弄复杂，使变得世故，曲解；*adj*.老于世故的
revisit /ˌriː'vɪzɪt/ *vt*.重游，再访，重临；*n*.再访问
hit rate *n*.［计］命中率
miniaturization /ˌmɪnətʃəraɪ'zeɪʃn/ *n*.小型化，微型化
radioligand /ræd'ɪɒlɪgənd/ *n*.放射性配体
fluorescence /flə'resns/ *n*.荧光，荧光性
fermentation /ˌfɜːmen'teɪʃn/ *n*.发酵
chemoinformatician 化学信息学家
pharmacophore /'fɑːməkəˌfɔː/ *n*.药效团，药效基因
isosterism /ˌaɪsəu'sterɪzəm/［物］同电子排列性
bioisosterism /'biːəuaɪsɒstərɪzəm/ 生物电子等排
resistance /rɪ'zɪstəns/ *n*.阻力，电阻，抵抗，反抗，抵抗力
affinity /ə'fɪnəti/ *n*.密切关系，吸引力，姻亲关系，类同
cyclopentene /saikləu'pentiːn/ *n*.［有化］环戊烯
pyrethrins Ⅰ /'paɪriːθrɪnz/ *n*.除虫菊素Ⅰ
empenthrin 右旋烯炔菊酯
scaffold /'skæfəuld/ *n*.脚手架，鹰架，绞刑台；*vt*.给……搭脚手架，用支架支撑
scaffold hopping 骨架跃迁
fenvalerate /fenvələ'reɪt/ *n*.［农药］氰戊菊酯
thiofluoximate 硫氟肟醚
receptor /rɪ'septər/ *n*.［生化］受体，接受器，感觉器官
enantiomer /ɪ'næntiəmɚ/ *n*.［有化］对映体，［数］［有化］对映异构体
chiral asymmetric synthesis 手性不对称合成

Notes

[1] virtual screening：虚拟筛选，基于小分子数据库开展的活性化合物筛选。利用小分子化合物与药物靶标间的分子对接运算，虚拟筛选可快速从几十万至上百万分子中，遴选出具有成药性的活性化合物，大大降低实验筛选化合物数量，缩短研究周期，降低药物研发的成本。

[2] High-Throughput Screening（HTS）：高通量筛选，指以分子水平和细胞水平的实验方法为基础，以微板形式作为实验工具载体，以自动化操作系统执行试验过程，以灵敏快速的检测仪器采集实验结果数据，以计算机分析处理实验数据，在同一时间检测数以千万的样品。

Exercises

Ⅰ Answer the following questions according to the text.

1. What are analogues? What are the three categories of analogues?
2. What screening methods are included in systematic screening?
3. What is virtual screening?
4. What synthetic methods are available for developing pesticides? Please list their detailed synthesis methods.

Ⅱ Translate the following English phrases into Chinese.

analogue design	targeted synthesis	isosteric replacement	basic scaffold
lead compound	structural analogue	asymmetric synthesis	chiral pesticide
biological assay	functional analogue	extensive screening	natural origin

Ⅲ Translate the following Chinese phrases into English.

高通量筛选	专利类似物	取代效应	表型筛选	计算机模拟设计
位置异构物	化学相似性	杀虫活性	性信息素	生物电子等排
光学异构体	聚集信息素	随机筛选	仿生合成	化学信息学家

Ⅳ Choose the best answer for each of the following questions according to the text.

1. What are the three categories of analogue design?
   A. analogues possessing chemical and active similarities
   B. analogues possessing only chemical similarities
   C. chemically different compounds displaying similar activities
   D. all of the above
2. If chemical and active are similar，what analogue is belong to?
   A. direct analogues　　B. functional analogues
   C. chemical similarities　　D. all of the above
3. If only biological properties are similar，what analogue is belong to?
   A. direct analogues　　B. functional analogues
   C. chemical similarities　　D. all of the above
4. What are the three systematic screening methods?

A. Extensive screening B. Random screening

C. High-Throughput Screening D. all of the above

5. The oxygen atom of an analogue is successively replaced by the NH and CH_2 groups, What kind of synthesis method is it?

A. Isosteric replacement B. Bioisosterism

C. Biomimetic synthesis D. Targeted synthesis

6. Replacements of fragments of the original molecule, What kind of synthesis method is it?

A. Isosteric replacement B. Bioisosterism

C. Biomimetic synthesis D. Targeted synthesis

7. What are the synthetic methods for designing and synthesizing pesticide molecules using these natural substances and their metabolites as lead compounds?

A. Isosteric replacement B. Bioisosterism

C. Biomimetic synthesis D. Targeted synthesis

8. What kind of synthesis method is based on the known three-dimensional structure of the receptor to synthesize the compounds using computersimulation design?

A. Isosteric replacement B. Bioisosterism

C. Biomimetic synthesis D. Targeted synthesis

Ⅴ Translate the following short passages into Chinese.

1. Direct analogue design involves straightforward molecular modifications, such as the synthesis of homologues, vinylogues, isosteres, positional isomers, optical isomers, modified ring systems and homodimers.

2. This method consists of screening new molecules, whether they are synthetic or of natural origin, on an animal model or on any biological test without having in mind hypotheses about its activites potential.

3. Random screening is one of the common approaches for pesticide discovery. This method screens many compounds to identify the compounds with relatively high biological activity.

4. Isosteric replacement requires substitution of one atom or group of atoms in the parent compound for another with similar electronic and steric configuration.

5. Biomimetic synthesis is the synthesis and modification of the unique substances of specific plants, animals, and microorganisms, such as the unique substances generated in plants to avoid predation and competition, and the sex pheromones and aggregation pheromones generated in insects for attraction and aggregation.

Combinatorial Chemistry

The probability that a directly isolated natural product will become a drug for a given disease is relatively low, except perhaps in the realm of antibiotics. However, the strength of the natural products approach is that these "base molecules" can serve both as leads to new active structures and as probes for new mechanisms of action. In a similar fashion, combinatorial biosynthesis can be utilized to produce what are now being called "unnatural natural products", where the biosynthetic machinery of a microbial cell is dissected and the relevant genes are "mixed and matched", followed by expression in a suitable heterologous host. Such compounds may be used in their own right or could be the starting materials for further synthetic modifications. In addition, novel methods of chemical syntheses that have the potential to produce base "natural product" molecules that can be optimized for specific medicinal chemistry purposes are now being reported.

Detailed analyses of active natural product skeletons have led to the identification of relatively simple key precursor molecules that form the building blocks for use in combinatorial synthetic schemes that have produced numbers of potent molecules, thereby enabling structure-activity relationships to be probed. Thus, in the study of the structure-activity relationships of the epothilones, solid-phase synthesis of combinatorial libraries was used to probe regions of the molecule important to retention or improvement of activity.

The approach above was further developed by the use of an active natural product as the central scaffold in the combinatorial approach in order to generate large numbers of analogues for structure-activity studies, the so-called "parallel synthetic approach". This embodies the concept of "privileged structures", originally proposed by Evans *et al.* and advanced further by Nicolaou *et al.* using benzopyrans as scaffolds to produce "libraries from libraries" that had very significantly different biological activities than the starting base structure. The Waldmann group developed the so-called "Biology-Oriented Synthesis of Natural Product-Inspired Libraries (BIOS)" approach. In the BIOS approach, a natural product inhibiting a protein target through interaction with a specific protein topology represents a biologically validated starting point for the development of closely related structures. These products may inhibit proteins with similar folds, since there are approximately a 1000 different topological folds in proteins. This approach is similar to the privileged structure concept, but with the added dimension of using protein topology patterns as the basis for subsequent screens.

A somewhat similar approach to protein-ligand interactions has been developed by the Quinn group, who hypothesized that the interior Protein Fold Topology (PFT) characteristics of key enzymes involved in the biosynthetic pathways of a particular class of natural products should mimic the PFT characteristics of the active site of proteins targeted by members of that natural product class. This was found to be the case with flavonoids, where the PFT characteristics of the relevant biosynthetic enzymes shared PFT characteristics of the phosphoinositol-3-kinase (PI3K) active site that is targeted by flavonoids. Thus, a comparison of the biosynthetic PFT characteristics of natural products with the PFT characteristics of various targets should permit the identification of potential targets for the natural product classes, and vice versa. Quinn's process can be thought of as "inside to target" whereas Waldmann's can be from "outside to target".

In the early days of combinatorial chemistry, the emphasis was on production of massive libraries of what turned out to be "flat" compounds. These did not generate many leads, and later enthusiasm waned for this approach. The use of natural product-like compounds as scaffolds, referred to as "Diversity Oriented Synthesis", led to the generation of smaller, more meaningful combinatorial libraries. This move was exemplified by the work of the Schreiber group, who combined the simultaneous reaction of maximal combinations of sets of natural-product-like core structures ("latent intermediates") with peripheral groups ("skeletal information elements") in the synthesis of libraries of over 1000 compounds bearing significant structural and chiral diversity.

Selected from: Newman D J, Cragg G M, Kingston D G I. Natural Products as Pharmaceuticals and Sources for Lead Structures. The Practice of Medicinal Chemistry, 2015: 128-130.

Words and Expressions

antibiotics /ˌæntɪbaɪˈɒtɪks/ *n*.［药］抗生素，抗生学
combinatorial biosynthesis 组合生物合成
heterologous /ˌhetəˈrɒləgəs/ *adj*. 不同的，不齐的，不等的，异种的
combinatorial chemistry 组合化学
precursor /priˈkɜːsə(r)/ *n*. 先驱，前导，前质，前兆
structure-activity relationship 构效关系
parallel synthetic approach 平行合成方法
privileged structure 优势结构
benzopyran 苯并吡喃
topology /təˈpɒlədʒi/ *n*. 拓扑学
flavonoid /ˈfleɪvənɔɪd/ *n*. 黄酮类，［有化］类黄酮
phosphoinositol-3-kinase 磷酸肌醇-3-激酶
peripheral /pəˈrɪfərəl/ *adj*. 外围的，次要的；*n*. 外部设备

Unit 30 Computer-Aided Design

Computer-aided drug design (CADD) is the coalescence of information on chemical structures, their properties, and their interactions with biological macromolecules. Further, these data are transformed into knowledge intended to aid in making better decisions for drug discovery and development. Because most of drugs and pesticides are small chemical molecules, the methods and technology based on the computer are same.

Historical Evolution

Assistance from computational chemistry and bioinformatics is necessary to handle the vast and ever-increasing relevant data to be analyzed. With the advent of computers and the ability to store and retrieve chemical information, serious efforts to compile relevant databases and construct information retrieval systems began. Collecting crystal structure information for small molecules was one of the first efforts that had a substantial long-term impact. The Cambridge Structural Database (CSD) stores crystal structures of small molecules and provides a fertile resource for geometrical data on molecular fragments for calibration of force fields and validation of results from computational chemistry. As protein crystallography gained momentum, the need for a common repository of macromolecular structural data led to the Protein Data Bank (PDB), originally located at Brookhaven National Laboratories and presently at Rutgers University. These efforts focused on the accumulation and organization of experimental results on the three-dimensional structure of molecules, both large and small.

As more chemical data accumulated with its implicit information content, a multitude of approaches began to extract useful information. Certainly, the shape and variability in geometry of molecular fragments from CSD was mined to provide fragments of functional groups for a variety of purposes. As compounds were tested for biological activity in a given assay, the desire to distill the essence of the chemical requirements for such activity to guide optimization by concepts, such as bioisosteres and pharmacophores, was generated. Initially, efforts focused on congeneric series as the common scaffold presumably eliminating the molecular alignment problem with the assumption that all molecules bound with a common orientation of the scaffold. This was the intellectual basis of the Hansch approach (quantitative structure-activity relationship, QSAR), in which substituent parameters from physical chemistry were used to correlate chemical properties with biological activity for a series of compounds with the same substitution pattern on the congeneric scaffold.

Considerable literature developed around the ability of numerical indices, derived from graph-theoretical considerations, to correlate with SAR data. The ability of various indices proved to be useful parameters in QSAR equations, such as surface area and molecular volume. Since computational time was at a premium during the early days of QSAR and such numerical indices could be calculated with minimal computations, they played a useful role and continue to be used predictively today. Work using QSAR led to the development of three-dimensional pharmacophores as descriptors molecular-recognition motifs both for medicinal chemistry, as well as for discerning protein-ligand interactions.

QSAR

An understanding of the many and diverse interactions of various chemicals with biological macromolecules as determined by their intermolecular forces, i. e. , hydrophobic, electrostatic, polar, and steric, was critical to the formulation and development of QSAR paradigm. The availability of rapidly expanding libraries of compounds, and the increased number of biological targets, created a need for a more organized approach to molecular design and development. The use of correlation analysis was useful in helping to mathematically delineate the importance of certain structural attributes of chemicals to their biological activities.

Pharmacophores

The word pharmacophore takes on two separate definitions: The first, attributed to Ehrlich, is defined as the set of atoms that a compound must possess for it to be active in a particular biological test. Even today, this usage of the word is common with medicinal chemists. However, here three-dimensional (3D) pharmacophores will be explored in detail. A 3D pharmacophore is defined as that set of properties and their arrangement in 3D that a compound must possess for it to be active in a particular biological test.

Pharmacophores form a key strategy in the ligand-based design of biologically active molecules. On the one hand, they typically are a part of 3D QSAR models derived to forecast potency. The pharmacophore identification can be performed either as a prelude to the actual QSAR analysis or with pharmacophore discovery coupled with the QSAR into one computer program. Additionally, pharmacophores form the query for 3D database searches that have the goal of discovering novel compounds with the desired biological properties. Lastly, one may use a pharmacophore model to superimpose compounds from two quite different series so as to use information from the better-explored series to propose where modifications in the less-explored series will be optimal and where they will not be tolerated.

Molecular Docking

The docking program is used to place many different computer-generated representations of a small molecule in a target structure (or in a user-defined part thereof, e. g. , the active site of an enzyme) in a variety of positions and orientations. Each such docking mode is called a "pose". In order to identify the energetically most favorable pose, each pose is evaluated (scored) based on its complementarity to the target in terms of shape and properties such as electrostatics. A good score indicates that molecule is potentially a good binder. This process is repeated for all molecules in the collection, which are subsequently rank-ordered by their scores (i. e. , their predicted affinities). This rank-ordered list is then used to select for purchase, synthesis, or biological investigation only those compounds that are predicted to be most active. Assuming that both the poses and the associated affinity scores have been predicted accurately, this selection will contain a relatively large proportion of active molecules, i. e. , it will be "enriched" with actives compared to a random selection.

Virtual Screening

With the increasing number of therapeutic targets, the need for a rapid search for small molecules that may bind to these targets is of crucial importance in the drug discovery process. One way of achieving this is the in silico or virtual screening (VS) of large compound collections to identify a subset of compounds that contains relatively many hits against the target, compared to a randomly selected subset. The compounds that are virtually screened can stem from commercial compound collections or virtual compound libraries. If a structure of the small molecule is known, ligand-based virtual screening could be considered. If a 3D structure or model of the target is available, structure-based virtual screening would be used.

De novo Design

Automated de novo drug design is a computational process whereby structural coordinate data of the site, together with a design strategy, are used as input to a de novo design algorithm. The algorithm then designs molecular structures to fit the site optimally without further human intervention. Potential candidate ligand structures are delivered as output from the algorithm. Depending on the type of de novo design algorithm, numerous different ligand structures can be generated from the same input and can be postprocessed for synthetic tractability and druglike properties. This approach offers the designer a set of alternative chemotypes from which to choose a number of different ligand series to explore. De novo design methods share a lot of characteristics with methods for virtual screening and library design.

If a fast throughput of protein structural data is achieved, the drug design process will need to be accelerated to cope with the expected avalanche of structural data. Automated de novo design methods have the capability of handling this mass of data to generate structures to fit crystallographically determined binding sites. The huge advantage of de novo design methods is that they can be used to assess in silico large numbers of potential structures for their fit to the site before any synthesis is embarked upon. In that way de novo design offers an overwhelming advantage to pharmaceutical companies seeking to seize a patent estate for the most promising lead compounds.

Selected from: John B T, David J T. Comprehensive Medicinal Chemistry Ⅱ Volume 4: Computer-Assisted Drug Design. Elsevier, 2006: 13-284.

Words and Expressions

coalescence /ˌkəuəˈlesns/ *n*. 合并，联合，接合

macromolecule /ˌmækrəˈmɒlɪkjuːl/ *n*. [高分子] 高分子，[化学] 大分子

bioinformatics /biːəuɪnˈfɔːmətɪks/ *n*. 生物信息学，生物资讯

crystallography /ˌkrɪstəˈlɒgrəfi/ *n*. 晶体学
geometrical *adj*. 几何学的，几何图案的；*n*. 几何形状
calibration /. kæləˈbreɪʃən/ *n*. 校准，校正，定标
bioisosteres *n*. 生物电子等排体，生物电子等排体原理
pharmacophore *n*. 药效团
congeneric /ˌkɒndʒɪˈnerɪk/ *adj*. 同源的，同属的，同族的，同种类的
scaffold /ˈskæfəʊld/ *n*. 脚手架，断头台，绞刑架，建筑架
substituent /sʌbˈstɪtʃʊənt/ *n*. 代替者，[化] 取代基
premium /ˈpriːmiəm/ *n*. 保险费，奖金，津贴；*adj*. 优质的，昂贵的
hydrophobic /ˌhaɪdrəˈfəʊbɪk/ *adj*. 疏水性的，疏水的
electrostatic /ɪˌlektrəʊˈstætɪk/ *adj*. 静电的
steric /ˈsterɪk/ *adj*. 空间的；*n*. 立体，空间位阻
forecast /ˈfɔːrkɑːst/ *n*. 预测，预报；*v*. 预测，预报
identification /aɪˌdentɪfɪˈkeɪʃən/ *n*. 鉴定，确定，辨认，确认
prelude /ˈpreljuːd/ *n*. 前奏，序幕；*v*. 开头，为……作序
superimpose /. suːpərɪmˈpəʊz/ *v*. 使重叠，使叠加，使附加于
enzyme /ˈenzaɪm/ *n*. 酶，酵素
de novo 重新；更始
crystallographically /ˌkrɪstəˈlɒgrəfi/ *n*. 结晶学；晶体学
enrich /ɪnˈrɪtʃt/ *v*. 丰富，充实，改进，使得到发展
druglike *n*. 类药物，类药
momentum /məˈmentəm/ *n*. 势头，[物] 动量，动力，冲力
in silico 计算机模拟的

Notes

[1] Cambridge Structural Database (CSD)：剑桥晶体结构数据库。

[2] Protein Data Bank (PDB)：蛋白质晶体结构数据库。

[3] Brookhaven National Laboratories：布鲁克海文国家实验室（BNL），位于纽约长岛萨福尔克县（Suffolk county）中部，隶属美国能源部，由石溪大学和Battelle成立的公司布鲁克海文科学学会负责管理。该实验室成立于1947年，历史上该实验室所获得的发现曾5次获得过诺贝尔奖。

[4] quantitative structure-activity relationship (QSAR)：定量构效关系。

Exercises

Ⅰ Answer the following questions according to the text.

1. What is the definition of CADD?
2. How many methods are discussed during the CADD? Please list.
3. Which methods could be used for ligand-based virtual screening?
4. Which methods could be used for structure-based virtual screening?

Ⅱ **Please write the full names of the following abbreviations.**

CADD　　CSD　　PDB　　QSAR　　VS

Ⅲ **Translate the following English phrases into Chinese.**

computational chemistry	molecular-recognition	molecular fragments
biological macromolecules	computer-aided drug design	de novo design
protein-ligand interactions	three-dimensional structure	lead compound

Ⅳ **Translate the following Chinese phrases into English.**

官能团　分子排列　分子设计　分子对接　力场

表面积　分子体积　药物化学　生物靶标　随机选择

Ⅴ **Choose the best answer for each of the following questions according to the text.**

1. Which of the following is not the method of CADD according to the text?

A. QSAR　　B. pharmacophore　　C. VS　　D. synthesis

2. Which method could not be used, if only a structure of the target is known?

A. QSAR　　B. pharmacophore　　C. VS　　D. molecular docking

3. Which of the following statements is true about CADD?

A. CADD doesn't need experimental data.

B. Compared to CADD, the traditional approach to drug discovery is really time-consuming and cost-intensive.

C. CADD could replace the traditional expiment approach.

D. CADD could not simulate the interactions between ligands and protein.

4. The intermolecular forces between various chemicals and biological macromolecules include ________.

A. hydrophobic　　B. electrostatic　　C. steric　　D. all of the above

Ⅵ **Translate the following sentences into Chinese.**

1. Computer-aided drug design (CADD) is the coalescence of information on chemical structures, their properties, and their interactions with biological macromolecules. Further, these data are transformed into knowledge intended to aid in making better decisions for drug discovery and development.

2. In order to identify the most favorable pose, each pose is scored based on its complementarity to the target in terms of shape and properties such as electrostatics. A good score indicates that molecule is potentially a good binder.

3. The huge advantage of de novo design methods is that they can be used to assess in silico large numbers of potential structures for their fit to the site before any synthesis is embarked upon. In that way de novo design offers an overwhelming advantage to pharmaceutical companies seeking to seize a patent estate for the most promising lead compounds.

Virtual Screening

The aim of all R&D activities in crop protection companies is to discover new active ingredients with ideal properties at lowest cost in the quickest time. This simple principle led to the introduction of a multitude of new technologies designed to accelerate the classical research process, e. g. , combinatorial chemistry and high-throughput screening. However, not each chemical needs to be purchased or synthesized to test its potency in expensive field trials. Very often, knowledge derived from previous experiments and an expert's intuition are sufficient to judge the potential of compounds with reasonable quality. Increased capacity of chemicals, however, resulting from new high-throughput techniques, prevents a rating by hand. Therefore, virtual screening strategies are increasingly applied to separate "good from bad" compounds at a very early stage.

1D and 2D Descriptors

Only a few compounds screened in early lead identification phases are synthesized in-house. More flexible and cost effective is to purchase chemicals from external suppliers. Most vendors provide lists of some ten to hundred thousand chemicals on compact discs and guarantee delivery within days to weeks. To explore this huge amount of data with the aid of computers, chemical information is transformed to computer-readable strings, e. g. , smiles code, and different descriptors are determined. 1-dimensional (1D) descriptors encode chemical composition and physicochemical properties, e. g. , molecular weight, stoichiometry, hydrophobicity, etc. 2D descriptors reflect chemical topology, e. g. , connectivity indices, degree of branching, number of aromatic bonds, etc. 3D descriptors consider 3D shape, volume or surface area.

Akin to Lipinski's "rule of five" that predicts a poor absorption of orally administered pharmaceuticals in case of exceedingly more than one of four particular molecular properties, i. e. , mass, clogP, number of hydrogen bond donors and acceptors, Briggs presented his "rule of 3" for agrochemical compounds. Thus, bioavailability is likely to be poor when three or more of his limits are exceeded.

3D Descriptors

1D and 2D descriptors are fast and inexpensive to calculate and therefore ideal for screening of large libraries with millions of compounds. Sometimes, however, geometrical features like shape or a particular localization of functional groups in three-dimensional (3D) space are necessary to describe molecules unambiguously. In this situation, it is necessary to use conformers of active (and non-active) compounds as reference templates for a ligand-based screening strategy. A second, even more promising concept may be applied in the presence of the 3-D coordinates of the molecular target protein. This so-called structure-based ligand design uses the binding site as a lock in order to find virtually the best fitting key, i. e. , new ligand.

Ligand-based Screening Strategies

Output from biochemical high-throughput screenings is a list of active compounds with indicated activity values, e. g. , IC_{50}. On 3D level, relevant conformers of active molecules may be superimposed in order to locate similar molecular functions crucial for biochemical activity in the same 3D space. The resulting agrophore model, synonymous with the pharmacophore model for drugs, should be able to differentiate active and non-active compounds by the presence or absence of particular functions in specific regions. Based on the spatial distribution of molecular key functions, 3D pharmacophore fingerprints may be extracted and used to detect similar molecules in 3D compound libraries.

While 3D pharmacophore/agrophore searches indicate best matches of relevant key functions, i. e. , (semi) qualitative result, a linear regression derived from a statistical analysis of spatial features and measured activities yields quantitative structure-activity relationships (3D QSAR). This may be extremely helpful for a detailed interpretation of existing results and the activity prediction of new or hypothetical compounds.

Structure-based Screening Strategies

Virtual screening concepts are most successful if coordinates of the molecular target protein are available. In this situation, the binding site of the target protein (the structure) is used as a lock in order to find the best fitting keys (ligands). Since this approach is independent of any further ligand information, it allows an unbiased probing of new scaffolds.

A structure-based virtual screening starts with the docking of each compound into the binding site pocket. Based on complementary features, best fitting poses are searched for each ligand. In general, three principal algorithmic approaches are used to dock small molecules into macromolecular binding sites. The first one separates the conformer generation of the ligands from the placement in the binding site. Subsequently, all relevant low-energy conformations are rigidly placed in the binding site [e. g. , GLIDE (Schrodinger, Inc.) and FRED (OpenEye Scientific Software)]. Only the remaining six rotational and translational degrees of freedom of the rigid conformer have to be considered. A second class of algorithms aims at optimizing the conformation and orientation of the molecule in the binding pocket simultaneously. The tremendous complexity of this combined optimization problem needs stochastic algorithms such as genetic algorithms or Monte Carlo simulations. The third class of docking algorithms exploits the fact that most molecules contain at least one small, rigid fragment that forms specific, directed interactions with the receptor site. Such so-called base fragments are docked rigidly at various initial positions. An incremental construction process explores the torsional conformational space adding now new fragments (e. g. , Flexx). Some programs apply further force field minimizations in order to optimize the actual pose of the fragments (e. g. , Glide XP, eHits, ICM).

Subsequent to the detection of appropriate binding poses, scoring functions are used to estimate the free energy of ligand binding for each conformer. Commonly used scoring functions may be divided into three general categories. Most important are empirical scoring functions. They approximate the free energy of binding as a weighted sum of terms, each term being a function of the ligand and protein coordinates. Each term describes different types of interactions such as lipophilic contacts or hydrogen bonds between receptor and ligand. The weighting factors are derived by multiple linear regression to experimental binding affinities or by approximate first principle considerations. A second class of scoring functions is based on molecular force fields, summing up the electrostatic and van der Waals interaction energies between receptor site and ligand. Knowledge-based scoring functions are derived from statistical analyses of experimentally determined protein-ligand X-ray structures. The underlying assumption is that inter-atomic contacts occurring more frequently than average are energetically favorable. Knowledge-based scoring functions are sums of many atom-pair contact distributions for protein and ligand atom type combinations.

The accuracy of docking procedures may be evaluated in two ways. First, by detecting known ligand-binding site poses of crystallized complexes, and second, by the enrichment of known active compounds of a mixed library, i. e. , enrichment factor.

Selected from: Ohkawa H, Miyagawa H, Lee P W. Pesticide Chemistry. Weinheim: WILEY-VCH Verlag GmbH, 2007: 77-87.

Words and Expressions

descriptor /dɪ'skrɪptər/ *n*. 描述符，描述子，描述器
bioavailability /ˌbaɪəʊəˌveɪlə'bɪlɪtɪ/ *n*. 生物利用度，生物有效率
fingerprint /'fɪŋgər. prɪnt/ *n*. 指纹，指印
conformation /ˌkɒnfɔː'meɪʃn/ *n*. 构象，构型
scoring function 打分函数，评分函数
free energy 自由能
hydrogen bond 氢键
enrichment factor 富集因子，浓缩系数

APPENDIXES

Appendix 1　专业词根、前缀、后缀

a-　从，自，无
　asymmetric　不均匀的，不对称的
　abiotic　无生命的，非生物的
-acetal　醛缩醇
acetal-　乙酰
acet-　醋；醋酸；乙酸
　acetylcholinesterase　乙酰胆碱酯酶
　acetylation　乙酰化作用
acid　酸，酸的
　acidify　使酸化
　shikimic acid　莽草酸
　boric acid　硼酸
　ascorbic acid　抗坏血酸，维生素 C
　indoleacetic acid　吲哚乙酸
　cyclopropanecarboxylic acid　环丙基甲酸
　oxalic acid　草酸
aer　空气
　aerial　航空的，空气的
　aerosol　浮质，气雾剂
　anaerobic　厌氧性的
ag　行动
　agency　代理，中介，作用
　agitate　搅动，煽动，激动
agr　农田
　agriculture　农业
　agrarian　田地的
-al　醛
-aldehyde　醛
　aldehyde　醛，乙醛
　cinnamaldehyde　肉桂醛
alkali-　碱的，碱性的
　alkaline　碱的，碱性的
　alkaloid　生物碱，植物碱基
allyl　丙烯基
alkoxy-　烷氧基
-amide　酰胺
-amidine　脒
-amine　胺
　glucosamine　葡（萄）糖胺，氨基葡（萄）糖
　octopamine　真蛸胺，章（鱼）胺
amino-　氨基
　amino acid　氨基酸
　aminomethylphosphonic acid　氨甲基膦酸
　ammonia　氨，氨水
-ane　烷
　hexane　己烷
　borane　硼烷
anhydride　酐
anilino-　苯胺基
　dinitroaniline　硝基苯胺
ant-　反对
　antagonism　对抗（状态），对抗性
　antagonist　敌手，对手
　antibiotics　抗生素，抗生学
　antibody　抗体
　antifeedant　拒食素
　antioxidant　抗氧化剂，硬化防止剂
aquo-　含水的
　aquatic　水的，水生的，水栖的
　aqueous　水的，水成的
aryl-　芳（香）基
　aryloxyphenoxypropionate　芳氧苯氧丙酸酯
-ase　酶
　carboxylesterase　羧酸酯酶
　hydrolase　水解酶
　acarboxylase　乙酰辅酶
　dehydrochlorinase　脱氯化氢酶

esterase 酯酶
monooxygenase 单（加）氧酶
oxidase 氧化酶
phosphorylase 磷酸化酶
-ate 盐，酯
glutamate 谷氨酸盐，谷氨酸酯
organophosphate 有机磷酸酯
carbamate 氨基甲酸盐
carbohydrate 碳水化合物，醣类
auto 自动，自己
autobiography 自传
autoclave 高压锅，高压灭菌器
azo- 偶氮
bacteri 细菌
bacterium 细菌
bacteriostatic 细菌抑制的
bio 生命，生物
biodiversity 生物多样性
bioinsecticide 生物杀虫剂
bioactivity 生物活性
bioassay 生物测定，生物鉴定
bor- 硼
boric acid 硼酸
botone 植物
botanic 植物的
bromo- 溴
dibromochloropropane 二溴氯丙烷
bronch 支气管
bronchus 支气管
bronchiole 细支气管
butyl 丁基
centi- 中心
centrifugation 离心法；离心过滤
carbonyl 羰基
-caboxylic acid 羧酸
chem 化学的
chemical 化学的
chlor- 氯
chlorinate 氯化
chloro- 氯代
chlorobenzene 氯苯
chlorofluorocarbon 氯氟烃
choline 胆碱
cholinesterase 胆碱脂酶
chromato 颜色
chromatography 色谱法
chromosome 染色体
cide 杀，切
pesticide 农药
biocide 生物灭绝
ecocide 生态灭绝
insecticide 杀虫剂
bactericide 杀菌剂
fungicide 杀真菌剂
herbicide 除草剂
nematicide 杀线虫剂
rodenticide 灭鼠剂
cis- 顺式
counter，contra 反，对应
counteracte 抵消，中和，阻碍
counterintuitive 违反直觉的
co 共同
covalent 共有原子价的，共价的
coverage 覆盖
cyclo- 环
cyclodiene 环戊二烯类杀虫剂
cyclohexanedione 环己二酮
de- 去，脱
defoliant 脱叶剂，落叶剂
dehydrate （使）脱水
deca- 十
dec- 十，癸
decimal 十进制的，小数的；小数
dehydro- 去氢；去水
dehydrochlorinase 脱氯化氢酶
des- 去，脱
desiccant 干燥剂
detoxicate 使解毒
detoxication 解毒；解毒作用
detoxify 使解毒
di- 二
2,4-dichlorophenol 2,4-二氯苯酚
4,4′-dichlorodiphenyldichloroethane 二氯二苯二氯乙烷

dikaryotic　双核的，双核细胞的
dinitrophenol　二硝基酚
diploid　双重的，倍数的，双倍的
dipyridyl　联吡啶
dis-　分离，否定，相反
discard　丢弃，抛弃，放弃
discordant　不调和的，不和谐的
discrepancy　相差，差异，矛盾
disharmony　不调和
disintegration　瓦解
disperse　（使）分散，（使）散开，疏散
dissipation　消散，挥霍，浪费，狂饮
distributor　分发者，销售者，批发商
eco-　生态的
eco-awareness　环保意识
ecocatastrophe　生态灾难
ecologist　生态学家
ecology　生态学，生态环境
electro-　电的
electrophoresis　电泳
electrostatic　静电的，静电学的
en-　在内
enantiomer　对映（结构）体
encompass　包围，环绕
endo-　内
endocrine　内分泌
endoparasitic　内部寄生物，体内寄生虫
-ene　烯
ethylene　乙烯
polyethylene　聚乙烯
hydroprene　烯虫乙酯
enzyme　酶
enzymology　酶学
epi-　表；差向
epidemic　流行的
epidermal　表皮的，外皮的
epoxy-　环氧
erythro-　红，赤
erythrocyte　红血球
-ester　酯
esterase　酯酶
esterify　（使）酯化
-ether　醚
ethoxy-　乙氧基
ethyl　乙基
ex-　外
exogenous　外生的，外因的
exoskeleton　外骨骼
exotoxin　外毒素
excretion　（动植物的）排泄，排泄物
exogenously　外生的，外因的
expel　驱逐
exceed　超越，胜过；超过其他
extra-　超过
extracellular　细胞外的
extraction　抽出，萃取
extrachromosomal　染色体外的
fluoro-　氟代
fluoride　氟化物
fluorocarbon　碳氟化合物
fumus　烟
fumigant　薰剂
fumigation　烟薰法，薰烟消毒法
fungi　真菌类
fungicide　杀真菌剂
fus　流，泻
confuse　使混乱
diffuse　散播；四散的
effuse　流出，吐露
infuse　注入，灌输
profuse　丰富的，浪费的
refuse　拒绝
suffuse　充满，染遍
transfuse　输血，倾注
gen　产生
generate　产生
genetics　遗传学
genesis　起源，创始
hemi-　半
hemisphere　半球
hemicycle　半圆形
hepta-　七
heptagon　七角形
heptaglot　使用七种语言的

herb　草
herbal　草本植物的
herbicide　除草剂
heter-　其他
heteroauxin　吲哚乙酸
hetero-　杂
heterokaryotic　异核体的
heterozygote　异质接合体，异形接合体
hexa-　六
hexagon　六角形
hexachlorobenzene　六氯苯
homo-　同类的，相同的
homogeneous　同类的，同族的
homogenize　使一致
homogenate　均匀混合物
homozygous　同型结合的，纯合子的
hydro-　氢或水
hydrolase　水解酶
hydrolysis　水解
hydroelectric　水电的
hyper　在上，超
hyperactivate　超活化
hypersonic　超声的
hypo-　低级的，次
-ide　无氧酸的盐，酐
anhydride　酐
-imine　亚胺
inter-　互相
interchange　交换
interlock　连锁
intra，intro-　在内，内部
intracellular　细胞内的
intravenous　静脉内的
iodo-　碘代
iso-　异，等，同
isomer　异构体
isothiocyanate　异硫氰酸盐（或酯）
-ite　亚酸盐
chlorite　亚氯酸盐
keto-　酮
ketone　酮
-lactone　内酯
leuco-　白
leukaemia　白血病
-lock　构成植物名称
charlock　田芥菜
mal-　坏，恶
malnutrition　营养不良
malformation　畸形
meta-　间，偏
methoxy-　甲氧基
methyl-　甲基
methyl　甲基
methyl bromide　溴甲烷
methyl parathion　甲基对硫磷
micro-　微，小
microbiology　微生物学
microflora　微生物区系
microcapsule　微胶囊，微囊体
microencapsulate　把……装入微胶囊
microflora　微生物区系，微生物丛
microscopical　显微镜的，精微的
mono-　（mon-）一，单
monoterpene　单萜
monooxygenase　单（加）氧酶；
mort　死亡
mortal　终有一死的
mortality　死亡率
mutant　突变异种
mutagenesis　突变形成，变异发生
mutagenic　诱变的，致突变的
mutagenicity　诱变（性）
mutation　变化，转变，（生物物种的）突变
neur-　神经
neural　神经的
neuro-　神经
neurotransmitter　神经传递素
nitro-　硝基
nona-　九
oct-　八，辛
octadecyl　十八（烷）基
octyl　辛基
-ol　醇，酚

ergosterol　麦角固醇
eugenol　丁子香酚
carvacrol　香芹酚，香荆芥酚
-oma　表示动作的结果；肿等名称
rhizoma　根茎
-ome/-omal　组/系（的）
cytochrome　细胞色素
chromosome　染色体
genome　基因组，染色体组
plastome　质体系
chromosomal　染色体的
ribosome　核糖体
-one　酮
cinerolone　瓜菊醇酮
acetone　丙酮
organ　器官，机体
organochlorine　有机氯（的）
organophosphate　有机磷酸酯（的）
organelle　细胞器官
organic　器官的，有机体的
organism　生物，有机体
organometallic　有机金属的
organosilicon　有机硅（化合物）
ortho-　邻，正，原
orthodox　正统的
orthopaedy　矫正术
-ous　亚酸的，低价金属
over-　过度，过分
overdose　药物过量
overgraze　过度放牧
-oxide　氧化物
oxidant　氧化剂
oxidase　氧化酶
oxidation　氧化（作用）
oxide　氧化物
oxo-　氧［代］
oxytetracycline　土霉素，氧四环素
para-　对位，仲
paraoxon　对氧磷
paraquat　百草枯
parathion　对硫磷
paradichlorobenzene　对二氯苯
penta-　五
pentachlorophenol　五氯苯酚
pentavalent　五价的，五种价的
petro-　石油
phen-　苯
phenol　苯酚，石炭酸
phenoxy　含苯氧基的
phenoxy-　苯氧（基）
phenyl-　苯基
phenylalanine　苯基丙氨酸
phenylpropene　苯基丙烯
phosphor-　磷；磷酸
phosphorus　磷
phosphorylase　磷酸化酶
photo-　光
photodegradation　光降解（作用）
photosynthesis　光合作用
phyllo-　叶
phylloplane　叶面
phyllosphere　叶圈，叶际
phyto　植物的
phytotoxic　植物性毒素的
phytochemical　植物化学的
poly-　聚，多
polyethylene　聚乙烯
polychlorovinyl　聚氯乙烯
polychlorinated　多氯化的
polychlorovinyl resin　聚氯乙烯树脂
polygenic　多基因的
polymorphous　多形的，多形态的
polysaccharide　多醣，聚糖，多聚糖
polyurethane　聚氨酯
post-　后来
postecdysis　蜕皮后期
post-emergence　出苗后至成熟前的
pre-　先
preempt　先占
pre-emergence　出土前的，出土前施用的
premise　前提，提论，假定；作出前提
preparatory　准（预，筹）备的，初步的
prephenate　预苯酸

proto- 原
proton 质子
pseudo- 假，伪，拟
pseudoreplication 假重复法
pur 清，纯，净
purify 使纯净
depurate 使净化
retro- 向后
retroact 倒行
retrospective 回顾的
semi- 半
semidominant 半显性的
semirefined 半精炼的
silico- 硅
silicone 硅树脂
stereo 立体的
stereochemistry 立体化学
stereoisomer 立体异构术
strepto- 链
streptomycin 链霉素
sulfo- 硫［代］，磺基
sulfone 砜
sulfoxide 亚砜
sulpho- 硫［代］，磺基
sulphonylurea 磺脲
super- 超过
superficial 表面的，肤浅的，浅薄的
sym- 对称
symmetry 对称
syn- 顺式，共，合
synthesis 合成
syndrome 综合症状
synchronously 同时地，同步地
synonymous 同义的
tert- 特，叔
tertiary 第三的；第三色；第三纪
tetra- 四
tetracycline 四环素
tetrahedral 有四面的，四面体的
therm 热
thermal 热的
thermometer 温度计
thermostable 耐热的，热稳定的
thio- 硫
thiophosphoric 硫代磷酸的
thiophanate 硫菌灵
tox- 毒，有毒的
toxalbumin 毒白蛋白
toxic 毒的，有毒的；中毒的
toxicant 有毒性的
toxication 中毒
toxicity 毒性，毒力
toxicological 毒物学的
toxicologist 毒物学者；毒理学家
toxicology 毒物学
toxigenic 产毒素的
toxin 毒素；毒质
toxine 毒素
toxoid 变性毒素；类毒素
toxoplasmosis 住血原虫病
-toxin 毒
exotoxin 外毒素
protoxin 原毒素，强亲和毒素
trans- 反（式）；转
transgenic 基因改造的
translocate 改变……的位置
translocation 迁移，移动，置换
transparence 透明，透明度
tri- 三
triazine 三嗪
trivalent 三价的
trichome 毛状体
trimethylsilylation 三甲硅烷基化
ultra- 超出，超过
ultralow 超低的
ultrafilteration 超滤
under- 在……下
underground 地下的
undersea 水下的，海面下的，海底的
-yne 炔
-zole 咪唑
fuberidazole 呋喃基苯并咪唑
thiabendazole 噻苯咪唑
benzimidazole 苯并咪唑，间（二）氮茚

Appendix 2　农药剂型名称及代码（GB/T 19378—2017）

代码	英文名称	剂型名称	代码	英文名称	剂型名称
AE	aerosol dispenser	气雾剂	OP	oil dispersible powder	油分散粉剂
CB	bait concentrate	浓饵剂	PA	paste	糊剂
CS	capsule suspension	微囊悬浮剂	PC	insect-proof cover	防虫罩
DC	dispersible concentrate	可分散液剂	PM	proof mat	防蚊片
DP	dustable powder	粉剂	PN	insect-proof net	防蚊网
DR	dispensor	挥散芯	PR	plant rodlet	条剂
DS	powder for dry seed treatment	种子处理干粉剂	PT	pellet	球剂
EC	emulsifiable concentrate	乳油	RB	bait	饵剂
EG	emulsifiable granule	乳粒剂	RK	repellent milk	驱蚊乳
EO	emulsion, water in oil	油乳剂	RP	repellent wipe	驱蚊巾
EP	Emulsifiable powder	乳粉剂	RQ	repellent liquid	驱蚊液
ES	emulsion for seed treatment	种子处理乳剂	RW	repellent floral water	驱蚊花露水
EW	emulsion, oil in water	水乳剂	SC	aqueous suspension concentrate	悬浮剂
FS	suspension concentrate for seed treatment (flowable concentrate for seed treatment)	种子处理悬浮剂	SE	suspo-emulsion	悬乳剂
FU	smoke generator	烟剂	SG	water soluble granule	可溶粒剂
GA	gas	气体制剂	SL	soluble concentrate	可溶液剂
GE	gas generating product	发气剂	SO	spreading oil	展膜油剂
GL	emulsifiable gel	乳胶	SP	water soluble powder	可溶粉剂
GR	granule	颗粒剂	ST	water soluble tablet	可溶片剂
GS	grease	脂剂	TB	tablet	片剂
GW	water soluble gel	可溶胶剂	TC	technical material	原药
HN	hot fogging concentrate	热雾剂	TK	technical concentrate	母药
LN	long-lasting insecticidal net	长效防蚊帐	UL	ultra low volume concentrate	超低容量液剂
LS	solution for seed treatment	种子处理液剂	WG	water dispersible granule	水分散粒剂
LV	liquid vaporizer	电热蚊香液	WP	wettable powder	可湿性粉剂
MC	mosquito coil	蚊香	WS	water dispersible powder for slurry seed treatment	种子处理可分散粉剂
ME	micro-emulsion	微乳剂	WT	water dispersible tablet	可分散片剂
MV	vaporizing mat	电热蚊香片	ZC	mixed formulations of CS and SC	微囊悬浮-悬浮剂
OD	oil dispersion	可分散油悬浮剂	ZE	mixed formulations of CS and SE	微囊悬浮-悬乳剂
OF	oil miscible flowable concentrate	油悬浮剂	ZW	mixed formulations of CS and EW	微囊悬浮-水乳剂
OL	oil miscible liquid	油剂			

Appendix 3 常见农药中英文名称

A

abamectin 阿维菌素
abscisic acid *S*-诱抗素
acephate 乙酰甲胺磷
acequinocyl 灭螨醌
acetamiprid 啶虫脒
acethion 家蝇磷
acetochlor 乙草胺
acetophos 乙酯磷
acetoprole 乙酰虫腈
acibenzolar-*S*-methyl 活化酯
acifluorfen 三氟羧草醚
aclonifen 苯草醚
acrinathrin 氟丙菊酯
acrylonitrile 丙烯腈
actidione 放线菌酮
adenine 腺嘌呤
Adoxophyes orana GV 棉褐带卷蛾颗粒体病毒
agricultural antibiotic 120 农抗 120
Agrotis segetum GV 黄地老虎颗粒体病毒
akton 硫虫畏
alachlor 甲草胺
L-alanine 双丙氨酰膦
alanycarb 棉铃威
albendazole 丙硫多菌灵
aldicarb 涕灭威
aldoxycarb 涕灭砜威
aldrin 艾氏剂
allantoin 尿囊素
allethrin 烯丙菊酯
d-allethrin 右旋烯丙菊酯
allidochlor 二丙烯草胺
alloxydim 禾草灭
2-allylphenol 邻烯丙基苯酚
allyxycarb 除害威
aluminium phosphide 磷化铝
ametryn 莠灭净
amicarbazone 氨唑草酮
amicarthiazol 拌种灵
amidithion 赛硫磷
amidosulfuron 酰嘧磺隆
amidothionate 果满磷
p-aminobenzen sulfonic acid 敌锈酸
aminocarb 灭害威
amiprophos 胺草磷
amiprophos-methyl 甲基胺草磷
amitraz 双甲脒
amitrole 杀草强
ammonium 2-nitrophenolate 邻硝基苯酚铵
ammonium 4-nitrophenolate 对硝基苯酚铵
ammonium sulfamate 氨基磺酸铵
amobam 代森铵
anabasine 新烟碱
ancymidol 环丙嘧啶醇
anilazine 敌菌灵
anilofos 莎稗磷
Anticarsia gemmatalis NPV 梨豆夜蛾核型多角体病毒
antu 安妥
apholate 不育特
aramite 杀螨特
arsenious acid 亚砷酸
ascomycin 长川霉素
asomate 福美胂
aspon 丙硫特普
asulam 磺草灵
atrazine 莠去津
aureonucleomycin 金核霉素
Autographa californica NPV 苜蓿银纹夜蛾核型多角体病毒
azaconazole 氧环唑
azadirachtin 印楝素
azafenidin 唑啶草酮
azamethiphos 甲基吡噁磷
azimsulfuron 四唑嘧磺隆

azinphos-ethyl　益棉磷
azinphos-methyl　保棉磷
aziprotryne　叠氮净
azocyclotin　三唑锡
azoxystrobin　嘧菌酯

B

Bacillus cereus　蜡质芽孢杆菌
Bacillus popillae　金龟子芽孢杆菌
Bacillus sphearicus H5a5b　球形芽孢杆菌
Bacillus thuringiensis　苏云金杆菌
baicalin　黄芩苷
barban　燕麦灵
barium carbonate　碳酸钡
barthrin　熏菊酯
Beauveria bassiana　球孢白僵菌
beflubutamid　氟丁酰草胺
belvitan　抑芽醚
benalaxyl　苯霜灵
benalaxyl-*M*　高效苯霜灵
benazolin　草除灵
benazolin-ethyl　草除灵乙酯
bendiocarb　噁虫威
benfluralin　乙丁氟灵
benfuracarb　丙硫克百威
benfuresate　呋草磺
benocacor　解草酮
benodanil　麦锈灵
benomyl　苯菌灵
benoxacor　解草嗪
benquinox　醌肟腙
bensulfuron-methyl　苄嘧磺隆
bensulide　地散磷
bensultap　杀虫磺
bentazone　灭草松
benthiavalicarb-isopropyl　苯噻菌胺
benthiazole　苯噻硫氰
bentranil　苯草灭
benzadox　苯草多克死
benzfendizone　双苯嘧草酮
benziothiazolinone　噻霉酮
benzipram　苄草胺
benzobicylon　双环磺草酮
benzofenap　吡草酮
benzoicacid　嘧磺隆
benzoximate　苯螨特
benzoylprop-ethyl　新燕灵
benzthiazuron　苯噻隆
6-benzylaminopurine　苄氨基嘌呤
berberine　小檗碱
beta-cypermethrin　高效氯氰菊酯
bialaphos-sodium　双丙氨膦
bifenazate　联苯肼酯
bifenox　甲羧除草醚
bifenthrin　联苯菊酯
bilanafos　双丙氨酰膦
binapacryl　乐杀螨
S-bioallethrin　*S*-生物烯丙菊酯
bioallethrin　生物烯丙菊酯
biopermethrin　生物氯菊酯
bioresmethrin　生物苄呋菊酯
biphenyl　联苯
bismerthiazol　叶枯唑
bispyribac-sodium　双草醚
bitertanol　联苯三唑醇
bitolylacinone　杀鼠新
blasticidin *S*　灭瘟素
borax　硼砂
bordeaux mixture　波尔多液
boscalid　啶酰菌胺
botulin type C　C型肉毒素
BPO1 *Bacillus cereus* sprain　BPO1蜡状芽孢杆菌
brassinolide　芸苔素内酯
brevibacterium　枯草芽孢杆菌
brodifacoum　溴鼠灵
brofenvalerate　溴灭菊酯
brofluthrinate　溴氟菊酯
bromacil　除草定
bromadiolone　溴敌隆
bromethalin　溴鼠胺
bromethrin　溴苄呋菊酯
bromoacetamide　溴乙酰胺

bromobutide　溴丁酰草胺
bromocyclen　溴西杀
bromofenoxim　溴酚肟
bromophos　溴硫磷
bromophos-ethyl　乙基溴硫磷
bromopropylate　溴螨酯
bromothalonil　溴菌腈
bromoxynil octanoate　辛酰溴苯腈
bromoxynil　溴苯腈
brompyrazon　溴莠敏
bromuconazole　糠菌唑
bronopol　溴硝醇
brornothalonil　溴菌腈
bufencarb　合杀威
bupirimate　乙嘧酚磺酸酯
buprofezin　噻嗪酮
busoxinone　羟草酮
butacarb　畜虫威
butachlor　丁草胺
butafenacil　氟丙嘧草酯
butamifos　抑草磷
butenachlor acetanilide　丁烯草胺
butenachlor　丁烯草胺
butethrin　苄烯菊酯
buthiobate　丁赛特
buthiuron　丁噻隆
butocarboxim　丁酮威
butonate　丁酯膦
butopyronoxyl　避蚊酮
butoxycarboxim　丁酮砜威
butralin　仲丁灵
butrizol　叶锈特
butroxydim　丁苯草酮
buturon　炔草隆
butylate　丁草敌
Buzura suppressaria NPV　油桐尺蠖核型多角体病毒

C

cadusafos　硫线磷
cafenstrole　唑草胺
calcium arsenate　砷酸钙
calvinphos　敌敌钙
camphechlor　毒杀芬
capsaicin　辣椒碱
captafol　敌菌丹
captan　克菌丹
carbamorph　硫酰吗啉
carbanolate　氯灭杀威
carbaryl　甲萘威
carbendazim　多菌灵
carbetamide　双酰草胺
carbofuran　克百威
carbonothioic acid　哒草特
carbophenothion　三硫磷
carbosulfan　丁硫克百威
carboxin　萎锈灵
carfentrazone-ethyl　唑酮草酯
carpropamid　环丙酰菌胺
cartap　杀螟丹
carvacrol　香芹酚
cellocidin　叶枯炔
chemagro　氯硝散
chinomethionate　灭螨猛
chinoso　18-羟基喹啉
chitin　壳多糖
chitosan　几丁聚糖
chlobenthiazone　灭瘟唑
chlomethoxyfen　甲氧除草醚
chloramben　草灭畏
chloramine phosphorus　氯胺磷
chloranil　四氯对醌
chloranocryl　丁酰草胺
chlorantraniliprole　氯虫酰胺
chlorbenside　杀螨醚
chlorbenzuron　灭幼脲
chlorbromuron　氯溴隆
chlorbufam　稗蓼灵
chlordane　氯丹
chlordecone　开蓬
chlordimeform　杀虫脒
chlorempenthrin　氯烯炔菊酯
chloretazate　玉雄杀

chlorethoxyfos 氯氧磷
chlorfenac 伐草克
chlorfenapyr 溴虫腈
chlorfenethol 杀螨醇
chlorfenson 杀螨酯
chlorfensulphide 敌螨特
chlorfenvinphos 毒虫畏
chlorfluazuron 氟啶脲
chlorflurenol 整形醇
chloridazon 氯草敏
chlorimuron-ethyl 氯嘧磺隆
chlorinitrofen 草枯醚
chlormephos 氯甲硫磷
chlormequat 矮壮素
chlormethiuron 灭虫脲
chlornidine 氯乙地乐灵
chlornitrofen 草枯醚
chloroazifop-propynyl 炔禾灵
chlorobenzilate 乙酯杀螨醇
chloroisobromine cyanuric acid 氯溴异氰尿酸
chloromethiuron 灭虫脲
chloromethylsilatrane 硅丰环
chloroneb 地茂散
chloronitrophen 氯硝酚
chloronitropropane 氯硝丙烷
chlorophacinone 氯鼠酮
chloropicrin 氯化苦
d-t-chloroprallethrin 右旋反式氯丙炔菊酯
chloropropylate 丙酯杀螨醇
chlorothalonil 百菌清
chlorotoluron 绿麦隆
chlorphenprop-methyl 燕麦酯
chlorphonium 矮形磷
chlorphoxim 氯辛硫磷
chlorpropham 氯苯胺灵
chlorpyrifos 毒死蜱
chlorpyrifos-methyl 甲基毒死蜱
chlorquinox 四氯喹噁啉
chlorsulfuron 氯磺隆
chlorthal-dimethyl 氯酞酸甲酯
chlorthiamide 氯硫酰草胺
chlorthion 氯硫磷
chlorthiophos 虫螨磷
chlorxuron 枯草隆
chlothizol 灭草荒
chlozolinate 乙菌利
choline chloride 氯化胆碱
choloroxuron 枯草隆
chrornafenozide 环虫酰肼
cinerin Ⅰ 瓜叶菊素Ⅰ
cinerin Ⅱ 瓜叶菊素Ⅱ
cinidon-ethyl 吲哚酮草酯
cinmethylin 环庚草醚
cinosulfuron 醚磺隆
cisanilide 咯草隆
citral 柠檬醛
citricacide-titanium chelate 柠檬酸钛
clefoxidim 环苯草酮
clethodim 烯草酮
climbazole 咪菌酮
clodinafop-propargyl 炔草酯
cloethocarb 除线威
clofencet-potassium 苯哒嗪钾
clofencet 苯哒嗪酸
clofentezine 四螨嗪
clomazone 异噁草酮
clomeprop 稗草胺
cloprop 调果酸
clopyralid 二氯吡啶酸
cloquitocet-mexyl 解草酯
cloransufam-methyl 氯酯磺草胺
clothianidin 噻虫胺
cloxyfonac 坐果酸
cnidiadin 蛇床子素
codlemone 苹果小卷叶蛾性信息素
colophonate 噻唑硫磷
Conidioblous thromboides 块状耳霉菌
copper abietate 松脂酸铜
copper acetate 乙酸铜
copper calcium sulphate 硫酸铜钙
copper carbonate 碱式碳酸铜
copper hydroxide 氢氧化铜

copper oxychloride　氧氯化铜
copper sulfate tribasic　碱式硫酸铜
copper sulfate　硫酸铜
copperchloride　王铜
coumachlor　氯灭鼠灵
coumafuryl　克鼠灵
coumaphos　蝇毒磷
coumatetralyl　杀鼠醚
coumithoate　畜虫磷
coumoxystrobin　丁香菌酯
CPTA　增色胺
credazine　醚草敏
crimidine　鼠立死
crotoxyphos　巴毒磷
crufomate　育畜磷
CTC　矮健素
cuaminosulfate　络氨铜
cuelure　诱蝇酮
cumyluron　苄草隆
cuppric nonyl phenolsulfonate　壬菌铜
cuprous oxide　氧化亚铜
curcumol　莪术醇
cyanamide　单氰胺
cyanatryn　氰草净
cyanazine　氰草津
cyanofenphos　苯腈膦
cyanophos　杀螟腈
cyanthoate　果虫磷
cyantraniliprole　溴氰虫酰胺
cyazofamid　氰霜唑
cyclafuramid　环菌胺
cyclanilide　环丙酰草胺
cyclethrin　环虫菊酯
cycloate　环草敌
cycloprate　环螨酯
cycloprothrin　乙氰菊酯
cyclosulfamuron　环丙嘧磺隆
cycloxydim　噻草酮
cycluron　环莠隆
Cydia plmnella GV　苹果小卷蛾颗粒体病毒
cyflufenamid　环氟菌胺
cyflurnetofen　丁氟螨酯
cyfluthrin　氟氯氰菊酯
β-cyfluthrin　高效氟氯氰菊酯
cyhalofop-butyl　氰氟草酯
gamma-cyhalothrin　精高效氯氟氰菊酯
lambda-cyhalothrin　高效氯氟氰菊酯
cyhalothrin　氯氟氰菊酯
cyhexatin　三环锡
cymiazole　螨蜱胺
cymoxanil　霜脲氰
cyometrinil　解草胺腈
cypendazole　氰菌灵
cypermethrin　氯氰菊酯
α-cypermethrin　顺式氯氰菊酯
β-cypermethrin　高效氯氰菊酯
cyphenothrin　苯醚氰菊酯
cyprazine　环丙津
cyprazole　三环赛草胺
cyproconazole　环丙唑醇
cyprodinil　嘧菌环胺
cyprofuram　酯菌胺
cypromid　环酰草胺
cyromazine　灭蝇胺
cythioate　畜蜱磷
cytosinpeptidemycin　博联生物菌素

D

DAEP　浸移磷
daimuron　杀草隆
dalapon　茅草枯
daminozide　丁酰肼
dayoutong　哒幼酮
dazomet　棉隆
DCPTA　增产胺
debacarb　咪菌威
dehydroacetic acid　脱氧乙酸
delachlor　异丁草胺
deltamethrin　溴氰菊酯
demephion　田乐磷
demeton　内吸磷
demeton-S-methyl　甲基内吸磷
demeton-S-methylsulphone　砜吸磷

Dendrolimus punctatus CPV　松毛虫质型多角体病毒
desmedipham　甜菜安
desmetryn　敌草净
diafenthiuron　丁醚脲
dialifos　氯亚胺硫磷
diallate　燕麦敌
diazinon　二嗪磷
dibromochloropropane　二溴氯丙烷
dibutyl phthalate　驱蚊叮
dibutyl succinate　驱虫特
dibutyladipate　驱虫威
dicamba　麦草畏
dicamba-methyl　增糖酯
dicapthon　异氯磷
dichlobenil　敌草腈
dichlofenthion　除线磷
dichlofluanid　苯磺菌胺
dichlone　二氯萘醌
dichlorbenzuron　除幼脲
dichlormate　苄胺灵
dichlormide　烯丙酰草胺
p-dichlorobene　对二氯苯
dichlorobenil　敌草腈
ortho-dichlorobenzene　邻二氯苯
para-dichlorobenzene　对二氯苯
dichlorophen　双氯酚
dichlorprop　2,4-滴丙酸
dichlorprop-*P*　精 2,4-滴丙酸
dichlorvos　敌敌畏
dichlozoline　菌核利
diclobutrazol　苄氯三唑醇
diclocymet　双氯氰菌胺
diclofop-methyl　禾草灵
diclomezine　哒菌酮
dicloran　氯硝胺
diclosulam　双氯磺草胺
dicofol　三氯杀螨醇
dicrotophos　百治磷
dicyclanil　环虫腈
dieldrin　狄氏剂
dienochlor　遍地克
diethatyl-ethyl　乙酰甲草胺
diethofencarb　乙霉威
diethyl aminoethylhexanoate　胺鲜酯
diethyltoluamide　避蚊胺
difenacoum　鼠得克
difenoconazole　苯醚甲环唑
difenolan　苯虫醚
difenoxuron　枯莠隆
difenzoquat　野燕枯
difethialone　噻鼠灵
diflovidazin　氟螨嗪
diflubenzuron　除虫脲
diflufenican　吡氟酰草胺
diflufenzopyr　氟吡草腙
diflumetorim　氟嘧菌胺
dikegulac　调呋酸
dimefluthrin　四氟甲醚菊酯
dimefox　甲氟磷
dimefuron　噁唑隆
dimepiperate　哌草丹
dimetachlone　菌核净
dimetan　地麦威
dimethacarb　混灭威
dimethachlor　二甲草胺
dimethametryn　异戊乙净
dimethenamid　二甲噻草胺
dimethenamid-*P*　高效二甲噻草胺
dimethipin　噻节因
dimethirimol　二甲嘧酚
dimethirn　苄菊酯
dimethoate　乐果
dimethomoph　烯酰吗啉
dimethrin　苄菊酯
dimethyl phthalate　避蚊酯
dimethylcarbate　驱蚊灵
dimethylvinphos　甲基毒虫畏
dimetilan　敌蝇威
dimexan　敌灭生
dimoxystrobin　醚菌胺
dimuron　杀草隆

dinex　消螨酚
diniconazole-*M*　高效烯唑醇
diniconazole　烯唑醇
dinitramine　氨基乙氟灵
dinobuton　敌螨通
dinocap　二硝巴豆酸酯
dinocton　二硝酯
dinopenton　消螨多
dinosam　戊硝酚
dinoseb acetate　地乐酯
dinoseb　地乐酚
dinotefuran　呋虫胺
dinoterb acetate　特乐酯
dinoterb　特乐酚
diofenolan　苯虫醚
dioxabenzofos　蔬果磷
dioxacarb　二氧威
dioxathion　敌噁磷
diphacinone　敌鼠
diphenamide　双苯酰草胺
diphenprophos　硫醚磷
diphenylacetonitrile　二苯乙腈
diphenylamine　二苯胺
dipropalin　地乐灵
dipropetryn　异丙净
dipropyl 2,5-pyridinedicarboxylate　驱蝇啶
diquat　敌草快
dirnethylvinphos　甲基杀螟威
disodium octaborate tetrahydrate　多硼酸钠
disosultap　杀虫双
disugran　麦草畏甲酯
disulfoton　乙拌磷
ditalimfos　灭菌磷
dithianon　二氰蒽醌
dithiopyr　氟硫草定
diuron　敌草隆
divostroside　异羊角拗苷
DMC　调节安
DMNP　二甲草醚
DNOC　二硝酚
dodemorph　十二环吗啉
dodine　多果定
DPU　二苯脲
drazoxolon　联氨噁唑酮
DTA　拒食胺
DT　琥胶肥酸铜

E

Ectropis oblique NPV　茶尺蠖核型多角体病毒
edifenphos　敌瘟磷
eluthiacet-methyl　嗪草酸甲酯
empenthrin　右旋烯炔菊酯
enadenine　烯腺嘌呤
ENA　萘乙酸乙酯
endosulfan　硫丹
endothal　茵多酸
endothion　因毒磷
endrin　异狄氏剂
enestroburin　烯肟菌酯
EPBP　伊比磷
EPN　苯硫磷
epofenonane　保幼醚
epoxiconazole　氟环唑
eradicane　菌达灭
erbon　抑草蓬
esfenvalerate　氰戊菊酯
esprocarb　禾草畏
etacelasil　乙烯硅
etaconazole　乙环唑
etem　代森硫
ethaboxam　噻唑菌胺
ethachlor　克草胺
ethalfluralin　乙丁烯氟灵
ethametsulfuron　胺苯磺隆
ethametsulfuron-methyl　胺苯磺隆
ethaprochlor　杀草胺
ethephon　乙烯利
ethidimuron　磺噻隆
ethiofencarb　乙硫苯威
ethiolate　硫草敌
ethion　乙硫磷
ethiozin　乙嗪草酮

ethiprole　乙硫虫腈
ethirimol　乙嘧酚
ETHN　肟螨酯
ethoate-methyl　益硫磷
ethobenzanid　乙氧苯草胺
ethofumesate　乙氧呋草磺
ethohexadiol　驱蚊醇
ethoprophos　灭线磷
ethoxyfen-ethyl　氯氟草醚
ethoxyquin　乙氧喹啉
ethoxysulfuron　嘧磺隆
ethychlozate　吲熟酯
ethyl hexanediol　驱蚊醇
ethyl-DDD　乙滴涕
ethylene dibromide　二溴乙烷
ethylene dichloride　二氯乙烷
ethylicin　乙蒜素
ethylmercury chloride　氯化乙基汞
etobenzanid　乙氧苯草胺
etofenprox　醚菊酯
etoxazole　乙螨唑
etoxinol　乙氧杀螨醇
etridiazole　土菌灵
etrimfos　乙嘧硫磷
etrofol　害扑威
eucalyptol　桉油精
eugenol　丁香酚
Euproctis pseudoconspersa NPV　茶毛虫核型多角体病毒
evermectin benzoate　甲氨基阿维菌素苯甲酸盐

F

FABA　氟蚧胺
FABB　氟螨胺
famoxadone　噁唑菌酮
famphur　伐灭磷
FDDT　氟滴滴滴
FDMA　苟丁酸胺
femclorim　解草啶
fenamidone　咪唑菌酮
fenaminosulf　敌磺钠
fenaminstrobin　烯肟菌胺
fenamiphos　苯线磷
fenapanil　咪菌腈
fenarimol　氯苯嘧啶醇
fenazaflor　抗螨唑
fenazaquin　喹螨醚
fenbuconazole　腈苯唑
fenbutatin oxide　苯丁锡
fenchlorazole　解草唑
fenchlorphos　皮蝇磷
fenclorim　解草啶
fenethocarb　双乙威
fenfluthrin　五氟苯菊酯
fenfuram　甲呋酰胺
fenhexamid　环酰菌胺
fenitropan　种衣酯
fenitrothion　杀螟硫磷
fenobucarb　仲丁威
fenoprop　2,4,5-涕丙酸
fenothiocarb　苯硫威
fenoxanil　稻瘟酰胺
fenoxaprop-ethyl　噁唑禾草灵
fenoxaprop-*P*-ethyl　精噁唑禾草灵
fenoxycarb　苯氧威
fenpiclonil　拌种咯
fenpirithrin　吡氯氰菊酯
fenpropathrin　甲氰菊酯
fenpropidin　苯锈啶
fenpropimorph　丁苯吗啉
fenpyroximate　唑螨酯
fenridazon-propyl　苯哒嗪丙酯
fenson　芬螨酯
fensulfothion　丰索磷
fenthiaprop-ethyl　噻唑禾草灵
fenthiaprop　噻唑禾草灵
fenthion　倍硫磷
fentin acetate　三苯基乙酸锡
fentin chloride　三苯基氯化锡
fentin hydroxide　三苯基氢氧化锡
fentin　三苯锡
fentrazamide　四唑酰草胺

fenuron 非草隆
fenvalerate 氰戊菊酯
ferbam 福美铁
ferimzone 嘧菌腙
fipronil 氟虫腈
flamprop-*M*-isopropyl 高效麦草伏丙酯
flamprop-*M*-metyl 高效麦草伏甲酯
flazasulfuron 啶嘧磺隆
flocoumafen 氟鼠灵
flonicamid 氟啶虫酰胺
florasulam 双氟磺草胺
floroacetamide 氟乙酰胺
fluacrypyrim 嘧螨酯
fluazifop 吡氟禾草灵
fluazifop-*P*-butyl 精吡氟禾草灵
fluazinam 氟啶胺
fluazolate 异丙吡草酯
fluazuron 啶蜱脲
flubendiamide 氟虫酰胺
flubenzimine 氟螨噻
flucarbawne-sodium 氟酮磺隆
fluchloraline 氯乙氟灵
flucycloxuron 氟环脲
flucythrinate 氟氰戊菊酯
fludioxonil 咯菌腈
flufenacet 氟噻草胺
flufenoxuron 氟虫脲
flufenprox 三氟醚菊酯
flufenpyr-ethyl 氟哒嗪草酯
flufiprole 丁烯氟虫腈
flumethrin 氟氯苯菊酯
flumetralin 氟节胺
flumetsulam 唑嘧磺草胺
flumiclorac 氟烯草酸
flumiclorac-pentyl 氟胺草酯
flumioxazin 丙炔氟草胺
flumorph 氟吗啉
fluometuron 氟草隆
fluopicolide 氟啶酰菌胺
fluopicolpkd 吡氟菌酯
fluopyrarn 氟吡菌酰胺
fluorbenside 氟杀螨
fluorethrin 氟苄呋菊酯
fluoridamid 增糖胺
fluoroacetamide 氟乙酰胺
fluoroacetanilide 氟乙酰苯胺
fluorochloridone 氟咯草酮
fluorodifen 三氟硝草醚
fluoroglycofen 乙羧氟草醚
fluoroglycofen-ethyl 乙羧氟草醚
fluoromide 氟氯菌核利
fluoronitrofen 氟除草醚
fluotrimazole 三氟苯唑
fluoxastrobin 氟嘧菌酯
flupoxam 氟胺草唑
flupropadine 氟鼠啶
flupyrsulfuron-methyl sodium 氟啶嘧磺隆
fluquinconazole 氟喹唑
flurazole 解草胺
flurenol 芴丁酯
fluridone 氟啶草酮
flurothiuron 氟硫隆
fluroxypyr 氟草烟
fluroxypyr-methyl 氯氟吡氧乙酸
flurprimidol 调嘧醇
flurtamone 呋草酮
flusilazole 氟硅唑
flusulfamide 磺菌胺
fluthiacet-methyl 氟噻乙草酯
flutolanil 氟酰胺
flutriafol 粉唑醇
tau-fluvalinate 氟胺氰菊酯
fluxofenim 氟草肟
folcisteine 代垅磷
folpet 灭菌丹
fomesafen 氟磺胺草醚
fondaren 壤虫威
fonofos 地虫硫膦
foramsulfuron 甲酰胺磺隆
forchlorfenuron 氯吡脲
formetanate 伐虫脒
formothion 安硫磷

fosamine ammonium　调节膦
fosetyl-aluminium　三乙膦酸铝
fosmethilan　丁苯硫磷
fosthiazate　噻唑膦
fosthietan　丁硫环磷
fpyridate　达草特
fuberidazole　麦穗宁
fuphenthiourea　呋苯硫脲
furalaxyl　呋霜灵
d-t-furamethrin　右旋炔呋菊酯
furamethrin　炔呋菊酯
furametpyr　呋吡菌胺
furan tebufenozide　呋喃虫酰肼
furathiocarb　呋线威
furcarbanil　二甲呋酰胺
cis-furconazole　呋醚唑
furconazole　呋菌唑
furethrin　糠醛菊酯
furilazole　解草噁唑
furmecyclox　拌种胺
furophanate　呋菌隆
furyloxyfen　呋氧草醚

G

gibberellic acide　赤霉酸
glenbar　敌草死
gliftor　鼠甘伏
glophosine　增甘膦
glufosinate　草铵膦
glyodin　果绿啶
glyoxime　乙二肟
glyphosate sesquisodium　增甘膦钠
glyphosate　草甘膦
glyphosate-trimesium　草硫膦
griseofulvin　灰黄霉素
guazatine acetate　双胍辛乙酸盐

H

halacrinate　丙烯酸喹啉酯
halfenprox　苄螨醚
halofenozide　氯虫酰肼
halosulfuron-methyl　氯吡嘧磺隆
haloxyfop-methyl　氟吡禾灵
haloxyfop-*P*-methyl　精吡氟氯禾灵
HCH　六六六
Helicoverpa zea NPV　谷实夜蛾核型多角体病毒
Heliothis armigera NPV　棉铃虫核型多角体病毒
heptachlor　七氯
heptenophos　庚烯磷
heptopargil　增产肟
herbisan　莠不生
heterophos　速杀硫磷
hexachlorobenzene　六氯苯
hexachlorophene　毒菌酚
hexaconazole　己唑醇
hexaflumuron　氟铃脲
hexazinone　环嗪酮
hexythiazox　噻螨酮
hopcide　害扑威
humic acid　腐植酸
hydramethylnon　氟蚁腙
2-hydrazinoethanol　羟基乙肼
hydroprene　烯虫乙酯
8-hydroxyquinoline sulfate　8-羟基喹啉硫酸盐
hymexazol　噁霉灵

I

imaxapic　甲基咪草烟
imazalil　抑霉唑
imazamethabenz　咪草酸
imazamox　甲氧咪草烟
imazapic　甲咪唑烟酸
imazapyr　咪唑烟酸
imazaquin　咪唑喹啉酸
imazethapyr　咪唑乙烟酸
imazosulfuron　唑吡嘧磺隆
imibenconazole　亚胺唑
imidacloprid　吡虫啉
imidaclothiz　氯噻啉
iminoctadine triacetate　双胍辛胺乙酸盐
iminoctadine　双胍辛胺

imiprothrin 炔咪菊酯
inabenfide 抗倒胺
indanofan 茚草酮
indol-3-yl acetic acid 吲哚乙酸
4-indol-3-yl butyric acid 吲哚丁酸
3-indol-3-yl propionie acid 吲哚丙酸
indoxacarb 茚虫威
iodfenphos 碘硫磷
4-iodophenoxyacetic acid 增产灵
iodosulfuron-methyl sodium 碘甲磺隆钠盐
ioxynil 碘苯腈
ioxynil-octanoate 辛酸碘苯腈
ipconazole 种菌唑
iprobenfos 异稻瘟净
iprodione 异菌脲
iprovalicarb 缬霉威
IPSP 丰丙磷
isazofos 氯唑磷
isobenzan 碳氯灵
isocarbamide 丁咪酰胺
isocarbophos 水胺硫磷
isodrin 异艾氏剂
isofenphos 异柳磷
isofenphos-methyl 甲基异柳磷
isolan 异索威
isomethiozin 丁嗪草酮
isonoruron 异草完隆
isopentenyladenine 异戊烯腺嘌呤
isoprocarb 异丙威
isopropalin 异丙乐灵
isoprothiolane 稻瘟灵
isoproturon 异丙隆
isothan 异喹丹
isothioate 异拌磷
isouron 异噁隆
isoxaben 异噁草胺
isoxachlortole 异噁氯草酮
isoxaflutole 异噁唑草酮
isoxapyrifop 异噁草醚
isoxathion 噁唑磷
ivermectin 伊维菌素
ixoxaben 异噁酰草胺
izopamfos 浸种磷

J

japothrins 喃烯菊酯
jasmolin Ⅰ 茉酮菊素Ⅰ
jasmolin Ⅱ 茉酮菊素Ⅱ
jasmonic acid 茉莉酸
jodfenphos 碘硫磷
juvenile hormone 保幼烯酯

K

kadethrin 噻恩菊酯
karbutilate 特胺灵
kasugamycin 春雷霉素
kinetin 糠氨基嘌呤
kinoprene 烯虫炔酯
kitazine 稻瘟净
kresoxim-methyl 醚菌酯
kumitox 灭螨胺

L

lactofen 乳氟禾草灵
Laphygma exigua NPV 甜菜夜蛾核型多角体病毒
lead arsenate 砷酸铅
lenacil 环草定
lepimectin 雷皮藤素
leptophos 溴苯磷
levamisole 保松噻
lime sulfur 石硫合剂
d-limonene *d*-柠檬烯
lindane 林丹
linuron 利谷隆
liuyangmycin 浏阳霉素
lufenuron 虱螨脲
Lymantria dispar NPV 舞毒蛾核型多角体病毒
lythidathion 除害磷

M

macbal 灭除威
magnesium chlorate 氯酸镁

malathion　马拉硫磷
maleic hydrazide　抑芽丹
malonoben　特螨腈
Mamestra brassicae NPV　甘蓝夜蛾核型多角体病毒
mancopper　代森锰铜
mancozeb　代森锰锌
mandipropamid　双炔酰菌胺
maneb　代森锰
matrine　苦参碱
MCPA　2甲4氯
MCPA-thioethyl　2甲4氯乙硫酯
MCPB　2甲4氯丁酸
mebenil　邻酰胺
mecarbam　灭蚜磷
mecarphon　四甲磷
mecoprop　2甲4氯丙酸
mecoprop-*P*　高2甲4氯丙酸
medimeform　杀螨脒
medinoterb　地乐施
mefenacet　苯噻酰草胺
mefenpyr-diethyl　吡唑解草酯
mefluidide　氟磺酰草胺
melitoxin　敌鼠灵
menazon　灭蚜硫磷
MENA　萘乙酸甲酯
mepanipyrim　嘧菌胺
meperfluthrin　氯氟醚菊酯
mephosfolan　地胺磷
mepiquate chloride　甲哌鎓
mepronil　灭锈胺
mesoprazine　灭莠津
mesosulfuron-methyl　甲磺胺磺隆
mesotrione　甲基磺草酮
mesulfan　灭菌方
meta-ammonium　安百亩
metaflumizone　氰氟虫腙
metalaxyl　甲霜灵
metaldehyde　四聚乙醛
metamifop　噁唑酰草胺
metamitron　苯嗪草酮
metam-sodium　威百亩
Metarhizium anisopliae　绿僵菌
metazachlor　吡唑草胺
metconazole　叶菌唑
metepa　不育胺
methabenzthiazuron　甲基苯噻隆
methacrifos　虫螨畏
methalaxyl-*M*　高效甲霜灵
methamidophos　甲胺磷
methane dithiocyanate　二硫氰基甲烷
methasulfocarb　磺菌威
methazole　灭草唑
methfuroxam　呋菌胺
methidathion　杀扑磷
methiocarb　灭梭威
methiopyrsulfuron　甲硫嘧磺隆
methomyl　灭多威
methoprene　烯虫酯
methoprotryne　甲氧丙净
methothrin　甲醚菊酯
methoxychlor　甲氧滴滴涕
methoxyfenozide　甲氧虫酰肼
methoxyphenone　苯草酮
cis-methrin　顺式苄呋菊酯
methuroxam　呋菌胺
methyl bromide　溴甲烷
methylacetophos　甲基乙酯磷
methylamineavermectin　富表甲氨基阿维菌素
methylarsenic sulphide　甲基硫化胂
1-methylcyclopropene　甲基环丙烯
methyldymron　甲基杀草隆
methyle bromide　溴甲烷
metobenzuron　吡喃隆
metobromuron　溴谷隆
metofluthrin　甲氧苄氟菊酯
S-metolachlor　高效异丙甲草胺
metolachlor　异丙甲草胺
metolcarb　速灭威
metomeclan　氯苯咯菌胺
metominostrobin　苯氧菌胺
metosulam　磺草唑胺

metoxadiazone　噁虫酮
metoxuron　甲氧隆
metrafenone　苯菌酮
metriam　代森联
metribuzin　嗪草酮
metsulfovax　噻菌胺
metsulfuron-methyl　甲磺隆
mevinphos　速灭磷
mexacarbate　兹克威
milbemycin　橘霉素
mildiomycin　灭粉霉素
milneb　代森环
mipafox　丙胺氟磷
mirex　灭蚁灵
MNFA　果乃胺
molinate　禾草敌
monalide　庚酰草胺
monoamitraz　单甲脒
monochloroacetic acid　一氯醋酸
monocrotophos　久效磷
monolinuron　绿谷隆
monosultap　杀虫单
monuron　灭草隆
moroxydine hydrochloride　盐酸吗啉胍
morphothion　茂硫磷
mtribuzin　嗪草酮
mucochloric anhydride　粘氯酸酐
muscalure　诱虫烯
mushroom polysaccharide　菇类蛋白多糖
myclobutanil　腈菌唑
myclozolin　甲菌利

N

nabam　代森钠
naftalofos　驱虫磷
naled　二溴磷
2-1-naphthy acetamide　萘乙酰胺
1-naphthyl acetic acide　萘乙酸
2-naphthyloxyacetic acid　萘氧乙酸
naproanilide　萘丙胺
napropamide　敌草胺
naptalam　萘草胺
n-decanol　正癸醇
neburon　草不隆
nemamort　二氯异丙醚
neochamaejasmin　狼毒素
Neodiprion leconter NPV　欧洲松木叶蜂核型多角体病毒
niclosamide　杀螺胺
nicosulfuron　烟嘧磺隆
nicotinamide　烟酰胺
nicotine　烟碱
nifluridide　氟蚁灵
nikkomycin　华光霉素
ningnanmycin　宁南霉素
nipyraclofen　吡氯草胺
nitenpyram　烯啶虫胺
nithiazine　噻虫醛
nitralin　甲磺乐灵
nitrapyrin　三氯甲基吡啶
nitrilacarb　戊氰威
nitrofen　除草醚
nitrostyrene　硫氰散
nitrothal-isopropyl　酞菌酯
NMT　甲苯酞氨酸
norbormide　鼠特灵
norflurazon　氟草敏
nornicotine　原烟碱
noruron　草完隆
Nosema locustae Canning　蝗虫微孢子虫
Nosema pyrausta　玉米螟微孢子虫
novaluron　氟酰脲
noviflumuron　多氟脲
nuarimol　氟苯嘧啶醇
nucleotide　核苷酸

O

OCA　缩水甘油酸
octhilinone　辛噻酮
ofurace　呋酰胺
oligosaccharins　氨基寡糖素
omadine　万亩定
omethoate　氧乐果
orbencarb　坪草丹

orysastrobin 肟醚菌胺
oryzalin 氨磺乐灵
oryzone 稻瘟清
K-othrin 苄呋烯菊酯
oxabetrinil 解草腈
oxadiargyl 丙炔噁草酮
oxadiazon 噁草酮
oxadixyl 噁霜灵
oxadizon 噁草酮
oxamyl 杀线威
oxasulfuron 环氧嘧磺隆
oxaziclomefone 噁嗪草酮
oxine citrate 8-羟基喹啉柠檬酸盐
oxine-copper 喹啉铜
oxolinic acide 喹菌酮
oxpoconazole fumarate 噁咪唑富马酸盐
oxpoconazole 噁咪唑
oxycarboxin 氧化萎锈灵
oxydemeton-methyl 亚砜磷
oxydeprofos 异亚砜磷
oxydisulfoton 砜拌磷
oxyenadenine 羟烯腺嘌呤
oxyfluorfen 乙氧氟草醚
oxymatrine 氧化苦参碱
oxytetracycline 土霉素
oxytetracyclini hydrochloridum 水合霉素

P

p,*p*′-DDT 滴滴涕
paclobutrazol 多效唑
Paecilomyces 拟青霉
paichongding 哌虫啶
paraoxon 对氧磷
paraquat 百草枯
parathion 对硫磷
parathion-methyl 甲基对硫磷
parinol 氯苯吡啶
PBPA 对溴苯氧乙酸
PCP-Na 五氯酚钠
pebulate 克草敌
pefurazoate 稻瘟酯
penconazole 戊菌唑
pencycuron 戊菌隆
pendimethalin 二甲戊乐灵
penfluron 氟幼脲
penoxsulam 五氟磺草胺
pentachlorophenol 五氯酚
pentanochlor 甲氯酰草胺
penthiopyrad 吡噻菌胺
pentmethrin 戊烯氰氯菊酯
pentoxazone 环戊噁草酮
perfluidone 黄草伏
cis-permethrin 顺式氯菊酯
permethrin 氯菊酯
pethoxamid 烯草胺
phenamacril 氰烯菌酯
9,10-phenanthraquinone 菲醌
phenazine oxide 叶枯净
phenazino-l-carboxylic acid 申嗪霉素
phenisopham 敌克草
phenkapton 芬硫磷
phenmedipham 甜菜宁
phenobenzuron 苯酰敌草隆
phenothiazine 吩噻嗪
phenothiol 酚硫杀
d-phenothrin 右旋苯醚菊酯
phenothrin 苯醚菊酯
phenproxide 苯螨醚
phenthoate ethyl 乙基稻丰散
phenthoate 稻丰散
2-phenylphenol 邻苯基苯酚
N-phenylphthalamic acid 苯肽胺酸
phorate 甲拌磷
phosalone 伏杀硫磷
phosazetim-bromo 溴代毒鼠磷
phosazetim 毒鼠磷
phosdiphen 氯瘟磷
phosfolan 硫环磷
phosfolan-methyl 甲基硫环磷
phosfolan 硫环磷
phosglycin 甘氨硫磷
phosmet 亚胺硫磷
phosphamidon 磷胺

phosphoramidothioicacid 抑草磷
phostex 四硫特普
phoxim 辛硫磷
phoxiom-methyl 甲基辛硫磷
phthalide 四氯苯酞
picloram 氨氯吡啶酸
picolinafen 氟吡草胺
picoxystrobin 啶氧菌酯
Pieris rapae granulosis virus，PrGV 菜青虫颗粒体病毒
pindone 杀鼠酮
pinolene 松脂二烯
piperalin 病花灵
piperlin 哌丙灵
piperophos 哌草磷
piproctanly 哌壮素
pirimicarb 抗蚜威
pirimioxyphos 嘧啶氧磷
pirimiphos-ethyl 嘧啶磷
pirimiphos-methyl 甲基嘧啶磷
plifenate 三氯杀虫酯
Plutella xylostella GV 小菜蛾颗粒体病毒
polycarbamate 代森福美锌
polyoxin 多抗霉素
polythialan 多噻烷
poppenate-methyl 羟戊禾灵
potasan 扑杀磷
potassium 4-CPA 对氯苯氧乙酸钾
potassium 4-nitrophenate 对硝基苯酚钾
prallethrin 右旋炔丙菊酯
pretilachlor 丙草胺
primisulfuron-methyl 氟嘧磺隆
pririmiphos-methyl 甲基嘧啶磷
probenazole 烯丙苯噻唑
procarbazone 丙苯磺隆
prochloraz manganese chloride complex 咪鲜胺锰络合物
prochloraz 咪鲜胺
procyazine 环丙青津
procymidone 腐霉利
prodiamine 氨氟乐灵
profenofos 丙溴磷
profluazol 氟唑草胺
profluralin 环丙氟灵
proglinazine 甘扑津
prohexadione 调环酸
prohexadione-calcium 调环酸钙
promacyl 蜱虱威
promecarb 猛杀威
prometon 扑灭通
prometryn 扑草净
promurit 灭鼠肼
propachlor 毒草胺
propamidine 丙烷脒
propamocarb 霜霉威
propanil 敌稗
propaphos 丙虫磷
propaquizafop 恶草酯
propargite 炔螨特
proparthrin 甲呋炔菊酯
propazine 扑灭津
propetamphos 胺丙畏
propham 苯胺灵
propiconazole 丙环唑
propineb 丙森锌
propionyl brassinolide 丙酰芸苔素内酯
propisochlor 异丙草胺
propoxur 残杀威
propoxycarbazone 丙苯磺隆
propylene dichloride 二氯丙烷
propyzamide 炔苯酰草胺
prornetryn 扑草净
prosulfocarb 苄草丹
prosulfuron 氟磺隆
protboate 发硫磷
prothiocarb 硫菌威
prothioconazole 丙硫菌唑
prothiofos 丙硫磷
prothoate 发硫磷
prynachlor 丙炔草胺
Pseudomonas chlororaphis strain 63-28 铜绿假单胞菌 63-28 菌株

pymetrozine 吡蚜酮
pyracarbolid 吡喃灵
pyraclofos 吡唑硫磷
pyraclonil 双唑草腈
pyraclostrobin 唑菌胺酚
pyraflufen-ethyl 吡草醚
pyramat 嘧啶威
pyraoxystrobin 唑菌酯
pyrazolate 吡唑特
pyrazophos 吡菌磷
pyrazosulfuron 吡嘧磺隆
pyrazothion 吡硫磷
pyrazoxon 吡氧磷
pyrazoxyfen 苄草唑
pyrethrins 除虫菊素
pyribenzoxim 嘧啶肟草醚
pyributicarb 稗草丹
pyridaben 哒螨灵
pyridalyl 三氟甲吡醚
pyridaphenthione 哒嗪硫磷
pyridate 哒草特
pyridinitril 啶菌腈
pyridyl phenylcarbamates 灭鼠安
pyridyl propanol 丰啶醇
pyrifenox 啶斑肟
pyriftalid 环酯草醚
pyrimethanil 嘧霉胺
pyrimidifen 嘧螨醚
pyriminobac-methyl 嘧草醚
pyrinuron 灭鼠优
pyripropanol 吡啶醇
pyriproxyfen 吡丙醚
pyrithiobac-sodium 嘧草硫醚
pyrolan 吡唑威
pyroquilon 咯喹酮
pyroxyfur 氯吡呋醚

Q

qingfengmycin 庆丰霉素
quinacetol sulfate 喹啉盐
quinalphos 喹硫磷
quinclorac 二氯喹啉酸
quinmerac 氯甲喹啉酸
quinoclamine 灭藻醌
quinonamid 氯藻胺
quinoxyfen 苯氧喹啉
quintozene 五氯硝基苯
quizalofop 喹禾灵
quizalofop-ethyl 喹禾灵
quizalofop-*P*-ethyl 精喹禾灵
quizalofop-*P*-tefuryl 喹禾糠酯
quizalofop-tefuryl 糠草酯
quyingding 驱蝇定

R

d-resmethrin 右旋苄呋菊酯
resmethrin 苄呋菊酯
RH-5849 抑食肼
rhodan 硫氰苯胺
ribavirin 三氮唑核苷
rich-*d*-transallethrin 富右旋反式烯丙菊酯
rimsulfuron 砜嘧磺隆
rotenone 鱼藤酮
ryanodine 鱼尼丁

S

safener 茵草敌
safmonella 沙门氏菌
saijunmao 噻菌茂
salicylanilide 防霉胺
salicylic acid 水杨酸
salifluofen 氟硅菊酯
salithion 蔬果磷
santonin 茴篙素
schradan 八甲磷
scilliroside 海葱素
sebuthylazine 另丁津
secbumeton 仲丁通
serniamitraz 单甲脒
sethoxydim 烯禾啶
shacaoan 杀草胺
siduron 环草隆
silafluofen 氟硅菊酯
silaid 调节硅

silatrane　毒鼠硅
silthiopham　硅噻菌胺
simazine　西玛津
simeconazole　硅氟唑
simetryn　西草净
sintofen　杀雄啉
SN 106279　氟萘草酯
sodium 4-CPA　对氯苯氧乙酸钠
sodium 5-nitroguaiacolate　5-硝基愈创木酚钠
sodium chlorate　氯酸钠
sodium dichloroisocyanurate　二氯异氰尿酸钠
sodium diphacinone　敌鼠钠
sodium ethylxanthate　促叶黄
sodium fluoroacetate　氟乙酸钠
sodium fluosilicate　氟硅酸钠
sodium nitrophenolate　复硝酚钠
sodium *p*-aminobenzen sulfonate　敌锈钠
sophamide　苏硫磷
spinetorarn　乙基多杀菌素
spinosad　多杀菌素
spirodiclofen　螺螨酯
spiromesifen　螺甲螨酯
spiroxamine　螺环菌胺
Spodoptera litura NPV　斜纹夜蛾核型多角体病毒
streptomycin　链霉素
strychnine　毒鼠碱
succinic acid　琥珀酸
sulcotrione　磺草酮
sulfallate　草克死
sulfentrazone　甲磺草胺
sulfluramid　氟虫胺
sulfometuron-methyl　甲嘧磺隆
sulfosulfuron　磺酰磺隆
sulfotep　治螟磷
sulfoxaflor　氟啶虫胺腈
sulfur　硫黄
sulfuryl fluoride　硫酰氟
sulphenone　一氯杀螨砜
sulphosate　草硫膦
sulprofos　硫丙磷
swep　灭草灵

T

tabatrex　驱虫特
tavron　稗草烯
tazimcarb　噻螨威
tea saponin　茶皂素
tebuconazole　戊唑醇
tebufenozide　虫酰肼
tebufenpyrad　吡螨胺
tebupirimfos　丁基嘧啶磷
tebutam　牧草胺
tebuthiuron　丁噻隆
tecloftalam　叶枯酞
tecnazene　四氯硝基苯
teflubenzuron　氟苯脲
tefluthrin　七氟菊酯
temephos　双硫磷
temivinphos　灭虫畏
tepraloxydim　吡喃草酮
terallethrin　环戊烯丙菊酯
terbacil　特草定
terbucarb　特草灵
terbuchlor　特丁草胺
terbufos　特丁硫磷
terbumeton　特丁通
terbuthylazine　特丁津
terbutryn　特丁净
tertraconazole　四氟醚唑
tetcyclacis　四环唑
tetrachlorvinphos　杀虫畏
tetraconazole　氟醚唑
tetradifon　三氯杀螨砜
d-tetramethrin　右旋胺菊酯
tetramethrin　胺菊酯
tetramethylfluthrin　四氟醚菊酯
tetramine　毒鼠强
tetramycin　梧宁霉素
tetrasul　杀螨好
tfatol　螨蜱胺
thallous sulphate　硫酸亚铊
thanite　杀那特

thenylchlor 甲氧噻草胺
thiabendazole 噻菌灵
thiacloprid 噻虫啉
thiamethoxam 噻虫嗪
thiapronil 噻丙腈
thiazfluron 噻氟隆
thiazopyr 噻草啶
thicyofen 噻菌腈
thidiazimin 噻二唑草胺
thidiazuron 噻苯隆
thifensulfuron-methyl 噻吩磺隆
thifluzamide 噻呋酰胺
thimet sulfoxide 保棉丰
thiobencarb 禾草丹
thiocarboxime 抗虫威
thiochlorfenphim 硫氯苯亚胺
thiocyclam 杀虫环
thiodiazole copper 噻菌铜
thiodicarb 硫双灭多威
thiofanox 久效威
thiometon 甲基乙拌磷
thionazin 虫线磷
thiophanate 硫菌灵
thiophanate-methyl 甲基硫菌灵
thioquinox 克杀螨
thiosemicarbazide 灭鼠特
thiram 福美双
thrizopyr 噻唑烟酸
thuringiensin 苏云金素
tiadinil 噻酰菌胺
TIBA 三碘苯甲酸
tiocarbazil 仲草丹
tirpate 环线威
TMPD 伊蚊避
tolclofos-methyl 甲基立枯磷
tolfenpyrad 唑虫酰胺
tolyfluanid 甲苯氟磺胺
toosedarin 楝素
toosendanin 川楝素
TOPE 甲草醚
toxisamate 多杀威
tralkoxydim 苯草酮
tralocythrin 溴氯氰菊酯
tralomethrin 四溴菊酯
tranid 棉果威
transfluthrin 四氟苯菊酯
triacontanol 三十烷醇
triadimefon 三唑酮
triadimenol 三唑醇
triallate 野麦畏
triamiphos 威菌磷
triapenthenol 抑芽唑
triarathene 苯螨噻
triasulfuron 醚苯磺隆
triaxamate 唑蚜威
triaziflam 三嗪氟草胺
triazophos 三唑磷
triazoxide 唑菌嗪
tribenuron-methyl 苯磺隆
tribufos 脱叶磷
S,*S*,*S*-tributyl phosphorotrithioate 脱叶磷
tricamba 杀草畏
trichlamide 水杨菌胺
trichlorfon 敌百虫
trichloronat 毒壤膦
triclopyr 绿草定
tricyclazole 三环唑
tridemorph 十三吗啉
tridiphane 灭草环
trietazine 草达津
trifenmorph 杀螺吗啉
trifloxystrobin 肟菌酯
trifloxysulfuron 三氟啶磺隆
triflumizole 氟菌唑
triflumuron 杀铃脲
trifluralin 氟乐灵
triflusulfuron-methyl 氟胺磺隆
triforine 嗪胺灵
trimethacarb 混杀威
trinexapac-ethyl 抗倒酯
triprene 烯虫硫酯
triptolide 雷公藤内酯醇

trirnexachlor 三甲环草胺
trithion-methyl 甲基三硫磷
triticonazole 灭菌唑
tritosulfuron 二氟甲磺隆
trizazinones 环嗪酮
tuberostemonine 百部碱

U

10-undecylenic acid 十一碳烯酸
uniconazole 烯效唑
urbacide 福美甲胂

V

valerate 戊菊酯
validamycin 井冈霉素
vamidothion 蚜灭磷
vcelangulin 苦皮藤素
veldrin 威尔磷
vernolate 灭草敌
Verticillium lecanii 蜡蚧轮枝菌
vertrine 藜芦碱
vinclozolin 乙烯菌核利
viniconazole 烯霜苄唑

W

warfarin 杀鼠灵
wuyiencin 武夷菌素

X

xylachlor 二甲苯草胺
xylylcarb 灭杀威

Z

zarilamid 氰菌胺
zhongshengmycin 中生霉素
zinc thiozole 噻唑锌
zine methanearsonate 甲基胂酸锌
zine phosphide 磷化锌
zineb 代森锌
ziram 福美锌
zoxamide 苯酰菌胺

Key to Exercises

Unit 1

Ⅰ 略。

Ⅱ 有机农药　神经毒剂　受体位点　筛选体系　杀虫活性　实际应用
物理性质　硫酸铜　基因修饰　空间排列　靶标生物　镜像

Ⅲ herbicide　insecticide　fungicide　nematicide　rodenticide　side effect
additive　diluents　antioxidant　combinatorial chemistry　repellent　active ingredient

Ⅳ 1. C　2. D　3. B　4. D　5. D

Ⅴ 1. 直到 1950 年，许多无机农药仍在使用，包括砷酸钙、硫酸铜、砷酸铅和硫黄。然而，在 1950 年以后，除了硫黄以外，这些无机农药已经完全被有机合成农药所代替。

2.“活性成分”是指能够预防、消灭、击退或减轻虫害的任何物质（或结构相似的物质组），或在《联邦杀虫剂、杀菌剂、灭鼠剂法案》中作为植物调节剂、干燥剂或脱叶剂功能的任何物质。

3. 配制剂型的目的是提高农药的药效，方便使用，并保证其在贮存过程中保持稳定。因此，添加剂包括惰性成分，如润湿剂、黏着剂、乳化剂、抗氧化剂和稀释剂。

Unit 2

Ⅰ 略。

Ⅱ 官能团　3 龄幼虫　结构变化　击倒作用　生物活性　代谢途径　神经传输
环境温度　哺乳动物毒性　光化学稳定性　在生物体外　环境降解

Ⅲ detoxication　aldehyde　isomerization　pheromone　carbon dioxide　juvenile pheromone　hydrolysis
degradation　phosphorylation　nucleophile　mode of action　organophosphate

Ⅳ 1. B　2. B　3. B　4. D　5. B

Ⅴ 1. 有机磷酸酯一词通常指含有磷原子的有机化合物。这些化合物被广泛用作杀虫剂，大量用于控制农业害虫和病媒。

2. 除虫菊提取物是从菊花中提取的。除虫菊酯具有快速击倒作用，能使昆虫麻痹，但对哺乳动物毒性低。

3. 新烟碱类杀虫剂由于对哺乳动物的低毒性和良好的内吸活性而被开发出来，理论上，它们远远优于以前开发的许多杀虫剂。

4. 昆虫生长调节剂（IGRs）对哺乳动物的毒性很低，而且通常作用缓慢，这让一些使用者担心他们是否能够控制害虫。

Unit 3

Ⅰ 略。

Ⅱ 植物源农药　触杀型杀虫剂　天然材料　神经毒剂　植物源杀虫剂　合成类似物　天敌
市况疲软　昆虫发育　预防剂　胃毒剂　粮食作物

Ⅲ biological pesticide　aerosol spray　insect population　crude material　mammalian toxicity
quick knockdown　synthetic insecticide　moderate toxicity　highly toxic　contact poison
oral toxicity　environmental stability　skin irritation　low toxicity　broad-spectrum activity

Ⅳ 1. D　2. B　3. D　4. A　5. C　6. D

Ⅴ 1. 大多数昆虫对低浓度的除虫菊酯非常敏感。毒素一接触昆虫就会立即击倒或瘫痪，但昆虫通常会代谢这些毒素并恢复。

2. 印楝杀虫剂是从印楝树的种子中提取的，印楝树生长在几个大洲干旱的热带和亚热带地区。这种植物在非洲和南亚长期被用作药物的来源，如伤口敷料和牙膏。

3. 鱼藤酮暴露在空气和阳光下会迅速降解。它不具有植物毒性，但对鱼类有极高毒性，对哺乳动物有中等毒性。它可以与除虫菊酯或胡椒酰丁醚混合，以提高其有效性。

4. 植物源农药对昆虫不产生击倒作用。喷洒后，昆虫数量减少。采用植物源农药作为预防措施将取得良好的效果。

Unit 4

Ⅰ 略。

Ⅱ 杀虫蛋白　杀虫毒素　紫外光解作用　半固体发酵　一级晶体蛋白　毒力因子
吸啜式昆虫　发病过程

Ⅲ insecticidal crystal　rasping insect　protein inclusion　microbial pesticide
pest-management　nontarget insects　water-soluble protein　midgut tissue

Ⅳ 1. B　2. D　3. A　4. B　5. A

Ⅴ 1. Bt 几乎只对不同种类的幼虫有活性，它通过破坏中肠组织来杀死昆虫，然后引起败血症，这可能不仅是 Bt 引起的，也可能是其他细菌引起的。

2. 杆状病毒是理想的，因为就目前所知，它们对非目标昆虫、人类和环境都是安全的。杆状病毒可能是唯一可用于控制昆虫物种的有效生物控制剂，它们为克服诸如耐药性等具体问题提供了途径。

3. 真菌比其他微生物更广泛地感染昆虫，鳞翅目昆虫（飞蛾和蝴蝶）、同翅目昆虫（蚜虫和介壳类昆虫）、膜翅目昆虫（蜜蜂和黄蜂）、鞘翅目昆虫（甲虫）和双翅目昆虫（苍蝇和蚊子）的感染非常普遍。

Unit 5

Ⅰ 略。

Ⅱ 油包水　疾病媒介物　喷洒剂型　液体制剂　惰性稀释剂　工业级　物理状态　气雾剂
商品名　水溶性　塑料容器　手动喷雾器

Ⅲ oil solutions　suspension concentrate　water-miscible liquids　economic threshold
fogging machine　emulsifiable concentrates　micro emulsion　sprayable suspension
physical condition　ultralow-volume concentrates

Ⅳ 1. D　2. C　3. D　4. D　5. A

Ⅴ 1. 制剂制备是指通过任何方法对农药化合物进行加工处理，以提高其贮存、运输、应用的效果或安全性。

2. 乳油，与乳化浓缩物同义，是指在工业级物料中加入足够的乳化剂，使乳油容易与水混合而喷洒的溶液。

3. 超低容量喷雾剂（ULV）仅用于商业用途，用于控制公共卫生、农业和森林害虫。它们通常是原液体形式的技术产品，或者，如果是固体，原产品溶解在最低限度的溶剂中。

4. 烟雾剂是严格为公共卫生用途而出售的剂型，用于控制害虫或病媒，如苍蝇和蚊子。

Unit 6

Ⅰ 略。

Ⅱ 60 目筛　毒剂　可湿性粉剂　水溶性粉剂　活性稀释剂　颗粒剂　残留活性　水分散粒剂
胃毒剂　手提式喷雾器　残留效应　土壤施用

Ⅲ solid formulation　sulfur dust　tracking powder　inert diluents　aerial application
systemic insecticide　porous material　ball mill　bait gun　slow-release insecticides

microencapsulated concentrate　spot application

Ⅳ 1. D　2. A　3. D　4. D　5. A

Ⅴ 1. 粉剂一直是最简单的农药加工剂型且最容易应用。未稀释的有毒粉剂的例子有硫黄粉、氟化钠。

2. 可湿性粉剂（WP）本质上是一种含有润湿剂的粉剂，可在喷洒前使可湿粉与水混合。除润湿剂或表面活性剂外，将原药加入惰性稀释剂（在这种情况下是磨细的滑石粉或黏土）中，在球磨机中充分混合。

3. 颗粒剂是由不同的惰性黏土制成的小颗粒，并混合一种有毒（原药）溶液以达到所需的浓度。溶剂蒸发后，颗粒被包装起来使用。

4. 饵剂是原药与食物或其他有吸引力的物质混合而成。饵剂要么吸引害虫，要么放置在害虫会发现它的地方。害虫取食含有杀虫剂的饵剂后被杀死。

Unit 7

Ⅰ 略。

Ⅱ 延长释药　光学效应　农药载体　电导　控制释放　表面面积　物理强度　农药靶向传递
疏水层　孔隙容积　化学反应性　可持续农业生态系统

Ⅲ nanobiopesticide　nanoparticle　nanoemulsion　nanocapsule　nanobiosensor
nanosilica　nanomaterial　biotic factor　host plant　cuticular lipid

Ⅳ 1. True　2. False　3. False　4. True　5. False

Ⅴ 1. "纳米生物技术"是纳米技术、生物技术、材料科学、化学工艺、系统工程、纳米晶体和纳米生物材料的多学科结合。

2. 经过精心设计的纳米颗粒可以变为"神奇的颗粒"，它可以包含杀虫剂、除草剂、化学物质或基因材料，还可以作为植物精确的靶标有针对性地释放内容物。

3. 现今的杀虫剂通常是通过喷洒在植物或土壤上作为"预防"治疗，或在疾病发生后使用。在此背景下，纳米技术为新型农药的开发提供了巨大的机遇。实际上，纳米技术提高了传统或新型杀虫剂的性能和可接受性，提高了它们的有效性、针对性、安全性、抗病能力，并最终降低了管理成本。

Unit 8

Ⅰ 略。

Ⅱ 熔融分散技术　乳化溶剂蒸发法　离子交联法　乳化溶剂扩散法　聚电解质复合法
高剪切　临界胶束浓度　两亲性三嵌段共聚物　阴离子乳化剂

Ⅲ nanocarrier　nanosphere　nanoliposome　nanodispersion
nanogel　nanoencapsulation　nanosuspension　nanoemulsion

Ⅳ 1. Not Given　2. True　3. False　4. False　5. Not Given

Ⅴ 1. 纳米制剂与其他农药或药物制剂类似，其设计目的是增加难溶活性成分的溶解度，并缓慢释放活性成分，或保护活性成分不过早降解。

2. 纳米胶囊是由一个被聚合物涂层或膜包围的内中心腔组成的囊状或储集型结构。限制活性成分（AIs）的内腔可能是亲水的，也可能是疏水的。

3. 在这个过程中，磷脂分子的疏水端彼此靠近，在双分子层内平行排列，从这个区域挤出水分子；而磷脂分子的极性端定位在超分子团聚体的边缘，与周围的极性水分子接触。

Unit 9

Ⅰ 略。

Ⅱ 中耕作物　喷雾容量　流失雾液　本征迁移率　喷雾漂移　旋转式喷雾机　雾液沉积
水压　液膜　预订目标　喷洒施用　雾滴大小分布

Ⅲ fine sprays　mobile target　post-emergence spray　Leaf Area Index selective application technique　coarse spray　spray runoff　static target

pre-emergence application　nonselective herbicide

Ⅳ 1. D　2. C　3. D　4. B　5. C　6. B　7. A　8. C　9. A

Ⅴ 1. 农药的最佳使用不仅需要正确的时间，而且还需要有效地将活性成分转移到害虫、杂草或疾病所在的作物区域。

2. 农药喷洒的简单变化可导致农药在作物内的分布发生巨大变化，从而对生物功效产生重大影响。

3. 许多现代喷雾器都配备了额外的功能，以减少喷雾漂移和改善沉积。其特点包括空气辅助，在喷雾器上的风扇产生的气流在喷雾器后面形成一个空气幕。

4. 优化农药应用的目的必须是最大限度地提高喷淋在目标部位的活性成分的比例。靶标位置必须是作物中农药生物活性最强的区域。可以看出，“靶标”不仅随着虫害的变化而变化，也随着农药的变化而变化。由于不同农药的作用方式不同，喷雾靶标的选择也会不同。

Unit 10

Ⅰ 略。

Ⅱ 光学传感器　空气协助系统　应用技术　手持喷雾器　直接注射　飞机喷雾器　封闭处理系统

手持喷雾枪　手动喷雾器　背负式喷雾器　施用设备　低压动力喷雾器

Ⅲ electrostatic spray　spot sprayer　precision farming　bulk-handling system　granular herbicide

high-capacity powered sprayer　compressed air sprayer　shielded sprayer

liquid and granular formulation

Ⅳ 1. C　2. B　3. B　4. C　5. D　6. C　7. A　8. D

Ⅴ 1. 在尽量减少农药对施药者和环境的有害影响方面，更好的施药设备和新技术使杀虫剂用量减少和漂移减少变得越来越重要。

2. 手动喷雾器，如压缩空气和背包喷雾器，专用作现场处理，不适合大面积的地区。

3. 静电喷雾系统可用于空中和地面应用。在地面应用系统中，采用了液体到达喷嘴前接触充液的原理。

4. 农药将继续发挥重要作用，帮助农民为全世界人民提供充足和安全的食品供应。应用行业将继续改变，使农药的使用尽可能安全。近年来，应用行业的技术进步速度非常快。随着科学家们继续关注未来的精准农业，设备行业将努力改进和开发所需的新设备，以实现应用更有效的目标。

Unit 11

Ⅰ 略。

Ⅱ 化学平衡　生态相互作用　种子科学　植物群落　植物区系组成　幼苗出苗

合成衍生物　化感作用

Ⅲ weed management　seed germination　seedling growth　invasive species　acid rain

allelopathic phenomenon　radicle growth　unsaturated lactone　adverse effect

Ⅳ 1. D　2. A　3. E　4. D　5. A

Ⅴ 1. 当一种植物通过其活的或腐烂的组织，通过化学抑制剂干扰另一种植物的生长时，就会发生化感作用。

2. 干扰是指植物对彼此生长产生的不利影响。竞争是干扰的一部分，由于一种或多种有限资源的耗尽或无法获得而发生。

3. 化感物质从简单的分子如氨，到醌、萜烯樟脑和桉油脑，复杂的共轭类黄酮如根皮苷，或杂环生物碱咖啡因。

4. 类似的假设应用于化感作用：（ⅰ）观察、描述和量化自然群落的干扰程度；（ⅱ）分离、描述和合成来自疑似化感植物的毒素；（ⅲ）在自然界中以适当的速度和时间使用毒素来重现症状；（ⅳ）监测释放、运动和吸收，并证明它们足以引起所观察到的效应。

Unit 12

Ⅰ 略。

Ⅱ 有机农业　作物轮作　根系分泌物　杂草综合治理　积极作用　杂草防除　作物品种　化感间作

Ⅲ seed bank　weed infestation　soil moisture　soil fertility　allelopathic cultivars
organic matter　crop productivity　cover crops　crop rotation　agro-ecosystem

Ⅳ 1. A　2. D　3. D　4. D　5. D

Ⅴ 1. 通过在离杂草很近的地方种植化感植物来控制这些化学物质的产生，或者将从死亡植物中获得的化感物质放置在杂草附近。

2. 种植覆盖作物的目的是保持农业生态系统的可持续性。种植覆盖作物的各种目标包括提高土壤肥力和土壤质量，以及抑制杂草和植物病原体。

3. 轮作可以抑制作物田的杂草危害，同时与其他方法结合使用可以提高杂草控制的效果。在轮作的情况下，如果没有来自邻近土地的杂草种子侵入田间，作物轮作就会更有效。

Unit 13

Ⅰ 略。

Ⅱ 农艺性状　耐除草剂　杀虫蛋白　杀虫基因　系统发育关系　转基因品种　鳞翅目害虫　敏感种

Ⅲ phenotypic characteristics　artificial diet　predation rate　insect-resistance　halo effect
host-plant　conventional breeding　indirect effects　food-webs　pest-resistant
genetically engineered crops　sweet corn

Ⅳ 1. D　2. A　3. D　4. D　5. C

Ⅴ 1. 20 世纪 80 年代末开发的第一批转基因作物从苏云金芽孢杆菌（*Bacillus thuringiensis* Berliner, Bt）中表达了杀虫蛋白，因为它们具有已知的特异性和优良的安全记录的微生物 Bt 配方。

2. 1996 年，首个转基因植物产业化，转基因品种种植面积稳步扩大。两个主要的特点是除草剂耐药性（HT）和抗虫性。

3. 综上所述，现有文献提供的证据表明，在商业化 Bt 作物中使用的杀虫蛋白对目标害虫以外的非目标物种没有直接的、不利的影响。

Unit 14

Ⅰ 略。

Ⅱ 双链 RNA　内源性基因　基因沉默　局部给药　刺吸式害虫　沉默复合物
病毒感染　模型微生物

Ⅲ chewing pest　target pest　susceptible strain　homeostasis　gene expression　reverse genetics
synergistic effects　crucial genes　environmental uptake　cellular uptake　transgenic plants
RNA interference　immunity system　expression level　target gene

Ⅳ 1. D　2. B　3. D　4. A　5. B

Ⅴ 1. 根据小 RNA 生物发生的类型和涉及的 Argonaute 蛋白的不同，RNAi 可以分为三种途径：微小 RNA（miRNA）、小干扰 RNA（siRNA）和 piwi-interaction RNA 途径。

2. 基于 RNA 的虫害控制依赖于沉默目标害虫的生存关键基因，最终导致害虫死亡或抑制害虫种群密度。

3. 环境 RNAi 方法可用于喷洒式 dsRNA 制剂的病虫害控制，类似于传统化学杀虫剂的应用。害虫对杀虫剂的传递和吸收主要是通过取食寄主植物过程或外皮完成的。

4. RNAi 可用于协助害虫控制策略，如植物产生的毒素、杀虫剂、病原体等，从环境中摄取 dsRNA 进入昆虫体内。这可能是由于从饮食中摄取的 dsRNA 或直接通过昆虫外皮摄取。

Unit 15

Ⅰ 略。

Ⅱ 对照组　实验单元　假重复　处理与对照　随机选择　实验室菌落　温度梯度
剂量响应曲线　定量反应　实验设计　解释变量　熏蒸活性

Ⅲ replication　treatment method　analysis of variance　growth rate　random number table
median lethal dosage　experimental bias　statistical analyses　logistical convenience
median lethal concentration

Ⅳ 1. F　2. F　3. T　4. F　5. T　6. T　7. F　8. F　9. T　10. F

Ⅴ 1. 毫无例外，所有的实验必须重复。重复是在与第一次试验相同的条件下（尽可能多）在不同的时间重复一次生物测定。重复测定的目的就是随机安排无法控制的过程和条件，其中包括工作人员和测定时间对测定结果的影响。重复测定另外一个目的是检定试验中的错误。鉴于此，确保重复测定真实性的最好办法就是从现配母液中进行系列剂量稀释。

2. 随机选择是从指定的试验单位到处理和对照过程是随机的。运用随机选择的目的是避免偏差，从而能真实地反映试验单位的实际反应。即使在试验中要记录更多的试验数据，即使试验的花费更多，但是最好还是采用完整的随机选择程序。

Unit 16

Ⅰ 略。

Ⅱ 琼脂凝胶　孢子活性　饲养手册　十字花科　田间测定　塑料海绵　试验标地
发酵残渣　人工饲料　缓冲溶液　饲料刺激　气象站

Ⅲ leaf discs　bioassay period　potency unit　lepidopterous insect　field plant bioassay
neonate larvae　target insect　microbial strains　leaf bioassay　potted-plant bioassay

Ⅳ 1. T　2. T　3. F　4. T　5. T　6. F　7. T　8. T　9. F　10. F

Ⅴ 1. 使用基于人工饲料的生物测定法的目的是为工作人员提供一种快速、标准化和简单的程序来估计微生物菌株的活性。

2. 标准化的生物测定饲料可以用于任何目标昆虫，只要将特定的噬菌体刺激剂加入到专门饲料中。此外，为了保持孢子活性，应避免在饲料中加入抗生素。

3. 让混合物在叶片上干燥。在距离叶 2cm 处切开叶柄。用镊子按住叶柄，插入琼脂层。放入 5 头初孵幼虫在每个小瓶中。用橡皮筋将棉布紧紧地封住小瓶。

4. 盆栽植物生物测定有几个用途，不能提供叶生物测定。这些用途包括：（ⅰ）用完整的植物器官测试苏云金杆菌的活动；（ⅱ）运用微生物产品喷雾或喷粉到植物上，其上剂量的精度超过了田间施用能达到的精度；（ⅲ）延长初孵幼虫接触到完整的叶片的生物测定的时间，超过 3 天；（ⅳ）检测微生物制剂的残余效应。

Unit 17

Ⅰ 略。

Ⅱ 交互抗性　致死毒物　多抗性　生化机制　第一代　杂交后代　分子遗传学　敏感个体
靶标密度　耐受剂量　遗传变异性　行为抗性

Ⅲ population density　resistant individual　fruit fly　heterozygote resistance
resistance development　biochemical or physiological property　tubulin protein　adult stage
host plant　penetration of the pesticide　detoxifying enzymes/detoxication enzymes
glutathione transferases

Ⅳ 1. B　2. D　3. D　4. C　5. D

Ⅴ 1.行为抗性这个术语描述了避免致命剂量的能力的发展。抗性不同于某些害虫表现出的自然耐受性。一种生化或生理特性使农药对大多数正常个体无效。

2.交叉抗性是指害虫种群由于只选择一种杀虫剂而对两种或两种以上的杀虫剂产生抗药性的现象。多重抗性是同时或连续接触两种或两种以上杀虫剂而诱发某些种产生抗性。

3.环境中的致命毒物会对种群产生巨大的影响。只有那些因为某种原因存活下来的个体才能繁衍后代。拥有等位基因或基因复制的个体对有毒环境的敏感度较低，将有更好的繁殖机会。因此，害虫的下一代将更高概率具有这些等位基因。如果农药或其他手段不能完全消灭害虫，迟早会出现抗药性。

4.这一战略意味着有些地区不使用农药，而是作为易感个体的庇护所。然后，这些昆虫可以在没有任何农药选择压力的情况下交配。存活下来的，或多或少有抗药性，在附近处理过田地里的个体将与来自保护区的个体交配，产生的后代将是杂合子，对杀虫剂敏感。

Unit 18

Ⅰ 略。

Ⅱ 结构基因　铜络合物　种群动态　分子遗传装置　抗性基因　抗性表达　代谢失活
代谢解毒　调控基因　多基因抗性　单克隆抗体　超氧化物歧化酶

Ⅲ genetic engineering　gene mutation　plastome mutator　minimum selection pressure
gene transcription　detoxication system　gene amplification　genetic adaptation
extrachromosomal gene　biosynthesis inhibitors

Ⅳ 1. B　2. D　3. A　4. C　5. B　6. C

Ⅴ 1.深入全面了解农药抗性的遗传、生化及生理机制，对于解决农药抗性问题至关重要。

2.植物利用与昆虫相同的一般抗性机制。对农作物有效使用除草剂之所以成为可能，是因为许多农作物能够快速代谢灭活这些化学物质，从而避免它们的毒性作用。靶标杂草明显缺乏这种能力。然而，很明显，将除草剂代谢成无害化合物的能力构成了杂草进化出抗除草剂能力的潜在重要基础。

3.合成农药很可能在一段时间内继续作为对付大多数害虫的主要武器，因为它们具有普遍的可靠性和迅速的作用，而且它们有能力维持今天城市消费者所要求的高质量农产品。虽然新农药提供了短期的解决办法，但仅用这种方法来控制害虫很难提供可行的长期战略。

Unit 19

Ⅰ 略。

Ⅱ 良好农业规范　残留试验　农药残留　累积残留　贸易争端　收获后　可检测水平
农药毒理学　农业商品　婴幼儿的敏感性

Ⅲ food safety　trace concentration　pesticide active ingredients　maximum residue limit (MRL)
acceptable daily intake (ADI)　mechanism of toxicity　risk assessment
globalization of the food chain　acute reference dose (ARfD)　integrated pest management (IPM)

Ⅳ 1. D　2. C　3. B　4. B　5. C　6. C

Ⅴ 1.消费者通常更倾向于没有农药残留的食物，但农药通常是IPM计划的组成部分。为了解决这一问题，人们在实践中达成了“食物链妥协”，以满足农民和消费者的需求。

2.食品法典标准的重要性在于，它们提供了全球统一、公正和权威的最大残留限量来源，考虑到特定农药商品的不同国家差距以及现有的残留试验数据。

3.来自主要粮食进口国家的实际农药监测方案表明，在许多情况下没有检测到农药残留，而绝大多数可检测的农药残留远远低于已确定的最大残留限量。

Unit 20

Ⅰ 略。

Ⅱ 方法敏感性　气相色谱　单残留分析法　旋转蒸发器　分析精确度　多残留分析法　分析方法　痕量残留分析　基体干扰　水溶性溶剂　反相分离　高效液相色谱　饱和食盐水

Ⅲ filter paper　mobile-phase　chromatographic column　sample process　development rate　mass　spectrometry　activated carbon　organic acids　sample analysis　retention time　standard operating procedure　reversed-phase solid-phase extraction

Ⅳ 1. C　2. C　3. D　4. B　5. B　6. D

Ⅴ 1. 单残留分析法（SRMs）描述了对特定作物中单一农药（或从该农药中衍生出的一组相关化合物）的分析，因为它们是为特定应用或作物注册特定农药而开发的。多残留分析法（MRMs）可用于多种不同类型的农药样品的分析。

2. 除非适用的政府法规要求提交的样品须保持原样不变，否则大多数实验室会对整个样品进行剪切和均质化。样品通常使用普通的商用食品加工机进行均质化，同时提供浸渍和混合。

3. 在高浓度的背景化学物质（通常称为基质干扰）存在的情况下，分离和检测农药残留是一项具有挑战性的任务。大多数粗提物在分析前需要进行一定的提纯。

4. 对监管实验室来说，农药水平的量化与农药的鉴别同样重要，因为农药使用的规则是基于可能存在的最大残留水平（MRL 或耐受性）。

Unit 21

Ⅰ 略。

Ⅱ 臭氧层　废物处理　光解反应　基质降解　低极性　持久性农药　低水溶性　不稳定中间体　拼凑　分散过程　降解产物　微生物代谢

Ⅲ aquatic organism　high water solubility　oxidation reaction　abiotic reaction　bioaccumulation　chemical degradation　biodegradation　environmental transformation　environmental toxicology　environmental analytical chemistry

Ⅳ 1. C　2. D　3. D　4. B　5. C　6. C　7. B

Ⅴ 1. 初始残留物以一个综合的速率消散，这个速率是单个过程的速率的总和，例如挥发、冲洗、浸出、水解、微生物降解等。

2. 根据化学品的物理性质、化学反应性、稳定性特征以及在使用或释放发生的环境中化学品的可用性和质量，残留物往往会重新分布，并优先于一个或多个组分或介质。

3. 有机磷可以在环境中水解成磷酸或硫代磷酸衍生物和取代苯酚或醇。

4. 有专家认为，微生物（细菌、真菌、藻类）降解占环境中所有降解反应的 90%以上，在大多数表层土壤、植物根部附近以及包括污水池和污水处理系统在内的富营养水域中几乎是唯一的降解途径。

Unit 22

Ⅰ 略。

Ⅱ 垃圾降解　环境胁迫　土壤污染　重组 DNA　碳利用率　无毒化合物　碳源　生物异源复合物

Ⅲ microbial community　mineral salt　endogenous microbe　nitrification　nutrient cycling　soil toxicant　enzymatic breakdown　parent compound　bioremediation　chromosomal DNA

Ⅳ 1. D　2. A　3. B　4. D　5. D

Ⅴ 1. 土壤微生物在垃圾降解、促进植物生长、养分循环以及污染物和农药的降解等方面发挥着重要作用。

2. 为了尽量减少农田的土壤污染和土壤毒物，可以利用各种处理方法（即污染土地的填埋、回收、热解和焚烧）。

3. 清除含有这些化合物的场所的物理化学方法既不经济，也不够充分。因此，生物法是利用微生物清除污染场所污染物的一种很有前景的方法。

4. 农药降解可通过生物转化、生物矿化、生物积累、生物降解、生物修复和共代谢等一种或多种方式进行。

5. 微生物介导农药在污染部位降解为简单的无机化学品；土壤、地下水、污泥、工业水系统和气体。

6. 这些酶催化代谢反应包括水解，氧化，氧双键的形成，氧化氨基（NH_2）为硝基，苯环上增加羟基，脱卤作用，还原硝基（NO_2）为氨基，硫替换成氧，侧链的代谢，环的开裂。

Unit 23

Ⅰ 略。

Ⅱ 标准差　肉毒杆菌毒素　剧毒　高毒　致死浓度　二项式定理　慢性剂量　急性剂量
剂量-反应曲线　慢性毒性　急性毒性　致死剂量

Ⅲ weakly toxic　subchronical dose　sublethal doses　normal distribution　confidence intervals
law of mass action　moderately toxic　environmental toxicology　subacute dose　effective dose
effective concentration　practically nontoxic

Ⅳ 1. B　2. D　3. C　4. C　5. A

Ⅴ 1. 毒理学一直被用作描述人类医学中毒药对人类影响的学科。毒物的概念包括毒物的吸收、排泄和代谢（毒物动力学），以及症状和其发展（毒物效应动力学）。

2. 一切物质都是毒药，没有一种不是毒药。正确的剂量能区别毒药和药物。

3. LD_{50} 是预期杀死一半暴露个体的剂量。有时我们更关注的是确定能杀死 90%或 10%个体的剂量，这些剂量分别称为 LD_{90} 和 LD_{10}。

4. 在许多情况下，急性或亚急性剂量可能在中毒多年后产生慢性症状或影响（吸烟和癌症）或在下一代产生影响。

Unit 24

Ⅰ 略。

Ⅱ 钠离子通道　免疫活性　急性症状　生殖毒性　基因表达　急性口服毒性　淋巴细胞　水生毒性

Ⅲ steroid receptor　aquatic organisms　leydig cell　reproductive hormones
toxicological symptoms　reactive oxygen species　endocrine disruption
priority target　cytotoxicity neurotoxicity　dietary intake　liver dysfunction

Ⅳ 1. A　2. B　3. D　4. C　5. D

Ⅴ 1. 拟除虫菊酯根据是否含有 α-氰基-3-苯氧苄基分为Ⅰ型和Ⅱ型。Ⅰ型拟除虫菊酯可引起震颤或抽搐，而Ⅱ型拟除虫菊酯暴露主要引起舞蹈手足徐动症。

2. 人接触拟除虫菊酯可导致诸如恶心和呕吐、呼吸困难、咳嗽、支气管痉挛、皮肤反应等急性症状。

3. 观测期越长，LD_{50} 越小。小鼠的年龄越大，LD_{50} 就越大。

4. 此外，拟除虫菊酯的水生毒性与温度有关。低温下毒性增加，这已被纳入毒性鉴定评估（TIE）程序，以确定环境毒性的原因。

Unit 25

Ⅰ 略。

Ⅱ 协同毒性　直接影响　风险评估　添加剂的毒性　种群下降　暴露时间
关键受体　植物生长调节剂

Ⅲ biological effect　toxicity assessment　toxic effect　central nervous system
antagonistic toxicity　negative impact　lethal toxicity　toxic action
homeostatic balance　toxic equivalent

Ⅳ 1. A　2. D　3. D　4. B　5. A

Ⅴ 1. 目前在我们星球的空气、土壤、水和沉积物中发现的大量化学污染物，主要是人类活动的结果，呼吁对其风险进行评估。

2.所有的化学品对哺乳动物的致命毒性基于大鼠或小鼠的生物测定，从口服 LD_{50} 和接触 LC_{50} 可以确定。

3.慢性毒性生物测定法用于检测重复或持续暴露于亚致死剂量或浓度的化学品后可能产生的生物效应。

4.除了确定混合物的总毒性外，重要的是评估化合物是否具有添加性、协同性或拮抗毒性，如果有，试着估计协同/拮抗比。

5.添加剂毒性发生在具有相同或类似作用模式的化合物中，如有机磷和氨基甲酸酯类杀虫剂、拟除虫菊酯、新烟碱类、三嗪类等。

Unit 26

Ⅰ 略。

Ⅱ 暴露评估　蒸汽压力　统计分布　低毒　离解常数　皮肤暴露　分配系数　浸出电位

Ⅲ soil adsorption　regulatory authority　degradation pathway　hydrophobic compound
soil erosion　aquatic environment　half-life　quantitative structure relationship

Ⅳ 1.T　2.F　3.T　4.T　5.T　6.T　7.T

Ⅴ 1.动物体内的降解途径也很重要，因为一些化学物质易于代谢分解并产生毒性较小的代谢物，这些代谢物可以通过尿液和粪便排出体外。

2.不同的地区和时间有大量的实际残留数据，因为综合监测调查进行了多年，并可从开放的文献获得。

3.合理的暴露评估不应过多地以准确的浓度水平为基础，而应以特定情景和情况下的浓度范围为基础。

4.在任何化学风险评估中，一个至关重要的方面是化学和毒性数据的准确性和可靠性。如果没有适当的数据，评估将无法确定可能情况下的实际风险。

Unit 27

Ⅰ 略。

Ⅱ 提高效率　可追溯系统　技术服务　害虫防治　临时登记　栽培制度　卫生杀虫剂
品质较差　农药管理　利用率　费用节省　十字花科蔬菜

Ⅲ zero growth　management regulations　residual contamination　upward trend　public health
agricultural products　business license　application technology　professional services
decentralized management

Ⅳ 1.A　2.A　3.A　4.D　5.C　6.C

Ⅴ 1.修订草案拟取消农药临时登记，并根据《农药经营许可管理办法》改为农药经营许可证制度。同时加入了农药追溯系统的要求，鼓励使用低毒生物农药，并强化企业提供农药使用指导的义务。

2.2015年，农业部制定并发布了到2020年的五年农药使用零增长行动计划。同年，农业部、国家发展和改革委员会、科技部、财政部、国土资源部、环境保护部、水利部、国家林业局，共同发布了“全国农业可持续发展规划（2015—2030）”。

3.未来我国农药管理的原则是实现下面五项要求：（1）改革农业经济结构的要求；（2）提高农药质量/安全和人类健康的要求；（3）促进农药产业发展的要求；（4）发展可持续农业的要求；（5）提高我国竞争力的要求。

Unit 28

Ⅰ 略。

Ⅱ 致癌效应　全膳食研究　残留监测　可接受水平　职业疾病　短持效期　累积接触　残留分析
非靶标生物　健康与环境风险

Ⅲ acute poisoning　chronic poisoning　secondary pest　pesticide tolerance　pesticide use restriction air pollution　violative residue　soil contamination　natural vegetation　natural pest population

Ⅳ 1. B　2. D　3. C　4. D　5. A　6. B　7. C

Ⅴ 1. 农业中用来控制有害物如害虫、杂草以及植物病害的农药，已经在过去的几十年内受制于众多的法律、法规及消费者的审核。农药从本质上来说是有毒性的化学品；因为许多农药可能在人类可消费的食品中残留，所以关于农药在人类饮食方面所构成的潜在健康危险引起了人们的极大关注。

2. 但是，应该强调的是，所确立的农药耐受水平依然要与生产商田间试验建立的最大残留量相等或稍高于这个水平。同样，这些耐受水平依然只代表执法工具而不应该和安全标准相混淆，尽管美国环保局在确立这些耐受水平之前的确考虑到可能的健康风险。

Unit 29

Ⅰ 略。

Ⅱ 类似物设计　靶向合成　电子等排取代　基本骨架　先导化合物　结构类拟物　不对称合成
手性农药　生物测定　官能团类似物　广泛筛选　天然来源

Ⅲ High-Throughput Screening　patentable analogue　substituent effect　phenotypic screening
computersimulation design　positional isomer　chemical similarity　insecticidal activity
sex pheromone　bioisosterism optical isomers　aggregation pheromone　random screening
biomimetic synthesis　chemoinformatician

Ⅳ 1. D　2. A　3. B　4. D　5. A　6. B　7. C　8. D

Ⅴ 1. 直接类似物设计包括直接的分子修饰，例如同系物、乙烯基化合物、同分异构体、位置异构体、光学异构体、修饰环体系和同型二聚体的合成。

2. 这种方法包括在动物模型或任何生物试验上筛选新分子，无论它们是合成的还是自然产生的，而不考虑其活性潜力的假设。

3. 随机筛选是农药发现的常用方法之一。该方法通过筛选多种化合物来鉴定具有较高生物活性的化合物。

4. 电子等排要求母体化合物中的一个原子或一组原子替换成另一个具有类似电子和空间构型的原子。

5. 仿生合成是对特定植物、动物和微生物的独特物质进行合成和修饰，如植物产生的独特物质以避免捕食和竞争，昆虫产生的性信息素和聚集性信息素以吸引和聚集。

Unit 30

Ⅰ 略。

Ⅱ Computer-aided drug design　Cambridge Structural Database　Protein Data Bank
quantitative structure-activity relationship　virtual screening

Ⅲ 计算化学　分子识别　分子碎片　生物大分子　计算机辅助药物设计　从头设计
蛋白质-配体互作　三维结构　先导化合物

Ⅳ functional group　molecular alignment　molecular design　molecular docking　force field
surface area　molecular volume　Medicinal Chemistry　biological targets　randomly selection

Ⅴ 1. D　2. A　3. B　4. D

Ⅵ 1. 计算机辅助药物设计（CADD）是有关化学结构、性质及其与生物大分子相互作用的信息的结合。此外，这些数据被转化为信息，旨在帮助药物发现和开发做出更好的决策。

2. 为了确定最有利的构型，每个构型都根据其形状和属性（如静电学）与目标的互补性进行打分。一个好的分数表明该分子可能是一个好的结合。

3. 从头设计方法的巨大优点是，在进行任何合成之前，它们可以用于在计算机模拟中评估大量适合实际情况的潜在结构。通过这种方式，从头设计为制药公司提供了压倒性的优势，这些公司试图为最有前途的先导化合物抢占专利。